The Collected Papers of
Albert Einstein

Volume 6

The Collected Papers of Albert Einstein

English Translations Published to Date

Volume 1: *The Early Years, 1879–1902.* Anna Beck, translator; Peter Havas, consultant (1987)

Volume 2: *The Swiss Years: Writings, 1900–1909.* Anna Beck, translator; Peter Havas, consultant (1989)

Volume 3: *The Swiss Years: Writings, 1909–1911.* Anna Beck, translator; Don Howard, consultant (1993)

Volume 4: *The Swiss Years: Writings, 1912–1914.* Anna Beck, translator; Don Howard, consultant (1996)

Volume 5: *The Swiss Years: Correspondence, 1902–1914.* Anna Beck, translator; Don Howard, consultant (1994)

Volume 6: *The Berlin Years: Writings, 1914–1917.* Alfred Engel, translator; Engelbert Schucking, consultant (1997)

The Collected Papers of
Albert Einstein

Volume 6

The Berlin Years: Writings, 1914–1917

English Translation of Selected Texts

Alfred Engel, Translator

Engelbert Schucking, Consultant

Princeton University Press
Princeton, New Jersey

Copyright © 1997 by the Hebrew University of Jerusalem

Published by Princeton University Press, 41 William Street, Princeton, NJ 08540
In the United Kingdom: Princeton University Press, Chichester, West Sussex

Princeton University Press books are printed on acid-free paper and meet the
guidelines for permanence and durability of the Committee on Production Guidelines
for Book Longevity of the Council of Library Resources

Printed in the United States of America

This book has been composed in Times Roman
Typeset by Michael Perlman

ISBN: 0—691—01734—4
Library of Congress catalog card number: 87—160800
10 9 8 7 6 5 4 3 2

ISBN-13: 978-0-691-01734-1 (cloth)

Contents

CONTENTS

Publisher's Foreword

We are pleased to be publishing this translation of Volume 6 of *The Collected Papers of Albert Einstein*. As with previous volumes, while every effort has been made to ensure the scientific accuracy of these translations, they are not intended for use without the documentary edition, which provides the extensive editorial commentary necessary for a full historical and scientific understanding of the source documents.

As the reader will note, this is the first volume to be translated on a more selective basis than the previous volumes. The editors of *The Collected Papers* have made this selection, and they have also recommended reprinting earlier translations of Documents 24, 30, 42, and 43 from other sources. We are doing so with the permission of the earlier publishers.

In addition, we welcome a new translator of Einstein's scientific writings: Alfred Engel, a mathematician; and a new consultant: Professor Engelbert Schucking, a physicist at New York University. We are pleased to be working with both of them and thank them for their dedication to this project.

Once again we thank Dr. Michael Perlman for his desk-top publishing skills and for conscientiously preparing the camera-ready copy for this volume.

Princeton University Press
May 1997

Translator's Preface

This volume contains several of Albert Einstein's most important papers on both quantum theory and the theory of general relativity. The translator's task was to try to preserve in English Einstein's peculiar style of writing, which reflects the kaleidoscope of his complex thought processes. Unlike many of his contemporaries, Einstein preferred to write more in the way he talked in the lecture hall, mentioning with didactic skill every thought to be considered at each step from one formula to the next. Occasionally this taxes even German grammar which, admittedly, is more tolerant of very long sentences than, for example, English. Thus, the translator put higher value on the tracing of the psychology of Einstein's thought processes historically and didactically than on linguistic elegance.

It is well known that Einstein was one of the first physicists to make extensive use of tensor calculus in his numerous papers. Understandably, this was an unfamiliar challenge to the typesetters of scientific journals—and typographical errors were numerous, as every reader of the originals has realized. In a text-critical edition these need to be annotated, as has been done by the editors, in square brackets [ ], and by the translator and consultant in curly brackets { }. The majority of typos are trivial mistakes in subscripted or superscripted indices; this also demonstrates that these errors did not originate with Einstein, no matter what his detractors said, especially those who also had other than truly scientific motives. Einstein, anxious like every author to see his important papers published as soon as possible, perhaps may have been negligent in proof reading, or he wanted to spare the typesetter the then laborious manual work of correcting a printing plate precisely at several cumbersome formulas. And so the typos are still there.

We felt, however, that the scientific community is better served if these distracting typos—while still pointed out in the margin and in annotations—are not dragged on forever through all translations. We tried to correct in the English translation all those that came to our attention in text or formulas, while being cognizant of the fact that many indices in the German originals are difficult to decipher.

For reasons of historical reference the page numbers of the original journals are usually also inserted in the margins at the appropriate places of the translation in brackets (e.g., [p. 123]).

Longer quotations in German which occur in the editors' notes sometimes without translation (e.g., in Documents 26 and 42) have also been translated here and appended as additional notes at the ends of these papers.

We would like to thank Prof. E. A. Spiegel for editorial help with Document 39.

We also owe thanks to Alice Calaprice, senior editor of the Einstein Project at

Princeton University Press, for lending her great experience and devotion to the finalization of the present volume. Michael Perlman deserves more than the usual credit for the preparation of the camera-ready text and, especially, for the setting of Einstein's numerous and frequently intricate formulas.

A. ENGEL, TRANSLATOR   E. L. SCHUCKING, CONSULTANT

Selected Documents

Doc. 1

On the Principle of Relativity

by Prof. Dr. A. Einstein

Member of the Royal Prussian Academy of Sciences

The editorial staff of the *Vossische Zeitung* has asked me to relate to their readers something about my field of work. I gladly honor this request. Although a deeper understanding of the theory of relativity is hardly possible without considerable effort, it may still be appealing for the nonscientist to hear something about the methods and results of this new branch of theoretical research. Even a cursory analysis of the processes we call motion already teaches us that we can perceive only the relative motion of things with respect to one another. We sit in a railway carriage and see (on the adjacent track) another carriage pass by. If we ignore the vibration of our carriage, we have no immediate means by which to decide whether the two carriages are moving "in reality." We find only that the relative position of the carriages changes in time. Even if we look at the telegraph poles alongside the track, nothing essential changes in this situation. For when we usually refer to telegraph poles (and the surface of the earth) "at rest," and every object moving relative to them "in motion," we merely use a customary and handy expression without deeper meaning. An observer in a "moving" railway carriage will not come into conflict with his perceptions if he states that the carriage is at rest and the ground and telegraph poles are in motion.

Physicists have found over the course of time that this characteristic of motion, to appear purely relative, is not merely attributable to primitive perception, but rather, that one is justified to call any single thing "at rest" among a multitude of things which are in relative (uniform) motion with respect to one another. Let's think again of a uniformly moving carriage on a straight track. Let the windows be closed airtight, with no light coming in; wheels and tracks are absolutely smooth. Inside the carriage is a physicist with all kinds of apparatus imaginable. We do know that all experiments done by this physicist will come out exactly the same as if the carriage were not not moving or, for that matter, if it were moving at a different velocity. This statement is essentially what physicists call the "principle of relativity." One can phrase this principle in a general fashion as: "The laws of nature perceived by an observer are *independent* of his state of motion."

This statement sounds harmless and self-evident. It would not have excited anybody but for the fact that the laws of the propagation of light, which, having emerged from the recent development in electrodynamics, seem to be incompatible with this principle. The phenomena of the optics of moving bodies lead to the interpretation that light in empty space always propagates with the same velocity, irrespective of the state of motion of the light source. Yet this result seems to be in contradiction with the aforementioned principle of relativity. After all, when a beam of light travels with a stated velocity relative to one observer, then—so it seems—a second observer who is himself traveling in the direction of the propagation of the light beam should find the light beam propagating at a lesser velocity than the first observer does. If this were really true, then the law of light propagation in vacuum would not be the same for two observers who are in relative, uniform motion to each other—in contradiction to the principle of relativity stated above.

This is where the theory of relativity comes in. This theory shows that the law of constancy of light propagation in vacuum can be satisfied simultaneously for two observers, in relative motion to each other, such that the same beam of light shows the same velocity to both of them.

The possibility for such an at first glance paradoxical interpretation can be understood from a more detailed analysis of the physical meaning of spatial and temporal statements. Of special importance to this question is the recognition of the relativity of the concept of simultaneity. Before the theory of relativity, it was believed that the statement "two events happening at two different places are simultaneous" had a clear meaning—clear without a special need to define the concept of simultaneity. A more detailed investigation, which does not skirt the issue of defining simultaneity, showed, however, that the simultaneity of two events is not absolute, but instead can only be defined relative to one observer of a given state of motion. It turns out that two events which are simultaneous with respect to one observer are, in general, not simultaneous with respect to a second observer who is moving relative to the first one. This signifies a fundamental change in our concept of time. (This is the most important, and also the most controversial theorem of the new theory of relativity. It is impossible to enter here into an in-depth discussion of the epistemological and "naturphilosophischen" assumptions and consequences which evolve from this basic principle.)* Those who want to familiarize themselves with a more detailed substantiation and justification can find sufficient instruction—without [1] difficult mathematical derivations—in E. Cohn's pamphlet, "Physikalisches über Raum und Zeit," and in an essay by Jos. Petzoldt, "Die Relativitätstheorie der [2] Physik," in the most recent issue of the *Zeitschrift für positivistische Philosophie*.

By combining the principle of relativity with the results of the constancy of light-velocity in vacuum, one arrives by a purely deductive manner at what is today called

---

Translator's note. The German "Naturphilosophie" embraces far more "metaphysics" than the (Baconian) English "natural philosophy"; thus, a literal translation would be very misleading.

"relativity theory." As an aid for the theoretical deduction of the laws of nature, this theory has already proven itself. Its significance lies in the fact that it provides conditions which every general law of nature must satisfy, for the theory teaches that phenomena in nature are such that the laws do not depend upon the state of motion of the observer onto whom the phenomena are spatially and temporally related.

Of the major results of the theory of relativity two should be mentioned here, because they are also of interest to the layman. First: the hypothesis of the existence of a space-filling medium for light propagation, the so-called light-ether, has to be abandoned. Light appears according to this theory no longer as a state of motion of an unknown carrier, but rather as a physical structure with a physical existence of its own. Second: the theory establishes that the inertia of a body is not an absolutely unchangeable constant, but rather that it grows with the energy content. The important conservation theorems of mass and of energy melt into a single theorem: the energy of a body also determines its mass.

Is the theory of relativity, sketched above, basically complete—or is it only a first step in a continuing development? On this question physicists who value the theory of relativity still hold differing opinions. Nevertheless, weighty arguments speak for the second alternative. We have stated above that the laws of nature are the same for a "uniformly moving" observer as they are for one "at rest." This means it is impossible for an observer to find criteria that would allow him to decide if he is at rest or in a state of uniform motion. "At rest" and "in uniform motion" are physically equivalent. This raises the question of whether the principle of relativity is limited to uniform motion. Could the laws of nature not be such that they are the same for two observers who are in nonuniform motion relative to each other? In recent years it has turned out that such an extension of relativity theory can be carried out, and that it leads to a general theory of gravitation which contains the Newtonian theory as a first approximation. According to this theory, light rays suffer a curvature in a gravitational field; though minute, it is just within the range of astronomical measurement. The future will teach us whether this generalized relativity theory, which from an epistemological aspect is very satisfying, conforms to reality.

Doc. 2

[p. 215]

Covariance Properties of the Field Equations of the Theory of Gravitation Based on the General Theory of Relativity

by Albert Einstein in Berlin and Marcel Grossmann in Zurich

In a paper[1] published in 1913 we based a generalized theory of relativity upon absolute differential calculus in a manner such that it also embraces the theory of gravitation. Two basically different kinds of systems of equations occur in this theory. For a gravitational field considered as given, we first established systems of equations for material (e.g., mechanical, electrical) processes. These equations are covariant under arbitrary substitutions of the space-time variables ("coordinates") and can be considered as generalizations of the corresponding equations of the original theory of relativity. Second, we established a system of equations that determines the gravitational field insofar as the quantities that determine the material processes are given; and this system can be considered a generalization of the Poisson equation of Newton's theory of gravitation. In the original theory of relativity, there is no corresponding system of equations for this. In contrast to the equations mentioned above, we could not demonstrate general covariance for those "gravitational equations." The reason is that their derivation was based (besides the conservation theorems) only upon the covariance with respect to *linear* transformations, and thus left it an open question as to whether or not there exist other substitutions that would transform the equations into themselves.

There are two reasons why the resolution of this question is of particular importance to the theory. The answer to this question gives, first, information on how far the basic idea of relativity theory can be developed; and this is of great import to the philosophy of space and time. And second, the judgment about the value of the theory from the point of view of physics depends to a high degree upon the answer to this question, as is shown by the following consideration.

The entire theory evolved from the conviction that all physical processes in a gravitational field occur just in the same way as they would without it, if an appropriately accelerated (three-dimensional) coordinate system would be introduced ("hypothesis of equivalence"). This hypothesis, which is based upon the experimental fact of the equality between gravitational and inertial mass, gains additional convincing force if the "apparent" gravitational field—which exists relative to the

[p. 216]

[3]

---

[1]   [1]"Entwurf einer verallgemeinerten Relativitätstheorie und einer Theorie der Gravitation" (Leipzig: B. G. Teubner, 1913). [In the following it is abbreviated as "Outline." The paper is
[2]   printed in this journal, *Zeitschrift für Mathematik und Physik,* vol. 62, pp. 225–261.]

accelerated three-dimensional coordinate system—can be viewed as a "real" gravitational field; in other words, if acceleration-transformations (i.e., nonlinear transformations) become permissible transformations in the theory.

At first glance it appears desirable to look for gravitational equations that are covariant toward arbitrary transformations. However, in §2 of the present paper[2] we will show by a simple consideration that the quantities $g_{\mu\nu}$ which characterize the gravitational field cannot completely be determined by generally-covariant equations.

In the following we shall demonstrate that the gravitational equations established by us are generally covariant just to the degree imaginable under the condition that the fundamental tensor $g_{\mu\nu}$ must be completely determined. It follows in particular that the gravitational equations are covariant with respect to quite varied acceleration transformations (i.e., nonlinear transformations).   [5]

§1. The Basic Equations of the Theory

We characterized the energetic response of a physical process by means of a covariant tensor $T_{\mu\nu}$ or its reciprocal contravariant tensor $\Theta_{\mu\nu}$, respectively. This tensor satisfies equations (10) of the "Outline," viz.,

$$\sum_{\mu\nu}\frac{\partial}{\partial x_\nu}(\sqrt{-g}\,\gamma_{\sigma\mu}T_{\mu\nu}) = \tfrac{1}{2}\sum_{\mu\nu}\sqrt{-g}\cdot\frac{\partial\gamma_{\mu\nu}}{\partial x_\sigma}T_{\mu\nu},$$ [6]

or respectively

$$\sum_{\mu\nu}\frac{\partial}{\partial x_\nu}(\sqrt{-g}\,g_{\sigma\mu}\Theta_{\mu\nu}) = \tfrac{1}{2}\sum_{\mu\nu}\sqrt{-g}\cdot\frac{\partial g_{\mu\nu}}{\partial x_\sigma}\Theta_{\mu\nu},$$

and they represent the energy-momentum equations of the material process. All [p. 217] equations of the theory take a particularly comprehensive form if one introduces the quantities

(1) $$\mathfrak{T}_{\sigma\nu} = \sum_\mu \sqrt{-g}\,\gamma_{\sigma\mu}T_{\mu\nu} = \sum_\mu \sqrt{-g}\,g_{\sigma\mu}\Theta_{\mu\nu}\,,$$

which differ from the components of a mixed tensor[3] only by a factor of $\sqrt{-g}$. Conceptually we call them the *complex of energy-density* of the physical process. Our equations above can now be rewritten as

---

[2]Compare also the remark in the appendix of the reprint in *Zeitschr. f. Math. u. Phys.*,   [4]
vol. 62.

[3]Compare §1 of part II of the "Outline."   [7]

[8]

(I)
$$\sum_{\nu}\frac{\partial \mathfrak{T}_{\sigma\nu}}{\partial x_{\nu}} = \tfrac{1}{2}\sum_{\mu\nu}\frac{\partial g_{\mu\nu}}{\partial x_{\sigma}}\gamma_{\mu\varrho}\mathfrak{T}_{\varrho\nu}.$$

If one introduces in place of the energy tensor of the gravitational field the "complex of the energy-density of the gravitational field," that is, the quantities

(2)
$$t_{\sigma\nu} = \sum_{\mu}\sqrt{-g}\cdot\gamma_{\sigma\mu}t_{\mu\nu} = \sum_{\mu}\sqrt{-g}\cdot g_{\sigma\mu}\vartheta_{\mu\nu},$$

then the "Outline" equations (14) and (13) respectively yield

(2a)
$$-2\varkappa t_{\sigma\nu} = \sqrt{-g}\left(\sum_{\beta\varrho\tau}\gamma_{\beta\nu}\frac{\partial g_{\varrho\tau}}{\partial x_{\sigma}}\frac{\partial \gamma_{\varrho\tau}}{\partial x_{\beta}} - \tfrac{1}{2}\sum_{\alpha\beta\varrho\tau}\delta_{\sigma\nu}\gamma_{\alpha\beta}\frac{\partial g_{\varrho\tau}}{\partial x_{\alpha}}\frac{\partial \gamma_{\varrho\tau}}{\partial x_{\beta}}\right),$$

where $\delta_{\sigma\nu} = 0$ or 1 depending on $\sigma \neq \nu$ or $\sigma = \nu$.

In place of the gravitational equations (21) and (18), respectively, of the "Outline" we now get the equations

(II)
$$\sum_{\alpha\beta\mu}\frac{\partial}{\partial x_{\alpha}}\left(\sqrt{-g}\gamma_{\alpha\beta}g_{\sigma\mu}\frac{\partial \gamma_{\mu\nu}}{\partial x_{\beta}}\right) = \varkappa(\mathfrak{T}_{\sigma\nu} + t_{\sigma\nu}).$$

In a manner analogous to the one used in §5 of the "Outline" one can now get from (I) and (II) the general conservation theorems, which can take the form

(III)
$$\sum_{\nu}\frac{\partial}{\partial x_{\nu}}(\mathfrak{T}_{\sigma\nu} + t_{\sigma\nu}) = 0.$$

§2. Remarks on the Choice of the Coordinate System

We want to show now that, completely independent of the gravitational equations we established, a complete determination of the fundamental tensor $\gamma_{\mu\nu}$ of a gravitational field with given $\Theta_{\mu\nu}$ by a generally-covariant system of equations is impossible.

[p. 218]

[9]

We can prove that if a solution for the $\gamma_{\mu\nu}$ for given $\Theta_{\mu\nu}$ is already known, then the general covariance of the equations allows for the existence of further solutions.

Assume a domain L within our four-dimensional manifold such that no "material process" shall exist within L, i.e., where the $\Theta_{\mu\nu}$ therefore vanish. By virtue of the given $\Theta_{\mu\nu}$ the $\gamma_{\mu\nu}$ are assumed determined everywhere outside of L and, therefore, also inside L (assumption a).

Instead of the original coordinates x_{ν} we now imagine new coordinates x_{ν}' introduced in the following manner. Everywhere outside of L we have $x_{\nu}' = x_{\nu}$, but inside L at least for part of it and at least for one index let there be $x_{\nu}' \neq x_{\nu}$.

[10]

Obviously, at least for part of L, this substitution achieves $\gamma_{\mu\nu}' \neq \gamma_{\mu\nu}$. On the other hand we have $\Theta_{\mu\nu}' = \Theta_{\mu\nu}$ everywhere, that is, outside of L, because for this

[11]

domain $x_{\nu}' = x_{\nu}$, and inside of L because for this domain $\Theta_{\mu\nu}' = 0 = \Theta_{\mu\nu}$.

Therefore, if all substitutions would be permitted, the same system of $\Theta_{\mu\nu}$ would have more than one system of the $\gamma_{\mu\nu}$ belonging to it, and this is a contradiction to assumption a).[4]

Once it is understood that an acceptable theory of gravitation implies necessarily a specialization of the coordinate system, it is also easily seen that the gravitational equations, given by us, are based upon a special coordinate system. An x_ν-differentiation of equations (II) and summation over ν, under simultaneous consideration of equations (III), yields the relations

(IV) $$\sum_{\alpha\beta\mu\nu} \frac{\partial^2}{\partial x_\nu \partial x_\alpha}\left(\sqrt{-g}\,\gamma_{\alpha\beta}g_{\sigma\mu}\frac{\partial\gamma_{\mu\nu}}{\partial x_\beta}\right) = 0,$$

and these are four differential conditions for the quantities $g_{\mu\nu}$. We want to write (IV) in the abbreviated form $\qquad B_\sigma = 0$.

These four quantities B_σ do not form a generally-covariant vector, as will be shown [p. 219] in §5. From this one can conclude that the equations $B_\sigma = 0$ represent a true condition for the choice of the coordinate system.[5]

§3. The Hamiltonian Form of the Gravitational Equations

In the following proof of covariance of the gravitational equations we will use the fact that these equations can be brought into the form of a variational principle.[6]

The gravitational equations (II) can be shown to be equivalent to the statement

(V) $$\int\left(\delta H - 2\varkappa\sum_{\mu\nu}\sqrt{-g}\,T_{\mu\nu}\delta\gamma_{\mu\nu}\right)d\tau = 0,$$ [15]

---

[4]This train of thought is already among the notes in the appendix of the reprint of the "Outline" in volume 62 of the *Zeitschr. f. Math. u. Phys.* The claim appended there about [12] the restriction on the coordinate systems, however, does not apply; the restriction to linear substitutions follows from (III) only if the quantities $t_{\sigma\nu}/\sqrt{-g}$ have tensorial character, and this turned out to be not justified. [13]

[5]The equations $B_\sigma = 0$ can also be obtained by imposing the divergence operator upon the gravitational equations in the manner of the absolute differential calculus, thereby using the conservation law of matter.

[6]We owe thanks to Mr. Paul Bernays in Zurich for suggesting the idea of simplifying [14] the proof by such a procedure.

where

(Va)
$$H = \tfrac{1}{2}\sqrt{-g}\sum_{\alpha\beta\tau\varrho}\gamma_{\alpha\beta}\frac{\partial g_{\tau\varrho}}{\partial x_\alpha}\frac{d\gamma_{\tau\varrho}}{dx_\beta}$$

and the $\gamma_{\mu\nu}$ are varied independently of each other such that the variation on the boundary of the four-dimensional domain of integration vanishes.

Utilizing for the calculation of δH the easily understood formulas

$$\delta(\sqrt{-g}) = -\tfrac{1}{2}\sum_{\mu\nu}\sqrt{-g}\,g_{\mu\nu}\delta\gamma_{\mu\nu},$$

$$\delta\left(\frac{\partial g_{\tau\varrho}}{\partial x_\alpha}\right) = \frac{\partial}{\partial x_\alpha}(\delta g_{\tau\varrho}) = -\sum_{\mu\nu}\frac{\partial}{\partial x_\alpha}(g_{\tau\mu}g_{\varrho\nu}\delta\gamma_{\mu\nu}),$$

$$\delta\left(\frac{\partial\gamma_{\tau\varrho}}{\partial x_\beta}\right) = \frac{\partial}{\partial x_\beta}(\delta\gamma_{\tau\varrho}),$$

and considering the fact that variations of surface integrals vanish, one finds

$$\int\delta H d\tau = \int\sum_{\mu\nu\alpha\beta\tau\varrho}\left(-\frac{\partial}{\partial x_\alpha}\left(\sqrt{-g}\,\gamma_{\alpha\beta}\frac{\partial g_{\mu\nu}}{\partial x_\beta}\right) + \sqrt{-g}\,\gamma_{\alpha\beta}\gamma_{\tau\varrho}\frac{\partial g_{\mu\tau}}{\partial x_\alpha}\frac{\partial g_{\nu\varrho}}{\partial x_\beta}\right.$$

[16]
$$\left. + \tfrac{1}{2}\sqrt{-g}\cdot\frac{\partial g_{\tau\varrho}}{\partial x_\mu}\frac{\partial g_{\tau\varrho}}{\partial x_\nu} - \tfrac{1}{4}g_{\mu\nu}\gamma_{\alpha\beta}\frac{\partial g_{\tau\varrho}}{\partial x_\alpha}\frac{\partial g_{\tau\varrho}}{\partial x_\beta}\right)\delta\gamma_{\mu\nu}\cdot d\tau.$$

Utilizing the definitions (14) and (16) of the "Outline," our condition (V) takes

[p. 220] the form

$$\int\sum_{\mu\nu}(D_{\mu\nu}(g) + \varkappa(t_{\mu\nu} + T_{\mu\nu}))\delta\gamma_{\mu\nu}\cdot\sqrt{-g}d\tau = 0.$$

As the $\delta\gamma_{\mu\nu}$ are supposed to be mutually independent, the equations (21) of the "Outline," i.e., our gravitational equations in covariant form, now become a consequence of this condition.

§4. Proof of a Lemma. Adapted Coordinate Systems

Our next task is the investigation of the covariance properties of equation (V). For this purpose we look first for the transformational properties of the integrals

[17]
$$J = \int H d\tau = \int\sqrt{-g}\sum_{\alpha\beta\tau\varrho}\gamma_{\alpha\beta}\frac{\partial g_{\tau\varrho}}{\partial x_\alpha}\frac{\partial\gamma_{\tau\varrho}}{\partial x_\beta}\cdot d\tau.$$

Let there be an arbitrary four-dimensional manifold M, referred to a coordinate system K of the x_ν. Furthermore, we refer the same manifold M to a second coordinate system K' of the x'_ν such that

$$dx_\nu = \sum_\mu \frac{\partial x_\nu}{\partial x'_\mu} dx'_\mu = \sum_\mu p_{\nu\mu} dx'_\mu$$

are the transformation formulas. J and J' shall be the values of the integral above relative to K and K', respectively. This gives

$$J' = \int \sqrt{-g'} \cdot \sum_{\alpha\beta\tau\varrho} \gamma'_{\alpha\beta} \frac{\partial g'_{\tau\varrho}}{\partial x'_\alpha} \frac{\partial \gamma'_{\tau\varrho}}{\partial x'_\beta} d\tau'.$$

Considering that $\sqrt{-g'} \cdot d\tau$ is a scalar, the transformation of J' in terms of the coordinate system K gives

$$J' = \int \sqrt{-g} \sum_{\substack{\mu\nu\alpha\beta \\ r s i k \\ m n \tau \varrho}} \left(\pi_{r\alpha} \pi_{s\beta} \gamma_{rs} \cdot p_{i\alpha} \frac{\partial}{\partial x_i} (p_{m\tau} p_{n\varrho} g_{mn}) p_{k\beta} \frac{\partial}{\partial x_k} (\pi_{\mu\tau} \pi_{\nu\varrho} \gamma_{\mu\nu}) \right) d\tau,$$

hence

$$J' = \int \sqrt{-g} \sum_{\mu\nu m n i k \varrho\tau} \left(\gamma_{ik} \frac{\partial}{\partial x_i} (p_{m\tau} p_{n\varrho} g_{mn}) \frac{\partial}{\partial x_k} (\pi_{\mu\tau} \pi_{\nu\varrho} \gamma_{\mu\nu}) \right) d\tau.$$

In further calculations we shall assume that the coordinate systems K and K' differ only by infinitesimals, i.e., the transformation is infinitesimal. We then have to set

$$x_\nu = x'_\nu - \varDelta x_\nu,$$

therefore

[p. 221]

$$p_{\nu\mu} = \frac{\partial x_\nu}{\partial x'_\mu} = \delta_{\nu\mu} - \frac{\partial(\varDelta x_\nu)}{\partial x'_\mu} = \delta_{\nu\mu} - \frac{\partial(\varDelta x_\nu)}{\partial x_\mu},$$

and

$$\pi_{\mu\nu} = \frac{\partial x'_\nu}{\partial x_\mu} = \delta_{\nu\mu} + \frac{\partial(\varDelta x_\nu)}{\partial x_\mu},$$

where the $\varDelta x_\nu$ are understood as infinitesimal quantities whose squares and products are negligible. This results in

$$J' - J = -4 \int \sqrt{-g} \sum_{m n i k \tau} \gamma_{ik} g_{mn} \frac{\partial \gamma_{\tau n}}{\partial x_k} \frac{\partial^2(\varDelta x_m)}{\partial x_\tau \partial x_i} \cdot d\tau.$$

Partial integration turns this into

(3) $$J' - J = -4 \int \sum_{m n i k \tau} \frac{\partial}{\partial x_\tau} \left(\sqrt{-g} \gamma_{ik} g_{mn} \frac{\partial \gamma_{\tau n}}{\partial x_k} \frac{\partial(\varDelta x_m)}{\partial x_i} \right) d\tau$$

$$+ 4 \int \sum_{m n i k \tau} \frac{\partial}{\partial x_i} \left(\sqrt{-g} \gamma_{ik} g_{mn} \frac{\partial \gamma_{\tau n}}{\partial x_k} \varDelta x_m \right) d\tau$$

[18]

$$- 4 \int \sum_{m n i k \tau} \frac{\partial^2}{\partial x_\tau \partial x_i} \left(\sqrt{-g} \gamma_{ik} g_{mn} \frac{\partial \gamma_{\tau n}}{\partial x_k} \right) \varDelta x_m \cdot d\tau.$$

We notice that the first two integrals can be written as surface integrals which we abbreviated as 0_1 and 0_2, respectively. The factor of Δx_μ in the third integral is readily recognized as B_m according to the notation introduced subsequent to equation (V). Equation (3) in abbreviated notation is now

(3a)
$$J' - J = 0_1 + 0_2 - 4\int \sum_m B_m \Delta x_m \cdot d\tau.$$

The reasons which led to a preference for coordinate systems in which the quantities $B_m = 0$ have been explicated in §2. We want to call these coordinate systems "adapted" to the manifold. It follows from equation (3a) that adapted coordinate systems are selected such that under fixed boundary values of the coordinates and their first derivatives (considered in an arbitrary coordinate system), the integral J becomes an extremum.

We now want to call a transformation between appropriate coordinate systems *admissible*.* When the transformation from K to K' is admissible, equation (3a) yields

$$J' - J = 0_1 + 0_2.$$

[p. 222]

§5. Proof of Covariance of the Gravitational Equations

In §4 we investigated a manifold M. We shall now consider a second manifold $\bar{M}$ which differs only infinitesimally from the former, and for which the quantities $g_{\mu\nu}$ and their first derivatives coincide on the boundary of the domain L with those of the corresponding manifold M. We impose the coordinate systems $\bar{K}$ and $\bar{K}'$ in the following manner:

a) Both coordinate systems be adapted ones for the manifold $\bar{M}$.
b) On the boundary of the domain L, let the coordinates $\bar{x}_\nu$ coincide with the x_ν and the $\bar{x}'_\nu$ with the x'_ν.
c) The coincidence of the coordinate systems shall not only apply on the boundary of the domain, but also for quantities of first order that are infinitesimally close to the boundary; this condition implies that the $\partial(\Delta x_\nu)/\partial x_\sigma$ coincide with the $\partial(\Delta \bar{x}_\nu)/\partial x_\sigma$.

Conditions (b) and (c) do not contradict each other, as can be seen in the following manner. Since manifold M is referred to as an adapted coordinate system, §4 shows the choice of coordinate system K is such as to make the integral J an

---

Translator's note. The word "berechtigt" in the German original is today mathematically understood as "admissible because of justification by previously stated conditions"; modern German texts also favor "zulässig" (= admissible) over the older "berechtigt" (= justified).

extremum under fixed boundary values of the coordinates and their first derivatives. It is then possible to put into the varied manifold $\overline{M}$ an adapted coordinate system $\overline{K}$ which coincides with the coordinate system K outside of L and deviates from K only inside of L; because an extremum of the integral J must also exist for the manifold $\overline{M}$ under unchanged boundary values—whereupon the satisfiability of the equations $B_m = 0$ follows also for the varied manifold.

Let us assume the coordinate systems K and K', which are used in the manifold M, are both adapted. According to (3b) the equations

$$J' - J = 0_1 + 0_2,$$
$$\overline{J}' - \overline{J} = \overline{0}_1 + \overline{0}_2,$$

or after subtraction

$$(\overline{J}' - J') - (\overline{J} - J) = (\overline{0}_1 - 0_1) + (\overline{0}_2 - 0_2).$$

are valid.

The specifications (b) and (c) and the relations between M and $\overline{M}$, together with (3), imply that both $\overline{0}_1 - 0_1$ and $\overline{0}_2 - 0_2$ vanish.

$\overline{M}$ can be called a manifold developed by variation of M. Therefore, we denote   [p. 223] analogously

$$\overline{J} - J = \delta_a J,$$
$$\overline{J}' - J' = \delta_a J',$$

and consequently get

(4) $\delta_a J' = \delta_a J.$

The index α is meant to express that, together with the manifold, the coordinate system is co-varied such that the varied coordinate system and the varied manifold are always adapted relative to each other, while on the boundary the coordinate system remains, of course, unvaried (so-called "adapted variation").

Our aim is to demonstrate that an equation

$$\delta J' = \delta J$$

is satisfied for *any* variation of the manifold, not just for an *adapted* variation as equation (4) says. However, we can let any variation of $g_{\mu\nu}$ evolve from an adapted one if we follow it with another variation of the coordinate system. It turns out that for a variation of $g_{\mu\nu}$, equivalent to just one variation of the coordinate system, the variation of J, denoted by $\delta_k J$, vanishes, provided we assume the variations δx_ν and their first derivations vanish on the boundary of the domain, and also provided furthermore that the coordinate system to be varied is an adapted system. The reason is that equation (3a) leads to the direct consequence

$$\delta_k J = O_1 + O_2 - 4\int' \sum_m B_m \delta x_m \cdot d\tau = 0.$$

Therefore, we can associate to equation (4) the equation

(5) $$\delta_k J' = \delta_k J = 0.$$

From these two equations—together with the fact that the superposition of an adapted variation and a mere coordinate variation is equivalent to any variation of the $\gamma_{\mu\nu}$—it follows that for such arbitrary variation one has

(6) $$\delta J' = \delta J.$$

From this equation, however, one can prove in a simple manner the covariance of equation (V). The $\delta\gamma_{\mu\nu}$ are after all contravariant, the $T_{\mu\nu}$ covariant, and thus [p. 224] $\sum\limits_{\mu\nu} T_{\mu\nu}\delta\gamma_{\mu\nu}$ is a scalar; and the same is true of $\sqrt{-g} \cdot d\tau$. Consequently,

{1} (7) $$\int \sqrt{-g'} \cdot \sum_{\mu\nu} T'_{\mu\nu}\delta\gamma'_{\mu\nu}\cdot d\tau' = \int \sqrt{-g} \cdot \sum_{\mu\nu} T_{\mu\nu}\delta\gamma_{\mu\nu}\cdot d\tau.$$

It follows from (6) and (7) that equation (V) is covariant toward all admissible transformations of the coordinate systems, provided the variations are chosen such that the $\delta\gamma_{\mu\nu}$ *and their first derivatives* vanish on the boundary of the domain. The variational theorem whose covariance has been proven in this manner is then a little less general than the one used in §3 for the derivation of the gravitational equations. However, a glance at the development of §3 shows that the derivation of those gravitational equations is not hampered by these restricting boundary conditions of the variation. With this it is proven that:

The gravitational equations (II) are covariant under all admissible transformations of the coordinate systems, i.e., under all transformations between coordinate systems which satisfy the conditions

(IV) $$B_\sigma = \sum_{\alpha\beta\mu\nu}\frac{\partial^2}{\partial x_\nu \partial x_\alpha}\left(\sqrt{-g}\cdot\gamma_{\alpha\beta}g_{\sigma\mu}\frac{\partial\gamma_{\mu\nu}}{\partial x_\beta}\right) = 0 .$$

We have claimed in §2 that the expressions B_σ do not form a covariant vector. We shall give the proof of it only now because it is especially simple when we utilize the results we have just obtained. All coordinate systems used in the foregoing (and called adapted systems) would be arbitrary coordinate systems if the B_σ were covariant. None of the steps in the proof would lose its convincing force due to this circumstance. The final result of the proof would be the completely general covariance of the gravitational equations. Then the following would be a general mixed tensor

{2} $$\mathfrak{A}_{\sigma\nu} = \frac{1}{\sqrt{-g}}\left(\sum_{\alpha\beta\mu}\frac{\partial}{\partial x_\alpha}\left(\sqrt{-g}\gamma_{\alpha\beta}g_{\sigma\mu}\frac{\partial\gamma_{\mu\nu}}{\partial x_\beta}\right) - \varkappa t_{\sigma\nu}\right) = \frac{\varkappa}{\sqrt{-g}}\cdot\mathfrak{T}_{\sigma\nu}$$

and consequently

$$\sum_{\sigma} \mathfrak{A}_{\sigma\sigma} = -\sum_{\alpha\beta\tau\varrho}\left(\frac{1}{\sqrt{-g}}\frac{\partial}{\partial x_\alpha}\left(\sqrt{-g}\,\gamma_{\alpha\beta}\frac{\partial \lg g}{\partial x_\beta}\right) - \tfrac{1}{2}\gamma_{\alpha\beta}\frac{\partial g_{\varrho\tau}}{\partial x_\alpha}\frac{\partial \gamma_{\varrho\tau}}{\partial x_\beta}\right)$$

would be a scalar under arbitrary transformations. However, as is known from the theory of differential invariants,[7] this quantity does not coincide with the only [p. 225] differential invariant of second order, viz.,

$$\sum_{imk}{}' \gamma_{im}\,\{ik,km\}.\tag{19}$$

This new theory of gravitation gains convincing power by the far-reaching covariance of the gravitational equations, even if the foregoing deliberations may not provide complete transparency of adapted coordinate systems and admissible transformations. We believe to have shown that the covariance of the equations is the optimum imaginable, since the conditions $B_\sigma = 0$, by which we restricted the coordinate systems, are a direct consequence of the gravitational equations.

––––––––––––

Additional notes by translator

{1}   "$\delta\gamma'_\mu$" on the left-hand side has been corrected to "$\delta\gamma'_{\mu\nu}$."
{2}   The parenthesis ")" was missing after $\partial\gamma_{\mu\nu}/\partial x_\beta$.

––––––––––––

[7] See §4 of part II of the "Outline."

Doc. 3

[p. 739] [The inaugural lectures of those members who newly entered the Academy since the Leibniz-Session of 1913 followed this lecture.]

Inaugural Lectures and Responses

Inaugural Lecture of Mr. Einstein

Most honored colleagues!

Please accept first my heartfelt gratitude for the greatest favor you could have bestowed upon a person such as I. By calling me into your Academy you have put me into a position that allows me to devote myself wholly to scientific studies and [1] frees me from the distractions and tribulations of a practical profession. I ask you to believe in my feelings of gratitude and the assiduousness of my striving, even if the fruits of my efforts may appear to you as meager ones.

Allow me to append a few general remarks about the position which my field of work, theoretical physics, takes vis-à-vis experimental physics. A mathematician friend recently said to me, half jokingly, "The mathematician knows some things, no [p. 740] doubt, but of course not those things one usually wants to get from him." The theoretical physicist is in a very similar position when the experimental physicist asks him for advice. So, what are the reasons for this strange lack of adaptability?

[2]   The methodology of the theoretician mandates implicitly that he use as his basis general assumptions, so-called principles, from which he can then deduce conclusions. His activity, therefore, has two parts: first, he has to ferret out these principles, and second, he has to develop the conclusions that can be deduced from these principles. His school provides him with excellent tools with which to fulfill the second-named task. Consequently, when the first one of his tasks is already solved for some area, or rather some complex of phenomena, then sufficient diligence and insight assure that success will not be denied to him. But the former task, namely to establish these principles which can serve as the basis of his deductions, is one of a completely different kind. Here there is no learnable, systematically applicable method which would led him to the objective. The researcher must rather eavesdrop on nature to become privy to these general principles, by recognizing in larger sets of experiential facts certain general traits that can then be strictly and precisely formulated.

Once this formulation is achieved, a chain of conclusions sets in, often with unforeseen connections, far transcending the domain of facts from which the principle has been wrested. However, as long as the principles that must serve as the basis for the deduction remain undiscovered, the individual experimental fact is of no help to the theoretician. In fact, he cannot even do much with individual empirically established general laws. Instead, he must rather remain in a state of helplessness vis-à-vis the individual results of empirical research until principles reveal themselves to him so that he can make them the basis of deductive developments.

Presently, theory is in just that position with respect to the laws of heat radiation and molecular movement at low temperatures. Just fifteen years ago, nobody doubted that GALILEI-NEWTONIAN mechanics and the MAXWELLIAN theory of the electromagnetic field as a basis of molecular movements would lead to a correct description of the electrical, optical, and thermal properties of matter. Just then PLANCK showed that the establishment of a law of heat radiation consistent with experience, required a method of calculation whose incompatibility with the principles of classical [p. 741] mechanics was becoming ever more apparent. With this method of calculation PLANCK introduced to physics the so-called quantum-hypothesis, which has since found outstanding confirmation. With this quantum hypothesis he toppled classical mechanics for those situations where we have sufficiently small masses with sufficiently small velocities and sufficiently large accelerations—so that today we can accept the laws of motion established by GALILEI and NEWTON only as limit-laws. But, despite the most strenuous efforts of theoreticians, we have not succeeded in replacing the principles of mechanics with those that are adequate for PLANCK's law of heat radiation or the quantum hypothesis. As incontrovertibly as it has been proven that heat derives from molecular movement, we still must concede today that the basic laws of this movement confront us in a similar manner as the planetary movements confronted the pre-NEWTONIAN astronomers.

I just pointed out a set of facts for whose theoretical treatment we lack the principles. But it can happen as well that clearly formulated principles lead to conclusions that are completely or almost completely outside the domain of facts that are presently accessible to our experience. In this case, protracted experimental research may be needed in order to find out whether or not the theoretical principles correspond to reality. Such a case offers itself with the theory of relativity.

An analysis of the temporal and spatial base concepts has shown that the theorem of a constant light velocity in vacuum, which follows from the optics of moving matter, does not at all force us toward a theory of a light-ether at rest. Instead, it has been possible to establish a general theory that takes cognizance of the fact that the translatory motion of the earth is never noticeable in the experiments conducted on earth. Thereby we use the relativity principle, which states: laws of nature do not change their form when one changes from the original (admissible) coordinate system to a new one that is in uniform translatory motion relative to the former. This theory found noteworthy confirmation in experience and has led to a simplification of the theoretical description of whole complexes of facts that were already connected.

From a theoretical point of view however, this theory does not grant full satisfaction because the relativity principle that has been stated above favors *uniform* motion. If it is true after all that *uniform* motion cannot have an absolute meaning from the physical point of view, we then have the obvious question whether or not this statement can be extended to nonuniform motions. It turned out that one arrives at a very distinct extension of relativity theory when one uses a relativity principle as a basis that is extended in this sense. In this manner one is led to a general theory of gravitation that includes its dynamics. At this moment, however, we do not have the experimental material necessary to test the justification of the introduction of this basic principle.

We have determined that inductive physics has questions for deductive physics and vice versa; and eliciting the answers will require the application of our utmost efforts. May we, by means of united efforts, soon succeed in advancing toward conclusive progress.

Doc. 4

Remarks on P. Harzer's Paper:
"On the Dragging of Light in Glass and on Aberration"

Doc. 4 is not translated for this volume.

Doc. 5

[p. 820]

Contributions to Quantum Theory

by A. Einstein

(Presented in the session of July 24, 1914)

(see p. 735 above)

Two considerations are presented here which—in some sense—belong together, as they show how far the most important newer results of the theory of heat, viz. [1] PLANCK's radiation formula and NERNST's theorem, can be derived in a purely thermodynamical manner, utilizing basic ideas of quantum theory but not enlisting [2] the help of the BOLTZMANN principle. Insofar as the following deductions correspond to reality, the theorem by NERNST is valid for chemically pure, crystallized substances, but not for mixed crystals. Nothing can be said about amorphous substances because of the still extant vagueness of the nature of the amorphous state.

In order to justify the attempt presented here to grasp the NERNST theorem theoretically, I must point to the fact that all efforts to theoretically derive the NERNST theorem in a thermodynamical manner, utilizing the experimental theorem [3] that the heat capacity vanishes at $T = 0$, have failed completely. I am very willing, [4] if colleagues so desire, to substantiate this claim against the individual attempts of proof.

§1. Thermodynamical Derivation of PLANCK's Radiation Formula. We consider a chemically uniform gas whose molecules carry one resonator[1] each. The energy of these resonators shall not assume every arbitrary value but only certain discrete values ε_σ (per mole). I take the liberty to consider two molecules chemically distinct, [p. 821] i.e., in principle separable by means of semipermeable walls, if their resonator energies ε_σ and ε_τ are different. By doing so I can view the gas that was originally seen as uniform as also a mixture of different gases whose constituents are characterized by distinct values ε_σ. By imposing the condition that this mixture is in thermodynamical equilibrium versus all changes in the ε-values of the molecules, I obtain the statistical law according to which the resonator energies of the molecules are partitioned. Then, retroactively treating the resonator energies again as "thermal energy," I get that portion of the specific heat of the gas which can be traced to the resonators on the molecules.

Let n_0, n_1, n_2 etc. be the moles of molecules and ε_0, ε_1, ε_2 etc. be their

---

[1]By "resonator" we mean here quite generally a carrier of inner molecular energy, without delineating its precise characteristics here in advance.

corresponding resonator energies. Energy U and entropy S of the mixture are then given by the expressions

$$U = \sum_\sigma n_\sigma \left\{ c\,T + u_0 + \varepsilon_\sigma \right\}$$

$$S = \sum_\sigma n_\sigma \left\{ c \lg T + R \lg V \right\} + \sum_\sigma n_\sigma \left\{ s_\sigma - R \lg n_\sigma \right\}. \qquad [5]$$

The specific heat c (at constant volume) per mole—following the idea sketched above—is to be taken at constant resonator energy ε_σ, which is the same for all components. s_σ is the entropy-constant of the gas type with resonator energy ε_σ; and this constant can a priori have different values for each σ. Next we have to form the free energy $F = U - TS$ and have to state the condition that for every reaction to be considered

[6]

$$\delta F = \delta\,(U - TS) = 0\,.$$

We take into account the totality of possible resonator reactions by considering for each ν the reaction

$$\delta n_0 = -1$$
$$\delta n_\sigma = +1\,.$$

In this manner one obtains the system of equations

$$\left(s_\sigma - \frac{\varepsilon_\sigma}{T} - R \lg n_\sigma \right) - \left(s_0 + \frac{\varepsilon_0}{T} - R \lg n_0 \right) = 0 \qquad [7]$$

or

$$\frac{n_\sigma}{n_0} = e^{(s'_\sigma - s'_0) - \frac{\varepsilon_\sigma - \varepsilon_0}{RT}}. \qquad 1)$$

This is the equilibrium distribution we were looking for, where $s'_\sigma = \dfrac{s_\sigma}{R}$.   [p. 822]

Let the resonator under consideration now be a monochromatic one with one degree of freedom and with frequency ν. In order to arrive at PLANCK's formula for the mean energy of such a structure, we have to introduce two hypotheses:

1. The entropy constants of all components of our mixture are equal even though the components differ in resonator energy; i.e., for all σ

$$s_\sigma = s_0\,.$$

This precondition corresponds to NERNST's theorem.   [8]

2. The resonator energy (per mole) is an integral multiple of $Nh\nu$:

$$\varepsilon_\sigma = \sigma N h\nu\,.$$

This is the quantum hypothesis for a monochromatic structure.

Based upon these hypotheses we get

$$n_\sigma = n_0 e^{-\frac{\sigma h\nu}{KT}}, \qquad 1\,a)$$

from which follows

[9]
$$\bar{\varepsilon} = \frac{\sum\limits_{0}^{\infty} \varepsilon_\sigma n_\sigma}{\sum\limits_{0}^{\infty} n_\sigma} = +RT^2 \frac{d}{dT}\{\lg \sum e^{-\frac{\varepsilon''}{RT}}\} = \frac{\sum \sigma N h\nu e^{-\frac{\sigma h\nu}{KT}}}{\sum e^{-\frac{\sigma h\nu}{KT}}}$$

$$= N\frac{h\nu}{e^{\frac{h\nu}{KT}} - 1}. \qquad 2)$$

[10]
This is PLANCK's formula for the mean energy of a one-dimensional monochromatic resonator.[2]

The fact that in this manner one arrives at PLANCK's formula is noteworthy in more than one respect. It appears, for one, that the concepts of physical and chemical [p. 823] change of a molecule lose their principal contrast. The quantum-like change of the physical state of a molecule seems not to be different in principle from the chemical change. One can go even further. The laws of BROWNIAN movement have led to blurring the principal contrast between a molecule and an arbitrarily extended physical system; DEBYE, on the other hand, has shown that arbitrarily extended [12] systems can be described with great success through quantum-theoretically different states. Even the quantum-like change of state of an extended system can be understood in a manner analogous to the chemical change of a molecule. In this sense, equations 1) and 2) can unhesitatingly be applied to proper vibrations of arbitrarily extended systems.

Furthermore, imagine the component of resonator energy ε_σ in the mixture to be separated from the others. Our derivation is based upon the assumption that this is possible, in principle, without change of the resonator energy. This assumption is analogous to the one in the theory of chemical equilibrium, namely that a chemical mixture can be separated into its simple constituents without chemical reactions taking place. One can now imagine changing the temperature of the isolated component while the resonator energy ε_σ stays constant. How far this is possible in practice depends upon the "reaction rate" with which the molecules change their ε. The component can be arbitrarily cooled without any loss of energy ε_σ if this rate is sufficiently small. In that case we have a structure similar to a radioactive one. For a principal understanding of the radioactive phenomena of diamagnetism it is, therefore, not necessary to assume the existence of a zero-point energy in the sense [13] of PLANCK. It is sufficient to assume the existence of a quantum-like partitioned energy that reaches its thermal equilibrium sufficiently slowly.

---

[2]It has been pointed out to me that BERNOULLI gave a similar derivation of PLANCK's [11] formula (ZS. f. Elektrochem. 20, 269, 1941). But BERNOULLI based his result on two erroneous formulas [4] and 5) of his paper].

On the other hand, this derivation is also useful for a better understanding of
NERNST's theorem, as can be concluded from the fact that the derivation of PLANCK's
formula required hypothesis 1. In order to get a better understanding of this
connection, we shall try to extend our analysis to structures of more than one degree [p. 824]
of freedom. How would he reason if the resonator of energy ε_σ would have two
degrees of freedom? In the derivation of 1), it was completely immaterial how the
structure of energy ε_σ was constituted; therefore, this equation still is to be kept. In
a similar manner one could keep hypothesis 2. If we also base ourselves upon
hypothesis 1, we again get equation 2) for the mean energy, that is to say, only half
of what is correct for the two-dimensional resonator. In order to obtain the correct
result here, the entropy constants of mixture components characterized by different ε_σ
can no longer be considered equal.

This is immediately understood if we replace the monochromatic resonator of two
degrees of freedom by two resonators of one degree of freedom each. The resonator
energy $\varepsilon_{\sigma\tau}$ is then to be taken as

$$\varepsilon_{\sigma\tau} = (\sigma + \tau)\, h\nu.$$

We obtain the correct value for mean energy if we always assume the two kinds of
molecules σ, τ and σ', τ' as separable unless $\sigma = \sigma'$, $\tau = \tau'$, and the components of
the mixture satisfy hypothesis 1. We then get

$$\frac{n_{\sigma\tau}}{n_{00}} = e^{-\frac{\varepsilon_{\sigma\tau} - \varepsilon_{00}}{RT}}$$

$$\bar{\varepsilon} = RT^2 \frac{d}{dT}\left\{ \lg \sum_\sigma \sum_\tau e^{-\frac{(\sigma+\tau) h\nu}{RT}} \right\} = 2N \frac{h\nu}{e^{\frac{h\nu}{RT}} - 1}. \qquad 2\,a)$$

That hypothesis 1 as an equivalent to NERNST's theorem must not be used as a basis
when the energy carrier ε_σ has two degrees of freedom and, if the state of the
molecule is characterized only by the energy ε_σ (without regard to how this energy
is distributed between the degrees of freedom), is probably connected with the
following: *Hypothesis 1 is permissible if and only if the state of the molecule—denoted by index σ in 3)—is completely characterized by quantum-theoretic*   [14]
standards such that it can be realized in only one way. In this case the correct   [15]
distribution law is   [p. 825]

$$\frac{n_\sigma}{n_0} = e^{-\frac{\varepsilon_\sigma - \varepsilon_0}{RT}}. \qquad 1\,a)$$

If we limit ourselves to the case of only discrete possibilities of realization of the
"inner state" of the molecule, to which ε_σ refers, then we have to stay with 1a);
assuming one chooses for each possibility of realization a separate index (resp. a

separate index system). Under this limitation, 1) remains valid[*] not only for a "molecule" in the ordinary sense, but also for a physical system looked at quantum-theoretically in the sense of JEANS-DEBYE. In this manner one stays safely within the bounds of experimentally verified quantum theory.

[16]

Energy ε_σ is referred to the gram-mole. The quantity $\dfrac{\varepsilon_\sigma}{N} = \varepsilon_\sigma^*$ for an individual molecule will always be introduced when the "molecule" is a system to be treated as a single structure accessible to experience. In this case one has to set

$$\frac{n_\sigma}{n_0} = w_\sigma = e^{-N\frac{\varepsilon_\sigma^* - \varepsilon_0^*}{RT}} . \qquad 1\,b)$$

§2. *Entropy. The* NERNST *Theorem.* We now think of a physical system to play the role of the "molecule" of the previous paragraph. For this purpose it is necessary to imagine the system not in isolation but rather linked with an infinitely large heat reservoir. We assume the system to be thermodynamically determined by temperature and one parameter λ (e.g., volume), or by several parameters. The possible states of the system and, therefore, its realizable energy values ε_σ^* will then depend upon the parameter values λ. Under constant λ we will have to accept equation 3b) as valid. The mean energy of the system is then given by

[17]

$$\overline{\varepsilon^*} = \frac{\sum \varepsilon_\sigma^* w_\sigma}{\sum w_\sigma} = \frac{\sum \varepsilon_\sigma^* e^{-\frac{N\varepsilon_\sigma^*}{RT}}}{\sum e^{-\frac{N\varepsilon_\sigma^*}{RT}}} = \frac{R}{N} T^2 \frac{d}{dT} \lg\{\sum e^{-\frac{N\varepsilon_\sigma^*}{RT}}\}. \qquad 3)$$

[p. 826]

From this follows the entropy at constant λ as a function of T:

$$S - S_0 = \int_{T_0}^{T} \frac{d\overline{\varepsilon^*}}{T} = \left|\frac{\overline{\varepsilon^*}}{T}\right|_{T_0}^{T} + \int_{T_0}^{T} \frac{\overline{\varepsilon^*}}{T^2}\, dT,$$

or with suitable choice of the value of S_0:

$$S = \frac{\overline{\varepsilon^*}}{T} + \frac{R}{N} \lg\{\sum e^{-\frac{N\varepsilon_\sigma^*}{RT}}\}. \qquad 4)$$

If the system has a large number of degrees of freedom, 1b) implies in a well-known manner that only those states of the system have to be considered which correspond to a small range of ε_σ^*. With the evaluation of the sum in 4) one can confine oneself to this narrow range within which ε_σ is set constant. One then gets

$$S = \frac{R}{N} \lg Z, \qquad 4\,a)$$

where Z is the number of elementary states possible under quantum theory; and the

[*]*Translator's note.* The word "gelten" = valid is missing in the German original.

energy value associated with Z is $\overline{\varepsilon^*}$.[3] Equation 4a) expresses BOLTZMANN's principle in the formulation of BOLTZMANN-PLANCK.

Up to now, we only considered changes of state under constant λ. The question hence arises whether or not 4a) remains valid under changes of state in the system when λ changes. This question cannot be answered without making special hypotheses. The most natural hypothesis which offers itself is EHRENFEST's adiabatic [18] hypothesis, which can be formulated thus:

With reversible adiabatic changes of λ every quantum-theoretically possible state changes over into another possible state.

It is a consequence of this hypothesis that the number Z of quantum-theoretically possible realizations does not change during adiabatic processes. Since the same is [19] true for S, we have to conclude from EHRENFEST's adiabatic hypothesis (which is a natural generalization of WIEN's displacement law) that the BOLTZMANN principle in [p. 827] formula 4a) has general validity. The entropy of a system has, therefore, for all [20] (thermodynamically defined) states of a system—provided they are quantum-theoretically realizable in the same number of ways—the same value.

We ask ourselves now if we can deduce some expectation with respect to the range of validity of NERNST's theorem. Let there be a physical system at absolute zero in two thermodynamically defined states A_1 and A_2. We can compare the entropy values of these states if we can find the number Z of quantum-theoretically possible realizations of the system.

We can consider the state of the system at absolute zero as quantum-theoretically and molecular-theoretically (i.e., in its micro-state) as completely described if the positions of the centers of gravity of the individual atoms which constitute the system are given (the atoms imagined are to be numbered). Z then is the number which says how many of these micro-states are possible without the system leaving its thermodynamically defined state.

If all phases of the system are chemically homogeneous and crystallized in spatial lattices such that it is determined in which positions the atoms of various kinds are situated, then I can change from one micro-state to another one, contained in Z, only in that manner that I exchange positions of atoms of the same kind. States caused by the exchanges of atoms of different kinds are, in contrast, not to be counted. If the system comprises of n_1 molecules of the first kind, n_2 of the second, etc., Z has the value

$$Z = n_1! \, n_2! \, \ldots$$

From this follows, under consideration of 4a), that the entropy in all these states has the same value. Therefore, *the validity of NERNST's theorem in PLANCK's*

---

[3]This is equivalent to a transition from a "canonical" to a "micro-canonical" ensemble.

formulation, i.e., for chemically simple, crystallized substances, follows.

[p. 828]   However, when two kinds of atoms form a mixture, we do not leave the thermodynamic state of the system if we exchange two atoms of different kinds. One then gets

$$Z = (n_1 + n_2)!,$$

while for substances in a nonmixed state

$$Z = n_1! \, n_2!$$

would be correct. The equality of entropies at absolute zero, therefore, does not hold in this case. Instead, one gets for the difference in entropies between the substances in mixed and in nonmixed states the value

$$\frac{R}{N} \lg \frac{(n_1 + n_2)!}{n_1! \, n_2!},$$

[21]   which, of course, reduces to $2R\lg 2$ if $n_1 = n_2 = N$.

Doc. 6

Response to Paul Harzer's Reply

Not translated for this volume.

Doc. 7

Lecture Notes for Course on Relativity at the University of Berlin, Winter Semester 1914–1915

Not translated for this volume.

Doc. 8

[2]
Manifesto to the Europeans

[1]
(mid-October 1914)

While technology and traffic clearly drive us toward a factual recognition of international relations, and thus toward a common world civilization, it is also true that no war has ever so intensively interrupted the cultural communalism of cooperative work as this present war does. Perhaps we have come to such a salient awareness only on account of the numerous erstwhile common bonds, whose interruption we now sense so painfully.

{1}
Even if this state of affairs should not surprise us, those whose heart is in the least concerned about common world civilization, would have a doubled obligation to fight for the upholding of those principles. Those, however, of whom one should expect such convictions—that is, principally scientists and artists—have thus far almost exclusively uttered statements which would suggest that their desire for the maintenance of these relations has evaporated concurrently with the interruption of the relations. They have spoken with explainable martial spirit—but spoken least of

{2}
all of peace.

Such a mood cannot be excused by any national passion; it is unworthy of all that which the world has to date understood by the name of culture. Should this mood achieve a certain universality among the educated, this would be a disaster.

It would not only be a disaster for civilization, but—and we are firmly convinced of this—a disaster for the national survival of individual states—the very cause for which, ultimately, all this barbarity has been unleashed.

Through technology the world has become *smaller*; the *states* of the large peninsula of Europe appear today as close to each other as the *cities* of each small Mediterranean peninsula appeared in ancient times. In the needs and experiences of every individual, based on his awareness of a manifold of relations, Europe—one could almost say the world—already outlines itself as an element of unity.

It would consequently be a duty of the educated and well-meaning Europeans to at least make the attempt to prevent Europe—on account of its deficient organization as a whole—from suffering the same tragic fate as ancient Greece once did. Should Europe too gradually exhaust itself and thus perish from fratricidal war?

The struggle raging today will likely produce no victor; it will leave probably only the vanquished. Therefore, it seems not only *good*, but rather bitterly *necessary, that educated men of all nations* marshall their influence such that—whatever the still uncertain end of the war may be—the *terms of peace shall not become the wellspring of future wars*. The evident fact that through this war all European relational conditions slipped into an *unstable and plasticized state* should rather be used to create an organic European whole. The technological and intellectual conditions for this are extant.

It need not be deliberated herein by which manner this (new) ordering in Europe is possible. We want merely to emphasize very fundamentally that we are firmly convinced that the time has come where *Europe must act as one in order to protect her soil, her inhabitants, and her culture.*

To this end, it seems first of all to be a necessity that all those who have a place in their hearts for European culture and civilization, in other words, those who can be called in *Goethe's* prescient words *"good Europeans,"* come together. For we must not, after all, give up the hope that their raised and collective voices—even beneath the din of arms—will not resound unheard, especially, if among these "good Europeans of tomorrow," we find all those who enjoy esteem and authority among their educated peers.

But it is necessary that the Europeans first come together, and if—as we hope—enough *Europeans in Europe* can be found, that it is to say, people to whom Europe is not merely a geographical concept, but rather, a dear affair of the heart, then we shall try to call together such a union of Europeans. Thereupon, such a union shall speak and decide.

To this end we only want to urge and appeal; and if you feel as we do, if you are likemindedly determined to *provide the European will the farthest-reaching possible resonance*, then we ask you to please send your (supporting) signature to us.

Additional notes by translator

{1} One may surmise that in 1914 all European nations understood "world civilization" as "western civilization."

{2} The German original is here syntactically faulty and not clear.

Doc. 9

[p. 1030] Plenary Session of Nov. 19, 1914—Communications of the phys.-math section of Oct. 29

The Formal Foundation of the General Theory of Relativity

A. Einstein

(Submitted on October 29, 1914 [see p. 965 above])

[1] In recent years I have worked, in part together with my friend Grossman, on a generalization of the theory of relativity. During these investigations, a kaleidoscopic mixture of postulates from physics and mathematics has been introduced and used as heuristical tools; as a consequence it is not easy to see through and characterize the theory from a formal mathematical point of view, that is, only based upon these papers. The primary objective of the present paper is to close this gap. In particular, it has been possible to obtain the equations of the gravitational field in a purely covariance-theoretical manner (section D). I also tried to give simple derivations of the basic laws of absolute differential calculus—in part, they are probably new ones (section B)—in order to allow the reader to get a complete grasp of the theory without having to read other, purely mathematical tracts. As an illustration of the mathematical methods, I derived the (Eulerian) equations of hydrodynamics and the field equations of the electrodynamics of moving bodies (section C). Section E shows that Newton's theory of gravitation follows from the general theory as an approximation. The most elementary features of the present theory are also derived inasfar as [2] they are characteristic of a Newtonian (static) gravitational field (curvature of light rays, shift of spectral lines).

A. The Basic Idea of the Theory

§1. Introductory Considerations

The original theory of relativity is based upon the premise that all coordinate systems in relative uniform translatory motion to each other are equally valid and equivalent [p. 1031] for the description of the laws of nature. When viewed from experience, this theory gains its main support from the fact that when we carry out experiments on earth we

---

Translator's note. Corrections have been made here to typographical errors that occurred in the original document. This applies to corrections mentioned in editorial notes [6], [7], [9]—[14], [19]—[22], [25]—[31], [38], [40], [41], [43]—[46], [49], [50], and to additional notes {5}—{20} at the end of this document.

do not have the slightest indication that the earth moves around the sun with a considerable velocity.

But the confidence we have in relativity theory also has another root. One cannot easily close one's mind to the following consideration. When K' and K are two coordinate systems in relative uniform translatory motion to each other, then these systems are completely equivalent from a kinematic point of view. We, therefore, look in vain to find a sufficient reason why one system should be more suitable as a system of reference for the formulation of laws of nature than another. Instead, we feel urged to postulate the equivalence of both systems.

But this argument immediately spawns a counterargument. The kinematic equivalence of these two coordinate systems is by no means limited to the case where the two systems K and K' are in *uniform translatory motion* to each other. From the kinematic aspect, this equivalence also obtains, i.e., just as well, when both systems rotate uniformly relative to each other. One feels pressed toward the idea that the existing theory of relativity is in need of a considerable generalization such that the apparently unjust preference for uniform translatory motion over other relative motions has to be eliminated from the theory. Everyone who studied this subject in detail must feel the desire for such an extension of the theory.   [3]

At first glance it appears that such extension of the theory of relativity would have to be rejected on physical grounds. Because: let there be a coordinate system K that is admissible in the sense of Galilei-Newton, and another system K' which is in uniform rotation relative to K; centrifugal forces will then act on masses which are at rest relative to K' while the masses which are at rest relative to K do not suffer such forces. Already Newton viewed this as proof that the rotation of K' had to be interpreted as "absolute"; in other words, that K' cannot claim the same right as K to be considered "at rest." This argument, however, is—as especially E. Mach has shown—not cogent. This is because we need not necessarily derive the existence of   [4] centrifugal forces from a motion of K'; instead, we can just as well derive them from the averaged rotational movement of distant ponderable masses in the environment, but relative to K', thereby treating K' as "at rest." If the Newtonian laws of [p. 1032] mechanics and gravitation do not allow for such interpretation, it may well be founded in deficiencies of this theory. The following important argument also speaks in favor of a relativistic interpretation. The centrifugal force which acts under given conditions upon a body is determined by precisely the same natural constant that also gives its action in a gravitational field. In fact, we have no means to distinguish a "centrifugal field" from a gravitational field. We thus always measure as the weight of a body on the surface of the earth the superposed action of both fields named above; and we cannot separate their actions. In this manner, the point of view to interpret the rotating system K' as *at rest* and the centrifugal field as a gravitational

field gains justification by all means. This interpretation is reminiscent of the one in   [5]
the original (more special) theory of relativity where the ponderomotively acting
force, upon an electrically charged mass which moves in a magnetic field, is the
action of that electric field which is found at the location of the mass as seen by a
reference system at rest with the moving mass.

From what has been said, one already sees that gravitation must play a
fundamental role in any theory of relativity that is extended along the lines we have
indicated above. After all, if one changes by a mere transformation from one system
of reference K to another one of K', then there exists a gravitational field relative to
K' which relative to K need not be there at all.

There arises the natural question: which systems of reference and which
transformations should be considered as "admissible" in a generalized theory of
relativity? This question can only be answered much later (section D). For the time
being we take the position to admit all coordinate systems and transformations that
are compatible with the conditions of continuity, as is always demanded for theories
in physics. It will turn out that the theory of relativity allows for a rather far-reaching
generalization that is free of almost any arbitrariness.

§2. The Gravitational Field

According to the original theory of relativity, a material point that is free of
gravitational and other forces moves in a straight line and uniformly according to the
formula

$$\delta\{\int ds\} = 0 \tag{1}$$

[p. 1033] where we set

$$ds^2 = -\sum_v dx_v^2. \tag{2}$$

[6]   We also set $x_1 = x$, $x_2 = y$, $x_3 = z$, $x_4 = it$. In this, ds is the "eigen-time"-
differential, i.e., this quantity gives the amount by how much the clock-time of a
clock (which is associated with the moving material point) progresses along the path-
element (dx, dy, dz). The variation in (1) has to be formed such that the coordinates x_v
at the end points of integration remain unvaried.

Equation (1) remains valid after any coordinate transformation, while (2) is
replaced by the more general form

$$ds^2 = \sum_{\mu v} g_{\mu v}\, dx_\mu\, dx_v. \tag{2a}$$

The ten quantities $g_{\mu v}$ are now functions of those x_v which are determined by the substitution that is applied. In the physical sense, the $g_{\mu\nu}$ determine the gravitational field which exists relative to the new coordinate system, as is also seen from the discussion in the previous paragraph. Equations (1) and (2a) determine the motion of a material point in a gravitational field which can be made to vanish with a suitable choice of the reference system. However, we want to assume in a generalizing manner that the movements of the material point in a gravitational field are determined by these equations.

The quantities $g_{\mu\nu}$ have a further, second significance. Because we can always set

$$ds^2 = \sum_{\mu\nu} g_{\mu\nu}\, dx_\mu\, dx_\nu = -\sum dX_\nu^2, \tag{2b}$$

where, however, the dX_ν are not complete differentials. But in the infinitesimally small, the quantities dX_ν can still be used as coordinates. Therefore, it is plausible to assume that the theory of relativity is valid in the infinitesimally small. The dX_ν are then the directly measurable coordinates in an infinitesimally small domain, and they are measured with unit-measuring rods and a suitably chosen unit-clock. In this sense, the quantity ds^2 can be called the naturally measured distance between two space-time points. In contrast, the dx_ν *cannot* be obtained in the same manner by measurements with clocks and rigid bodies. They are rather connected with the naturally measured distance ds via (2b) in a manner that is determined by the $g_{\mu\nu}$.

After what has been said, ds can be defined independent of the choice of the [p. 1034] coordinate system; it is thus a scalar. In the general theory of relativity, ds takes the same role the element of a worldline had in the original theory of relativity.

In the following, we want to derive the most important theorems of absolute differential calculus, since they take in our theory the role of the theorems of ordinary vector and tensor theory with its three- and four-dimensional vector calculus (with respect to the Euclidean element ds). The laws of general relativity theory which correspond to known laws of the original theory of relativity can easily be derived with the help of those theorems.

B. From the Theory of Covariants

§3. Four-Vectors

The covariant four-vector. Four functions A_ν of coordinates that are defined for any coordinate system are called a *covariant* four-vector or a covariant tensor of rank one if for any *arbitrarily chosen* line element with the components dx_ν the sum

$$\sum_\nu A_\nu dx_\nu = \phi \tag{3}$$

remains an invariant (scalar) with respect to arbitrary coordinate transformations. The quantities A_ν are called the "components" of a four-vector.

From this definition follows immediately the transformation law of these components. If the symbols A'_ν, dx'_ν refer to the same point of the continuum, but refer to another arbitrary coordinate system, we have

$$\sum_\nu A'_\nu dx'_\nu = \sum_\alpha A_\alpha dx_\alpha = \sum_{\alpha\nu} A_\alpha \frac{\partial x_\alpha}{\partial x'_\nu} dx'_\nu.$$

Since this equation is supposed to remain valid for arbitrarily chosen dx'_ν, the transformation law we are looking for follows as

[7]
$$A'_\nu = \sum_\alpha \frac{\partial x^\alpha}{\partial x'_\nu} A_\alpha. \tag{3a}$$

Vice versa, it is easy to show that the validity of this transformation law implies that A_ν is a covariant four-vector.

[p. 1035]   *The contravariant four-vector.* Four functions A^ν of coordinates which are defined for any coordinate system are called a *contravariant* four-vector or a contravariant tensor of rank one if the transformation law of the A^ν is the same as the one for the components dx_ν of the line element. From this follows the transformation law

$$A^{\nu\prime} = \sum_\alpha \frac{\partial x'_\nu}{\partial x_\alpha} A^\alpha. \tag{4}$$

Following Ricci and Levi-Civita, we denote the contravariant character by affixing

[8]   the index as a superscript. According to this definition, the dx_ν themselves are components of a contravariant four-vector; but complying with custom, we shall here leave the index as a subscript.

It follows immediately from the two definitions given above that the expression

$$\sum_\nu A_\nu A^\nu = \phi \tag{3b}$$

is a scalar (invariant). We call ϕ the inner product of the covariant vector (A_ν) and the contravariant vector (A^ν).

By adding (or subtracting) the components of two covariant or contravariant four-vectors, one obtains again a covariant or contravariant four-vector, respectively. This fact follows from the transformation equations (3a) and (4) because of their linearity in the vector components.

§4. Tensors of Second and Higher Ranks

Covariant tensors of second and higher ranks. Sixteen functions $A_{\mu\nu}$ of the coordinates are called the components of a covariant tensor of rank two if the sum

$$\sum_{\mu\nu} A_{\mu\nu} dx_\mu^{(1)} dx_\nu^{(2)} = \Phi \tag{5}$$

is a scalar; here $dx_\mu^{(1)}$ and $dx_\nu^{(2)}$ denote the components of two arbitrarily chosen line elements.

This yields the relation

$$\sum_{\mu\nu} A'_{\mu\nu} dx_\mu^{(1)'} dx_\nu^{(2)'} = \sum_{\alpha\beta} A_{\alpha\beta} dx_\alpha^{(1)} dx_\beta^{(2)} = \sum_{\alpha\beta\mu\nu} \frac{\partial x_\alpha}{\partial x'_\mu} \frac{\partial x_\beta}{\partial x'_\nu} A_{\alpha\beta} dx_\mu^{(1)'} dx_\nu^{(2)'} \qquad [9]$$

and considering that it is supposed to hold for arbitrarily selected $dx_\mu^{(1)'}$ and $dx_\nu^{(2)'}$, one also gets the sixteen equations

$$A'_{\mu\nu} = \sum_{\alpha\beta} \frac{\partial x_\alpha}{\partial x'_\mu} \frac{\partial x_\beta}{\partial x'_\nu} A_{\alpha\beta}. \tag{5a}$$

This equation is again equivalent to the definition above.

Obviously, covariant tensors of third and higher ranks can be defined in an [p. 1036] analogous manner.

The symmetric covariant tensor. If a covariant tensor satisfies for one coordinate system the condition that two of its components which differ by a mere exchange of indices are equal ($A_{\alpha\beta} = A_{\beta\alpha}$), then this holds—as a glance at (5a) shows—for every other coordinate system. In this case, the sixteen transformation equations of a covariant tensor of rank two reduce to ten. If $A_{\mu\nu} = A_{\nu\mu}$, the tensorial character of $(A_{\mu\nu})$ can be proven merely by showing that

$$\sum_{\mu\nu} A_{\mu\nu} dx_\mu dx_\nu = \Phi \tag{5c} \qquad \{1\}$$

is a scalar. Taking (5a) into account, this follows from the identity

$$\sum_{\mu\nu} A'_{\mu\nu} dx'_\mu dx'_\nu = \sum_{\alpha\beta} A_{\alpha\beta} dx_\alpha dx_\beta = \sum_{\alpha\beta\mu\nu} A_{\alpha\beta} \frac{\partial x_\alpha}{\partial x'_\mu} \frac{\partial x_\beta}{\partial x'_\nu} dx'_\mu dx'_\nu.$$

Symmetric covariant tensors of higher rank can be defined in complete analogy.

The covariant fundamental tensor. The quantity

$$ds^2 = \sum g_{\mu\nu} dx_\mu dx_\nu$$

has a special role in the developing theory; we shall call this quantity the square of the line element. From what has been previously said, it follows that $g_{\mu\nu}$ is a covariant (symmetric) tensor of rank two. We shall call it the "covariant fundamental tensor."

Note. We could have defined the covariant tensor alternatively as the total of sixteen quantities $A_{\mu\nu}$ which transform exactly like the sixteen products $A_{\mu}B_{\nu}$ of two covariant vectors (A_{μ}) and (B_{ν}). If one sets

$$A_{\mu\nu} = A_{\mu}B_{\nu} \tag{6}$$

one immediately obtains from (3a)

$$A'_{\mu\nu} = A'_{\mu}B'_{\nu} = \sum_{\alpha\beta} \frac{\partial x_{\alpha}}{\partial x'_{\mu}} \frac{\partial x_{\beta}}{\partial x'_{\nu}} A_{\alpha}B_{\beta} = \sum_{\alpha\beta} \frac{\partial x_{\alpha}}{\partial x'_{\mu}} \frac{\partial x_{\beta}}{\partial x'_{\nu}} A_{\alpha\beta},$$

and with regard to (5a) it follows that $A_{\mu\nu}$ is a covariant tensor. Analogous reasoning applies to tensors of higher ranks. However, not every covariant tensor can be represented in this form, because $(A_{\mu\nu})$ has sixteen components, but A_{μ} and B_{ν} [p. 1037] together have only eight components; therefore, there are—based upon (6)—algebraic relations between the $A_{\mu\nu}$ which tensor components in general do not satisfy. Yet one gets to *any* tensor by adding[1] several of them of type (6), simply setting

$$A_{\mu\nu} = A_{\mu}B_{\nu} + C_{\mu}D_{\nu} + \ldots \tag{6a}$$

The same is true, in analogy, for covariant tensors of higher ranks. This representation of tensors formed from four-vectors proves useful for many theorems. An analogous note applies to tensors of higher ranks.

Contravariant tensors. Just as covariant tensors can be built from covariant four-vectors according to (6) and (6a), resp., one can also form contravariant tensors from contravariant four-vectors with the equations

$$A^{\mu\nu} = A^{\mu}B^{\nu} \tag{7}$$

and

$$A^{\mu\nu} = A^{\mu}B^{\nu} + C^{\mu}D^{\nu} + \ldots \tag{7a}$$

respectively. From this definition, the transformation law follows immediately, according to (4), as

$$A^{\mu\nu'} = \sum_{\alpha\beta} \frac{\partial x'_{\mu}}{\partial x_{\alpha}} \frac{\partial x'_{\nu}}{\partial x_{\beta}} A^{\alpha\beta}. \tag{8}$$

Contravariant tensors of higher ranks are defined in analogy to those of rank two. Just as above, the special case of symmetric tensors deserves attention.

---

[1]It is obvious that the addition of corresponding components of a tensor again leads to components of a tensor, as has been shown for tensors of rank one (four-vectors). (Addition and subtraction of tensors.)

Mixed tensors. It is also possible to form tensors of second and higher ranks which are covariant with respect to some indices, contravariant with respect to others. They are called mixed tensors. A mixed tensor of rank two is, for example,

$$A_\mu^{\ \nu} = A_\mu B^\nu + C_\mu D^\nu. \tag{9}$$

Anti-symmetric tensors. Next to symmetric covariant and contravariant tensors, {2} so-called anti-symmetric covariant and contravariant tensors play an important role. It is characteristic of them that components whose pair of indices are exchanged are *oppositely* equal. For example, a contravariant tensor $A^{\mu\nu}$ which satisfies the condition $A^{\mu\nu} = -A^{\nu\mu}$ is called an anti-symmetric contravariant tensor of rank two, [p. 1038] or also a six-vector (because it has 12 non-zero components which, in pairs, have equal magnitudes). A contravariant tensor $A^{\mu\nu\lambda}$ of rank three is anti-symmetric if it satisfies the conditions

$$A^{\mu\nu\lambda} = -A^{\mu\lambda\nu} = -A^{\nu\mu\lambda} = A^{\nu\lambda\mu} = -A^{\lambda\nu\mu} = A^{\lambda\mu\nu}.$$

One realizes that (in a continuum of four dimensions) this anti-symmetric tensor has only four numerically non-zero components.

With formulas (5a) and (8), resp., one can easily show that this definition has a meaning which is independent of the choice of the system of reference.

According to (5a) we have, e.g.,

$$A'_{\nu\mu} = \sum_{\alpha\beta} \frac{\partial x_\alpha}{\partial x'_\nu} \frac{\partial x_\beta}{\partial x'_\mu} A_{\alpha\beta}.$$

Replacing $A_{\alpha\beta}$ with $-A_{\beta\alpha}$ (which is permissible by the hypothesis made) and then exchanging the summation indices α and β in the double-sum, one gets

$$A'_{\nu\mu} = \sum_{\alpha\beta} \frac{\partial x_\alpha}{\partial x'_\mu} \frac{\partial x_\beta}{\partial x'_\nu} A_{\alpha\beta} = -A'_{\mu\nu},$$

as has been claimed. The proof for contravariant tensors and tensors of rank three and four is analogous. Anti-symmetric tensors of higher than fourth rank cannot exist in a four-dimensional continuum because all components with two identical indices vanish.

§5. Multiplication of Tensors

The outer product of tensors. We have seen that by multiplication of the components {3} of a tensor of rank one, one obtains the components of tensors of higher rank (see equations 6, 8, and 9). In analogy we can always derive tensors of higher rank from those of lower rank by multiplying the components of one tensor with those of another. If, for example, $(A_{\alpha\beta})$ and $(B_{\lambda\mu\sigma})$ are covariant tensors, then $(A_{\alpha\beta} \cdot B_{\lambda\mu\sigma})$

is also a covariant tensor (of rank five). The proof follows immediately from the representability of tensors by sums of products of four-vectors:

$$A_{\alpha\beta} = \sum A_\alpha^{(1)} A_\beta^{(2)},$$

$$B_{\lambda\mu\sigma} = \sum B_\lambda^{(1)} B_\mu^{(2)} B_\sigma^{(3)},$$

that is,
$$A_{\alpha\beta} B_{\lambda\mu\sigma} = \sum A_\alpha^{(1)} A_\beta^{(2)} B_\lambda^{(1)} B_\mu^{(2)} B_\sigma^{(3)},$$

and, therefore, $(A_{\alpha\beta} B_{\lambda\mu\sigma})$ is a tensor of rank five.

[p. 1039]   This operation is called "outer multiplication," and the result the "outer product" of tensors. One sees that with this operation, the character and rank of tensors to be "multiplied" are immaterial. Furthermore, the commutative and the associative laws apply to the sequence of such operations.

{4}   *The inner product of tensors.* The operation shown in formula (3b), with the tensors of rank one A_ν and A^ν, is called "inner multiplication," and the result the "inner product." Because of the representability of tensors of higher rank by means of four-vectors, this operation can be extended easily to tensors in general.

If, for example, $A_{\alpha\beta\gamma}...$ is a covariant and $A^{\alpha\beta\gamma}...$ is a contravariant tensor, both of the same rank, then

$$\sum_{\alpha\beta\gamma} (A_{\alpha\beta\gamma}... A^{\alpha\beta\gamma}\cdots) = \Phi$$

is a scalar. The proof follows directly if one puts

$$A_{\alpha\beta\gamma}... = \sum A_\alpha B_\beta C_\gamma \cdots$$

$$A^{\alpha\beta\gamma}\cdots = \sum A^\alpha B^\beta C^\gamma ...,$$

and then multiplies them taking (3b) into account.

The mixed product of tensors. The most general multiplication of tensors results when some indices are used for outer, others for inner multiplication. The tensors A and B result in a tensor C according to the following scheme:

$$\sum_{\alpha\beta\gamma...\alpha'\beta'\gamma'} (A_{\alpha\beta\gamma...\rho\sigma\tau...}^{\alpha'\beta'\gamma'...\lambda\mu\nu...} B_{\alpha'\beta'\gamma'...rst...}^{\alpha\beta\gamma...lmn...}) = C_{\rho\sigma\tau...rst...}^{\lambda\mu\nu...lmn...}.$$

The proof that C is a tensor follows from combining the last two proofs we sketched above.

§6. On Some Relations Concerning the Fundamental Tensor $g_{\mu\nu}$

The contravariant fundamental tensor. If one forms in the determinantal scheme of the $g_{\mu\nu}$ the minors for every $g_{\mu\nu}$ and divides them by the determinant $g = |g_{\mu\nu}|$ of those $g_{\mu\nu}$, one obtains certain quantities $g^{\mu\nu}(= g^{\nu\mu})$ for which we want to prove that they form a contravariant symmetrical tensor.

From this definition it follows next that

$$\sum_{\sigma} g_{\mu\sigma} g^{\nu\sigma} = \delta_{\mu}^{\nu}, \tag{10}$$

where δ_{μ}^{ν} signifies the quantity 1 or 0 resp., depending upon $\mu = \nu$ or $\mu \neq \nu$.[2] [p. 1040]
Furthermore, {5}

$$\sum_{\alpha\beta} g_{\alpha\beta} dx_{\alpha} dx_{\beta}$$

is a scalar which we can equate—according to (10)—with

$$\sum_{\alpha\beta\mu} g_{\mu\beta} \delta_{\alpha}^{\mu} dx_{\alpha} dx_{\beta}$$

and also equal to

$$\sum_{\alpha\beta\mu\nu} g_{\mu\beta} g_{\nu\alpha} g^{\mu\nu} dx_{\alpha} dx_{\beta}.$$

However, according to the previous paragraph, the

$$d\xi_{\mu} = \sum_{\beta} g_{\mu\beta} dx_{\beta}$$

are the components of a covariant vector, and similarly of course also the

$$d\xi_{\nu} = \sum_{\alpha} g_{\nu\alpha} dx_{\alpha}.$$

Our scalar, therefore, takes the form

$$\sum_{\mu\nu} g^{\mu\nu} d\xi_{\mu} d\xi_{\nu}.$$

It can be easily proven that $g^{\mu\nu}$ is a contravariant tensor; this is based upon the facts that the sum above is a scalar, the $d\xi_{\mu}$ are by nature arbitrarily selectable components of a covariant four-vector, and $g^{\mu\nu} = g^{\nu\mu}$.

Note. According to the multiplication theorem of determinants,

$$\left| \sum_{\alpha} g_{\mu\alpha} g^{\alpha\nu} \right| = |g_{\mu\alpha}| \cdot |g^{\alpha\nu}|.$$

On the other hand,

---

[2]According to the previous paragraph, δ_{μ}^{ν} is a mixed tensor ("mixed fundamental tensor").

$$\left| \sum_{\alpha} (g_{\mu\alpha} g^{\alpha\nu}) \right| = |\delta_{\mu}^{\nu}| = 1.$$

From this follows

$$|g_{\mu\nu}| \cdot |g^{\mu\nu}| = 1. \tag{11}$$

The Invariant of Volume. For the immediate neighborhood of a point of our continuum we can always put, according to (2b),

$$ds^2 = \sum_{\mu\nu} g_{\mu\nu} \, dx_{\mu} \, dx_{\nu} = \sum_{\sigma} dX_{\sigma}^2, \tag{12}$$

provided we allow imaginary values for the dX_{σ}. For the choice of the system of the dX_{σ} there are still infinitely many possibilities; however, all these systems are connected [p. 1041] by linear orthogonal substitutions. From this follows that the integral

$$d\tau_0^* = \int dX_1 dX_2 dX_3 dX_4$$

over a volume element is an invariant, i.e., completely independent of the choice of the coordinates.

We want to find a second expression for this invariant. There are at any rate relations of the form

$$dX_{\sigma} = \sum_{\mu} \alpha_{\sigma\mu} dx_{\mu}, \tag{13}$$

from which it follows that

$$d\tau_0^* = |\alpha_{\sigma\mu}| d\tau \tag{14}$$

if $d\tau$ and $d\tau_0^*$ resp. denote the integrals

$$\int dx_1 ... dx_4 \quad \text{and} \quad \int dX_1 dX_2 dX_3 dX_4 \text{ resp.}$$

extended over the same elementary domain. Furthermore, according to (12) and (13),

$$g_{\mu\nu} = \sum_{\sigma} \alpha_{\sigma\mu} \alpha_{\sigma\nu}, \tag{15}$$

and consequently, according to the multiplication theorem of determinants,

{6}
$$|g_{\mu\nu}| = \left| \sum_{\sigma} (\alpha_{\sigma\mu} \alpha_{\sigma\nu}) \right| = |\alpha_{\mu\nu}|^2. \tag{16}$$

Considering (14), one now gets

$$\sqrt{g} \, d\tau = d\tau_0^*, \tag{17}$$

where we put $|g_{\mu\nu}| = g$ for short. With this we have found the invariant we were looking for.

Note. That the dX_σ correspond to the customary coordinates in the original theory of relativity is seen from (12). Three of these coordinates are real-valued, and one is imaginary (e.g., dX_4). Consequently, $d\tau_0$ is imaginary. On the other hand, the determinant g with real-valued time coordinates in the original theory of relativity is negative, since the $g_{\mu\nu}$ (under suitable choice of the time unit) get the values

$$\left.\begin{array}{rrrr} -1 & 0 & 0 & 0 \\ 0 & -1 & 0 & 0 \\ 0 & 0 & -1 & 0 \\ 0 & 0 & 0 & 1 \end{array}\right\}; \tag{18}$$

$\sqrt{g}$ is therefore also imaginary. This is quite generally the case, as we will show in §17. In order to avoid imaginaries, we put   [p. 1042]

$$d\tau_0 = \frac{1}{i}\int dX_1 dX_2 dX_3 dX_4$$

and instead of (17) we write

$$\sqrt{-g}\,d\tau = d\tau_0. \tag{17a}$$

The anti-symmetric fundamental tensor of RICCI and LEVI-CIVITA. We claim that

$$G_{iklm} = \sqrt{g}\,\delta_{iklm} \tag{19}$$

is a covariant tensor. δ_{iklm} denotes $+1$ or -1 depending upon 1234 results from $iklm$ by an even or odd permutation of the indices.

In order to prove this, we note first that the determinant

$$\sum_{iklm} \delta_{iklm} dx_i^{(1)} dx_k^{(2)} dx_l^{(3)} dx_m^{(4)} = V \tag{20}$$

is—aside from an irrelevant factor—equal to the volume of an elementary pentahedron whose corners are represented by one point of the continuum and the four end points of the arbitrary line elements $(dx_i^{(1)})$, $(dx_k^{(2)})$, $(dx_l^{(3)})$, $(dx_m^{(4)})$ extending from this point. According to (19) and (20), one has   [10]

$$\sum_{iklm} G_{iklm} dx_i^{(1)} dx_k^{(2)} dx_l^{(3)} dx_m^{(4)} = \sqrt{g}\,V.$$

Since from (17) the right-hand side is a scalar, (G_{iklm}) is a covariant tensor, precisely because δ_{iklm} is defined as an anti-symmetric covariant tensor.

From the tensor G_{iklm} one easily forms a contravariant tensor by mixed multiplication according to the scheme

$$\sum_{\alpha\beta\lambda\mu} G_{\alpha\beta\lambda\mu} g^{\alpha i} g^{\beta k} g^{\lambda l} g^{\mu m} = G_{iklm}. \tag{21}$$

The contravariant tensorial character now follows directly from §4. Due to (19), the left-hand side takes the form

$$\sqrt{g} \sum_{\alpha\beta\lambda\mu} \delta_{\alpha\beta\lambda\mu} g^{\alpha i} g^{\beta k} g^{\lambda l} g^{\mu m},$$

which, according to known theorems of the theory of determinants, equals

$$\sqrt{g} \delta_{iklm} \sum \delta_{\alpha\beta\lambda\mu} g^{1\alpha} g^{2\beta} g^{3\lambda} g^{4\mu},$$

which after (11),

$$\frac{1}{\sqrt{g}} \delta_{iklm}.$$

[p. 1043] With this we have proven that

$$G^{iklm} = \frac{1}{\sqrt{g}} \delta_{iklm} \tag{21a}$$

is a contravariant anti-symmetrical tensor.

In the theory of general anti-symmetric tensors, finally, a mixed tensor has an important role; it is formed from the fundamental tensor of the $g_{\mu\nu}$ and its components are

$$G_{ik}^{lm} = \sum_{\alpha\beta} \sqrt{g} \delta_{ik\alpha\beta} g^{\alpha l} g^{\beta m} = \sum_{\alpha\beta} \frac{1}{\sqrt{g}} \delta_{lm\alpha\beta} g_{\alpha i} g_{\beta k}. \tag{22}$$

The tensorial character of these two expressions is evident from what has been said above and in §4. One has to prove only that they are equal.

According to (21) and (19), the latter one can only be brought into the form

[11]
$$\sum_{\lambda\mu\rho\sigma\alpha\beta} \sqrt{g} \delta_{\lambda\mu\rho\sigma} g^{\lambda l} g^{\mu m} g^{\rho\alpha} g^{\sigma\beta} g_{\alpha i} g_{\beta k}$$

and after summation over α and β, considering (10), one gets

$$\sum_{\lambda\mu} \sqrt{g} \delta_{\lambda\mu ik} g^{\lambda l} g^{\mu m}.$$

This expression deviates from the first one in (22) only in the notation of the summation indices and in the (irrelevant) sequence of the index-pairs $\lambda\mu$ and ik in $\delta_{\lambda\mu ik}$.

The mixed tensor G_{ik}^{lm} is anti-symmetric in its indices i,k as well as in l,m, as can be seen from (22).

With the help of the fundamental tensor, and following the rules given in §5, we can produce from any tensor others of different character. For example, we can transform the covariant tensor $(T_{\mu\nu})$ into the contravariant tensor $(T^{\mu\nu})$ according to the rule

$$T^{\mu\nu} = \sum_{\alpha\beta} T_{\alpha\beta} g^{\alpha\mu} g^{\beta\nu} \tag{23}$$

while vice versa one gets

$$T_{\mu\nu} = \sum_{\alpha\beta} T^{\alpha\beta} g_{\alpha\mu} g_{\beta\nu}. \tag{23a}$$

The mutual equivalence of equations (23) and (23a) follows easily by using (10). The tensors $(T^{\mu\nu})$ and $(T_{\mu\nu})$ are called "reciprocal." If one of two reciprocal tensors is symmetrical or anti-symmetrical, then so is the other; this follows from (23) and/or (23a). It also applies to tensors of any rank.

Dual six-vectors. If, furthermore, $(F^{\mu\nu})$ is an anti-symmetric tensor (of rank two), [p. 1044] we can form a second anti-symmetric tensor $F^{\mu\nu*}$ with the equation

$$F^{\mu\nu*} = \frac{1}{2}\sum_{\alpha\beta} G^{\mu\nu}_{\alpha\beta} F^{\alpha\beta}. \tag{24}$$

$F^{\mu\nu*}$ is called the contravariant six-vector "dual" to $F^{\mu\nu}$. Vice versa, $F^{\mu\nu}$ is dual to $F^{\mu\nu*}$. Because, if we multiply (24) by $G^{\sigma\tau}_{\mu\nu}$ and sum over μ and ν, we get

$$\frac{1}{2}\sum G^{\sigma\tau}_{\mu\nu} F^{\mu\nu*} = \frac{1}{4}\sum_{\alpha\beta\mu\nu} G^{\sigma\tau}_{\mu\nu} G^{\mu\nu}_{\alpha\beta} F^{\alpha\beta}.$$

But, since according to (22)

$$\sum_{\mu\nu} G^{\sigma\tau}_{\mu\nu} G^{\mu\nu}_{\alpha\beta} = \sum_{\mu\nu\lambda\kappa\lambda'\kappa'} \sqrt{g}\,\delta_{\mu\nu\lambda\kappa} g^{\lambda\sigma} g^{\kappa\tau} \frac{1}{\sqrt{g}} \delta_{\mu\nu\lambda'\kappa'} g_{\lambda'\alpha} g_{\kappa'\beta} = 2(\delta^{\sigma}_{\alpha}\delta^{\tau}_{\beta} - \delta^{\sigma}_{\beta}\delta^{\tau}_{\alpha})$$

[see footnote 3] one finally gets

$$\frac{1}{4}\sum_{\alpha\beta\mu\nu} G^{\sigma\tau}_{\mu\nu} G^{\mu\nu}_{\alpha\beta} F^{\alpha\beta} = \frac{1}{2}(F^{\sigma\tau} - F^{\tau\sigma}) = F^{\sigma\tau} \qquad\qquad [13]$$

from which our claim follows.

An analogous result applies to covariant six-vectors. One can also easily prove that six-vectors that are reciprocal to two dual ones are dual themselves.

---

[3]The second formulation is based upon the fact that $\delta_{\mu\nu\lambda\kappa}$ differs from zero only when all indices are different. This leaves only two possibilities, $(\lambda = \lambda', \kappa = \kappa')$ and $(\lambda = \kappa', \kappa = \lambda')$. Considering this, one gets first, by summing over μ and ν,

$$2\sum_{\lambda\kappa} \{g^{\lambda\sigma} g^{\kappa\tau} g_{\lambda\alpha} g_{\alpha\beta} - g^{\lambda\sigma} g^{\kappa\tau} g_{\lambda\beta} g_{\kappa\alpha}\}, \qquad\qquad [12]$$

where further summation needs to consider only the index combinations $(\lambda\kappa)$ with $\lambda \neq \kappa$. But, since the parenthesis vanishes for $\lambda = \kappa$ anyway, one can go over *all* combinations. Considering (1), one then gets the expression given in the text.

§7. The Geodesic Line and Equations of Motions of Points

It has already been explained in §2 that the movement of a material point in a gravitational field is governed by

$$\delta\left\{\int ds\right\} = 0. \tag{1}$$

From a mathematical point of view, the movement of a point, therefore, corresponds
[14]   to a geodesic line in a four-dimensional manifold. We insert here the well-known
[p. 1045] derivation of the explicit equations of this line for the sake of completeness.

We are dealing here with a line between two points $P^{(1)}$ and $P^{(2)}$ relative to which all lines infinitely close and joining both points satisfy equation (1). If λ denotes a function of the coordinates x_v, then the "surface" of constant λ will cut just one point from each one of these infinitely adjacent lines; and the coordinates of this point can be looked at as a function of λ alone, provided the curve is given. If we put

$$w^2 = \sum_{\mu v} g_{\mu v} \frac{dx_\mu}{d\lambda} \frac{dx_v}{d\lambda},$$

we can rewrite (1) in the form

$$\int_{\lambda_x}^{\lambda_2} \delta w d\lambda = 0, \tag{1a}$$

because the limits of integration λ_1 and λ_2 are the same for all curves under consideration. If δx_v denotes the increases that have to be given to the x_v in order to go from a point of the desired geodesic line to a point of the same λ on one of the varied lines, then one gets

$$\delta w = \frac{1}{w}\left\{\frac{1}{2}\sum_{\mu v \sigma} \frac{\partial g_{\mu v}}{\partial x_\sigma} \frac{dx_\mu}{d\lambda} \frac{dx_v}{d\lambda} \delta x_\sigma + \sum_{\mu v} g_{\mu v} \frac{dx_\mu}{d\lambda} \delta\left(\frac{dx_v}{d\lambda}\right)\right\}.$$

Substituting this expression into (1a), partially integrating the last term and considering that the δx_v vanish for $\lambda = \lambda_1$ and $\lambda = \lambda_2$ one obtains

$$\int_{\lambda_1}^{\lambda_2} d\lambda \sum_\sigma (K_\sigma \delta x_\sigma) = 0,$$

where

$$K_\sigma = \sum_{\mu v}\left[\frac{d}{d\lambda}\left\{\frac{g_{\mu v}}{w} \frac{dx_\mu}{dl}\right\} - \frac{1}{2w} \frac{\partial g_{\mu v}}{\partial x_\sigma} \frac{dx_\mu}{d\lambda} \frac{dx_v}{d\lambda}\right]$$

is used as an abbreviation. From this, it follows that

$$K_\sigma = 0 \qquad (23) \qquad [15]$$

is the equation of the geodesic line.

In the original theory of relativity, those geodesic lines for which $ds^2 > 0$ correspond to the motion of material points, those where $ds = 0$ represent light rays. [p. 1046] This will also be the case in a generalized theory of relativity. Excluding the latter case ($ds = 0$) from our consideration, we can choose the "length of arc" s along a geodesic line as our parameter λ. The equation of a geodesic line then transforms into

$$\sum_\mu g_{\sigma\mu} \frac{d^2 x_\mu}{ds^2} + \sum_{\mu\nu} \begin{bmatrix} \mu\nu \\ \sigma \end{bmatrix} \frac{dx_\mu}{ds} \frac{dx_\nu}{ds} = 0, \qquad (23a)$$

where we introduced, after CHRISTOFFEL, the abbreviation   [16]

$$\begin{bmatrix} \mu\nu \\ \sigma \end{bmatrix} = \frac{1}{2}\left(\frac{\partial g_{\mu\sigma}}{\partial x_\nu} + \frac{\partial g_{\nu\sigma}}{\partial x_\mu} - \frac{\partial g_{\mu\nu}}{\partial x_\sigma} \right), \qquad (24)$$

This expression is symmetric in the indices μ and ν. Finally, (23a) is multiplied by $g^{\sigma\tau}$ and summed over σ. Considering (10) and using the well-known CHRISTOFFEL symbols   [17]

$$\begin{Bmatrix} \mu\nu \\ \tau \end{Bmatrix} = \sum_\sigma g^{\sigma\tau} \begin{bmatrix} \mu\nu \\ \sigma \end{bmatrix} \qquad (24a)$$

one gets in place of (23a)

$$\frac{d^2 x_\tau}{ds^2} + \sum_{\mu\nu} \begin{Bmatrix} \mu\nu \\ \tau \end{Bmatrix} \frac{dx_\mu}{ds} \frac{dx_\nu}{ds} = 0. \qquad (23b)$$

This is the equation of the geodesic line in its most comprehensive form. It expresses the second derivatives of the x_ν with respect to s by means of the first derivatives. Differentiating (23b) with respect to s would yield equations that would also allow to reduce higher differential quotients of coordinates with respect to s to their first derivatives. In this manner one would obtain the coordinates in a TAYLOR expansion of the variable s. Equation (23b) is equivalent to the equation of motion of a material point in its MINKOWSKI form where s denotes the "eigen-time."

§8. Forming Tensors by Differentiation

The fundamental significance of the concept of a tensor rests, as is well known, upon the fact that the transformation equations of the tensor components are linear and homogeneous. This in turn causes the tensor components to vanish relative to any coordinate system if they vanish relative to just *one* such system. Therefore, once a group of equations of a physical system has been brought into a form which shows the vanishing of all components of a tensor, then this equation system is independent [p. 1047] of the choice of the coordinate system. In order to establish such equations of tensors, one has to know the laws by which new tensors can be formed from given ones. It has already been discussed how this can be done in an algebraic manner. We still have to derive the laws by which new tensors can be formed from known ones through differentiation. The laws of these differential expressions have already been given by CHRISTOFFEL, RICCI, and LEVI-CIVITA. I give here a particularly simple derivation for this, which appears to be new.

All differential operations on tensors can be traced to the so-called "extension." In the case of the original theory of relativity, i.e., the case where only linear orthogonal substitutions are admitted as "admissible," the law can be phrased as follows. If $T_{\alpha_1...\alpha_l}$ is a tensor of rank l, then $\dfrac{\partial T_{\alpha_1...\alpha_l}}{\partial x_l}$ is a tensor of rank $l + 1$. The

so-called "divergence" of tensors follows easily now with equation (10) of §6 from the special tensor δ_μ^ν, which under restriction to linear orthogonal transformations—where the distinction between covariant and contravariant is void—is to be replaced by the notation $\delta_{\mu\nu}$. By means of inner multiplication of the tensor $\delta_{\mu\nu}$ with the tensor of rank $l + 1$, obtained by "extension," we get the tensor of rank $l - 1$

{7}
$$T_{\alpha_1...\alpha_{l-1}} = \sum_{\alpha_l s} \frac{\partial T_{\alpha_1...\alpha_l}}{\partial x_\sigma} \delta_{\alpha_l s} = \sum_{\alpha_l} \frac{\partial T_{\alpha_1...\alpha_l}}{\partial x_{\alpha_l}}.$$

This is the divergence of the tensor $T_{\alpha_1...\alpha_l}$ formed relative to the index α_l. It is now our task to find the generalization of this operation in case substitutions are not restricted to the previous conditions (linearity-orthogonality).

Extension of a covariant tensor. Let $\phi(x_1...x_4)$ be a scalar and S a given curve in our continuum. The "length of arc" s is measured from a point P on S in a distinct direction, as has been explained in §§1 and 8. The values of the function Φ at points on S in our continuum can then also be looked at as a function of s. Obviously, the [p. 1048] quantities $d\phi/ds$, $d^2\phi/ds^2$, etc., are then also scalars, i.e., quantities that are defined independently of the coordinate system. However, since

$$\frac{d\phi}{ds} = \sum_\mu \frac{\partial \phi}{\partial x_\mu} \frac{dx_\mu}{ds} \tag{25}$$

and since from every point the curves S can be drawn in any direction, the quantities

$$A_\mu = \frac{\partial \phi}{\partial x_\mu} \tag{26}$$

are, according to §3, components of a covariant four-vector (tensor of rank one), which we can suitably view as the "extension" of the scalar Φ (a tensor of rank zero).

Furthermore, we get from (25),

$$\frac{d^2\phi}{ds^2} = \sum_{\mu\nu} \frac{\partial^2\phi}{\partial x_\mu \partial x_\nu} \frac{dx_\mu}{ds} \frac{dx_\nu}{ds} + \sum_\tau \frac{\partial\phi}{\partial x_\tau} \frac{d^2 x_\tau}{ds^2}.$$

We now specialize our analysis by assuming the line S to be geodesic, a choice which is independent of the adopted system of reference. According to (23b) we then get

$$\frac{d^2\phi}{ds^2} = \sum_{\mu\nu} \left[\frac{\partial^2\phi}{\partial x_\mu \partial x_\nu} - \sum_\tau \begin{Bmatrix} \mu\nu \\ \tau \end{Bmatrix} \frac{\partial\phi}{\partial x_\tau} \right] \frac{dx_\mu}{ds} \frac{dx_\nu}{ds}. \tag{27}$$

Now we focus on the quantities

$$A_{\mu\nu} = \frac{\partial^2\phi}{\partial x_\mu \partial x_\nu} - \sum_\tau \begin{Bmatrix} \mu\nu \\ \tau \end{Bmatrix} \frac{\partial\phi}{\partial x_\tau}, \tag{28}$$

which satisfy the conditions of symmetry

$$A_{\mu\nu} = A_{\nu\mu}$$

according to (24) and (24a). Due to this and due to (5c), one derives from (27) and from the scalar character of $d^2\phi/ds^2$ that $A_{\mu\nu}$ is a (symmetric) covariant tensor of rank two. We can view $A_{\mu\nu}$ as the extension of the covariant tensor of rank one, $A_\mu = d\phi/dx_\mu$, and therefore write (28) in the form

$$A_{\mu\nu} = \frac{\partial A_\mu}{\partial x_\nu} - \sum_\tau \begin{Bmatrix} \mu\nu \\ \tau \end{Bmatrix} A_\tau. \tag{28a}$$

One can now expect that not only a four-vector of type (26) but rather *any* covariant four-vector will become a covariant tensor of rank two when differentiation (extension) is applied according to (28a). We will verify this next.

First, it is easy to see that the components A_μ of any covariant four-vector in a [p. 1049] four-dimensional continuum can be represented in the form

$$A_\mu = \psi_1 \frac{\partial\phi_1}{\partial x_\mu} + \psi_2 \frac{\partial\phi_2}{\partial x_\mu} + \psi_3 \frac{\partial\phi_3}{\partial \phi_\mu} + \psi_4 \frac{\partial\phi_4}{\partial x_\mu},$$

where the quantities ψ_λ and ϕ_λ are scalars. In order to satisfy the equation, we equate arbitrarily $\phi_\nu = x_\nu$ (in the special coordinate system under consideration) and, furthermore, $\psi_v = A_v$ (again in the special coordinate system under consideration). The tensorial character of the quantities $A_{\mu v}$ in (28a) can be understood if we show that $A_{\mu v}$ is a tensor, provided we set in (28a) $A_\mu = \psi \dfrac{d\phi}{dx_\mu}$, while ψ and ϕ are scalars. According to (28), the

$$\psi \left[\frac{\partial^2 \phi}{\partial x_\mu \partial x_v} - \sum_\tau \begin{Bmatrix} \mu v \\ \tau \end{Bmatrix} \frac{\partial \phi}{\partial x_\tau} \right]$$

are components of a tensor, due to (26) and (6); likewise are the

$$\frac{\partial \psi}{\partial x_\mu} \frac{\partial \phi}{\partial x_v}.$$

After addition, we have the tensorial character of

$$\frac{\partial}{\partial x_v} \left[\psi \frac{\partial \phi}{\partial x_\mu} \right] - \sum_\tau \begin{Bmatrix} \mu v \\ \tau \end{Bmatrix} \left(\psi \frac{\partial \phi}{\partial x_\tau} \right).$$

Equation (28a) makes a tensor also from the four-vector $d\phi/dx_\mu$ and therefore, as proven before, from any covariant four-vector A_μ. This concludes the desired proof.

It is easy to find the extension of covariant tensors of any rank after we derived already the extension of a tensor of rank one. According to (6) and (6a), we can represent any covariant tensor as a sum of tensors of the type

$$A_{\alpha_1 \dots \alpha_l} = A^{(1)}_{\alpha_1} A^{(2)}_{\alpha_2} \dots A^{(l)}_{\alpha_l},$$

where the $A^{(v)}_{\alpha_v}$ represent covariant four-vectors. Due to (28a) we first have

$$A^{(v)}_{\alpha_v s} = \frac{\partial A^{(v)}_{\alpha_v}}{\partial x_s} - \sum_\tau \begin{Bmatrix} \alpha_v s \\ \tau \end{Bmatrix} A^{(v)}_\tau$$

[p. 1050] as a covariant tensor of rank two. By the rules of outer multiplication we multiply it with all of $A_{\alpha_\mu}{}^{(\mu)}$, except the component $A^{(v)}_{\alpha_v}$, and thus get a tensor of rank $l + 1$ in whose formation the index v was privileged. In this manner one can form l tensors by merely privileging the indices $v = 1, v = 2, \dots v = l$ in sequence. Finally, adding them all together, one obtains the tensor of rank $(l + 1)$,

$$A_{\alpha_1 \dots \alpha_l s} = \frac{\partial A_{\alpha_\tau \dots \alpha_l}}{\partial x_s} - \sum_\tau \left[\begin{Bmatrix} \alpha_1 s \\ \tau \end{Bmatrix} A_{\tau \alpha_2 \dots \alpha_l} + \begin{Bmatrix} \alpha_2 s \\ \tau \end{Bmatrix} A_{\alpha_1 \tau \alpha_3 \dots \alpha_l} + \dots \right]. \tag{29}$$

[18]   This formula has already been found by CHRISTOFFEL and produces, as has been said, from any covariant tensor of rank l another one of rank $(l + 1)$, which we call the "extension" of the former. All differential operations on tensors can be derived from this operation.

Multiplying (29) by $g^{\alpha_1\beta_1} g^{\alpha_2\beta_2} \ldots g^{\alpha_l\beta_l}$ such that the multiplication is an inner one with respect to the α indices and an outer one with respect to the β indices, one obtains a tensor which is contravariant in $\beta_1 \ldots \beta_l$ and covariant in s. Writing again α instead of β, one gets

$$A_s^{\alpha_1\ldots\alpha_l} = \frac{\partial A^{\alpha_1\ldots\alpha_l}}{\partial x_s} + \sum_\tau \left[\begin{Bmatrix} ST \\ \alpha_1 \end{Bmatrix} A^{\tau\alpha_2\ldots\alpha_l} + \begin{Bmatrix} ST \\ \alpha_2 \end{Bmatrix} A^{\alpha_1\tau\alpha_3\ldots\alpha_l} + \ldots \right]. \tag{30}$$

This tensor can be called the *extension of the contravariant tensor*. A glance at (29) and (30) shows that the extension defined in this manner always produces one covariant index. It is also easy to give a general formula for the extension of a mixed tensor, and this would amount to an amalgamation of the formulas (29) and (30).

Divergence. The extension of a contravariant tensor of rank l produces a mixed tensor of rank $l + 1$. It can be transformed into a contravariant tensor of rank $l - 1$ by inner multiplication with the mixed fundamental tensor (10); and it can be done in l different ways. One can therefore distinguish usually l different divergences of a contravariant tensor. One of them is

$$A^{\alpha_1\ldots\alpha_{l-1}} = \sum_{\alpha_l s} A_s^{\alpha_1\ldots\alpha_l} \delta_{\alpha_l}^s. \tag{31} \qquad [19]$$

With symmetrical and anti-symmetrical tensors, the end result of the divergence is independent of the privileged α_v.

Some supplementary formulas. Before we apply the formulas we just derived to special cases, we want to derive a few differential properties of the fundamental tensor. Differentiating the determinant $|g_{\mu v}| = g$ with respect to x_α, one gets   [p. 1051]

$$\frac{1}{g} \frac{\partial g}{\partial x_\alpha} = \sum_{\mu v} \frac{\partial g_{\mu v}}{\partial x_\alpha} g^{\mu v} = \frac{2}{\sqrt{g}} \frac{\partial \sqrt{g}}{\partial x_\alpha}. \tag{32}$$

From (24a), (24), and (32) follows

$$\sum_\tau \begin{Bmatrix} \mu\tau \\ \tau \end{Bmatrix} = \sum_\tau \begin{Bmatrix} \tau\mu \\ \tau \end{Bmatrix} = \frac{1}{2} \sum_{\tau\alpha} g^{\tau\alpha} \frac{\partial g_{\tau\alpha}}{\partial x_\mu} = \frac{1}{\sqrt{g}} \frac{\partial \sqrt{g}}{\partial x_\mu}. \tag{33} \qquad \{8\}$$

Differentiating (10) yields

$$\sum_\sigma \frac{\partial g_{\mu\sigma}}{\partial x_\alpha} g^{v\sigma} = - \sum_\sigma \frac{\partial g_{v\sigma}}{\partial x_\alpha} g_{\mu\sigma}. \tag{34}$$

When one multiplies this equation by $g_{v\tau}$ and sums over v, or likewise multiplies by $g_{\mu\tau}$ and sums over μ, one obtains—taking (10) into account—two equations, respectively, which read after the appropriate change of indices

$$\frac{\partial g_{\mu\nu}}{\partial x_{\alpha}} = - \sum_{\sigma\tau} \frac{\partial g^{\sigma\tau}}{\partial x_{\alpha}} g_{\sigma\mu} g_{\tau\nu} \tag{35}$$

$$\frac{\partial g^{\mu\nu}}{\partial x_{\alpha}} = - \sum_{\sigma\tau} \frac{\partial g_{\sigma\tau}}{\partial x_{\alpha}} g^{\sigma\mu} g^{\tau\nu}. \tag{36}$$

Extension and divergence of a four-vector. The extension of the covariant four-vector is given by (28a). By exchanging the indices μ and ν and subtracting, one obtains the anti-symmetric tensor

$$A_{\mu\nu} - A_{\nu\mu} = \frac{\partial A_{\mu}}{\partial x_{\nu}} - \frac{\partial A_{\nu}}{\partial x_{\mu}}. \tag{28b}$$

The extension A_{ν}^{μ} of the contravariant four-vector A^{μ} follows from (30) as

$$A_{\nu}^{\mu} = \frac{\partial A^{\mu}}{\partial x_{\nu}} + \sum_{\tau} \begin{Bmatrix} \nu\tau \\ \mu \end{Bmatrix} A^{\tau}.$$

The divergence of this is

$$\Phi = \sum_{\mu\nu} A_{\nu}^{\mu} \delta_{\mu}^{\nu} = \sum_{\mu} \left(\frac{\partial A^{\mu}}{\partial x_{\mu}} + \begin{Bmatrix} \mu\tau \\ \mu \end{Bmatrix} A^{\tau} \right),$$

or, due to (33),

$$\Phi = \frac{1}{\sqrt{g}} \sum_{\mu} \frac{\partial}{\partial x_{\mu}} (\sqrt{g} A^{\mu}). \tag{37}$$

Now replacing A^{μ} with the contravariant vector $\sum g^{\mu\nu} \frac{\partial \phi}{\partial x_{\nu}}$, where ϕ denotes a scalar, one obtains the well-known generalization of the Laplacian $\Delta\phi$:

[p. 1052]

$$\Phi = \sum_{\mu\nu} \frac{1}{\sqrt{g}} \frac{\partial}{\partial x_{\mu}} \left(\sqrt{g} g^{\mu\nu} \frac{\partial \phi}{\partial x_{\nu}} \right). \tag{38}$$

Extension and divergence of a tensor of rank two. Applied to covariant and contravariant tensors of rank two, (29) and (30) resp. produce tensors of rank three.

$$A_{\mu\nu s} = \frac{\partial A_{\mu\nu}}{\partial x_{s}} - \sum_{\tau} \left(\begin{Bmatrix} \mu s \\ \tau \end{Bmatrix} A_{\tau\nu} + \begin{Bmatrix} \nu s \\ \tau \end{Bmatrix} A_{\mu\tau} \right) \tag{29a}$$

$$A_{s}^{\mu\nu} = \frac{\partial A^{\mu\nu}}{\partial x_{s}} + \sum_{\tau} \left(\begin{Bmatrix} s\tau \\ \mu \end{Bmatrix} A^{\tau\nu} + \begin{Bmatrix} s\tau \\ \nu \end{Bmatrix} A^{\mu\tau} \right). \tag{30a}$$

From these one can easily convince oneself that the "extensions" of the fundamental tensors $g_{\mu\nu}$ and $g^{\mu\nu}$, respectively, vanish.

The divergence of $A^{\mu\nu}$ with respect to the index ν follows from (31), (30a), and (33):

$$A^{\mu} = \sum_{s\nu} A_{s}^{\mu\nu} \delta_{\nu}^{s} = \frac{1}{\sqrt{g}}\left(\sum_{\nu} \frac{\partial(A^{\mu\nu}\sqrt{g})}{\partial x_{\nu}} + \sum_{\tau\nu} \begin{Bmatrix} \tau\nu \\ \mu \end{Bmatrix} A^{\tau\nu}\sqrt{g}\right). \tag{39}$$

For an *anti-symmetric* tensor (six-vector) this produces, due to the symmetry of $\begin{Bmatrix} \tau\nu \\ \mu \end{Bmatrix}$ in the indices τ and ν,

$$A^{\mu} = \frac{1}{\sqrt{g}}\sum_{\nu} \frac{\partial(A^{\mu\nu}\sqrt{g})}{\partial x_{\nu}}. \tag{40}$$

The case of a symmetrical $A^{\mu\nu}$ allows for a rearrangement of (39), which is of importance for what follows next. We form a covariant four-vector $\sum_{\mu} A^{\mu}g_{\mu\sigma} = A_{\sigma}$, which is reciprocal to (A^{μ}):

$$A_{\sigma} = \frac{1}{\sqrt{g}}\left(\sum_{\mu\nu} g_{\mu\sigma} \frac{\partial(A_{\mu\nu}\sqrt{g})}{\partial x_{\nu}} + \sqrt{g}\sum_{\tau\nu} \begin{bmatrix} \tau\nu \\ \sigma \end{bmatrix} A^{\tau\nu}\right) \tag{9}$$

$$= \frac{1}{\sqrt{g}}\left(\sum_{\mu\nu} \frac{\partial(g_{\mu\sigma}A^{\mu\nu}\sqrt{g})}{\partial x_{\nu}} + \frac{1}{2}\sqrt{g}\sum_{\mu\nu}\left(-\frac{\partial g_{\mu\sigma}}{\partial x_{\nu}} + \frac{\partial g_{\nu\sigma}}{\partial x_{\mu}} - \frac{\partial g_{\mu\nu}}{\partial x_{\sigma}}\right)A^{\mu\nu}\right). \tag{20}$$

And from this, provided $A^{\mu\nu}$ is symmetrical:

$$A_{\sigma} = \frac{1}{\sqrt{g}}\sum_{\mu\nu}\left(\frac{\partial(g_{\mu\sigma}A^{\mu\nu}\sqrt{g})}{\partial x_{\nu}} - \frac{1}{2}\frac{\partial g_{\mu\nu}}{\partial x_{\sigma}} A^{\mu\nu}\sqrt{g}\right), \tag{41}$$

for which one can also write

$$A_{\sigma} = \frac{1}{\sqrt{g}}\left(\sum_{\nu} \frac{\partial(A_{\sigma}^{\nu}\sqrt{g})}{\partial x_{\nu}} - \frac{1}{2}\sum_{\mu\nu\tau} g^{\tau\mu}\frac{\partial g_{\mu\nu}}{\partial x_{\sigma}} A_{\tau}^{\nu}\sqrt{g}\right) \tag{41a}$$

after the mixed tensor $\sum_{\mu} g_{\sigma\mu}A^{\mu\nu} = A_{\sigma}^{\nu}$ has been introduced.

The Riemann-Christoffel Tensor. Formula (29) allows for a very simple derivation [p. 1053] of the well-known criterion which allows us to decide whether or not a given continuum with a given line element is Euclidean—in other words, whether or not it is possible to select a suitably chosen substitution such that ds^2 is everywhere equal to the sum of the squares of the coordinate differentials.

By twofold extension, we form, according to (29), from the covariant four-vector A_μ the tensor of rank three $(A_{\mu\nu\lambda})$. One obtains

$$A_{\mu\nu\lambda} = \frac{\partial^2 A_\mu}{\partial x_\nu \partial x_\lambda} - \sum_\tau \left[\begin{Bmatrix} \mu\lambda \\ \tau \end{Bmatrix} \frac{\partial A_\tau}{\partial x_\nu} + \begin{Bmatrix} \mu\nu \\ \tau \end{Bmatrix} \frac{\partial A_\tau}{\partial x_\lambda} \right]$$

{10}
$$- \sum_\tau \begin{Bmatrix} \nu\lambda \\ \tau \end{Bmatrix} \frac{\partial A_\mu}{\partial x_\tau} + \sum_{\sigma\tau} \begin{Bmatrix} \nu\lambda \\ \tau \end{Bmatrix} \begin{Bmatrix} \tau\mu \\ \sigma \end{Bmatrix} A_\sigma$$

[21]
$$- \sum_\sigma \left[\frac{\partial}{\partial x_\lambda} \begin{Bmatrix} \mu\nu \\ \sigma \end{Bmatrix} - \sum_\tau \begin{Bmatrix} \mu\lambda \\ \tau \end{Bmatrix} \begin{Bmatrix} \nu\tau \\ \sigma \end{Bmatrix} \right] A_\sigma.$$

From this follows immediately that $(A_{\mu\lambda\nu} - A_{\mu\nu\lambda})$ is also a covariant tensor of rank three, and consequently

[22]
$$\sum_\sigma \left[\frac{\partial}{\partial x_\lambda} \begin{Bmatrix} \mu\nu \\ \sigma \end{Bmatrix} - \frac{\partial}{\partial x_\nu} \begin{Bmatrix} \mu\lambda \\ \sigma \end{Bmatrix} + \sum_\tau \left(\begin{Bmatrix} \mu\nu \\ \tau \end{Bmatrix} \begin{Bmatrix} \lambda\tau \\ \sigma \end{Bmatrix} - \begin{Bmatrix} \mu\lambda \\ \tau \end{Bmatrix} \begin{Bmatrix} \nu\tau \\ \sigma \end{Bmatrix} \right) \right] A_\sigma$$

is a covariant tensor σ of rank three; that is, the square bracket is a tensor of rank four $(K^\sigma_{\mu\nu\lambda})$ which is covariant in μ, ν, λ and contravariant in σ. All components of this tensor vanish when the $g_{\mu\nu}$ are constants. This vanishing happens always when it occurs relative to one suitably chosen coordinate system. The vanishing of the bracketed expression for all index combinations is therefore a *necessary condition* for the possibility that the line element can be brought into Euclidean form. However, it requires further proof to show that the condition is also *sufficient*.

V-Tensors. A glance at the formulas (37), (39), (40), (41), (41a) shows that tensor components are often encountered with the factor $\sqrt{g}$. For this reason we want to introduce a special notation for tensor components which are multiplied with $\sqrt{g}$ (or $\sqrt{-g}$ in case g is negative). We write these products with German (or Gothic) letters and set, for example

$$A_\sigma \sqrt{g} = \mathfrak{A}_\sigma$$

$$A_\sigma^\nu \sqrt{g} = \mathfrak{A}_\sigma.$$

We call (A_σ), (A_σ^v), *V*-tensors (volume tensors). When multiplied by $d\tau$, they [23]
represent tensors in the previously defined sense because $\sqrt{g}d\tau = \sqrt{g}dx_1dx_2dx_3dx_4$ is [p. 1054]
a scalar. For example, equation (41a) can be rewritten in this notation as

$$\mathfrak{A}_\sigma = \sum_v \frac{\partial \mathfrak{A}_\sigma^v}{\partial x_v} - \frac{1}{2}\sum_{\mu\tau v} g^{\tau\mu}\frac{\partial g_{\mu v}}{\partial x_\sigma}\mathfrak{A}_\sigma^v. \tag{41b}$$

C. The Equations of Physical Processes in a Given Gravitational Field

Every equation of the original theory of relativity has an equivalent, generally
covariant equation, in the sense of the previous section (B) which takes its place in
the generalized theory of relativity. For the establishment of these equations, the
fundamental tensor of the $g_{\mu v}$ has to be considered as given. In this manner, one
obtains generalizations of those physical laws which are already known in the original
theory of relativity. The generalized equations show to us the influence of the
gravitational field upon those processes to which these equations relate. Only the
differential relations of the gravitational field proper remain unknown for the time
being; they have to be found in a special manner. We want to subsume all other laws
(e.g., mechanical, electromagnetical) under the name "laws of material processes."

§9. Energy-Momentum Theorem for "Material Processes"

The energy-momentum theorem is the most general law concerning "material
processes." In the original theory of relativity and using the formulation of
MINKOWSKI-LAUE, it can be written as follows: [24]

$$\left.\begin{array}{l}
\dfrac{\partial p_{xx}}{\partial x} + \dfrac{\partial p_{xy}}{\partial y} + \dfrac{\partial p_{xz}}{\partial z} + \dfrac{\partial(i\,\mathfrak{i}_x)}{\partial l} = f_x \\[2mm]
\dfrac{\partial p_{yx}}{\partial x} + \dfrac{\partial p_{yy}}{\partial y} + \dfrac{\partial p_{yz}}{\partial z} + \dfrac{\partial(i\,\mathfrak{i}_y)}{\partial l} = f_y \\[2mm]
\dfrac{\partial p_{zx}}{\partial x} + \dfrac{\partial p_{zy}}{\partial y} + \dfrac{\partial p_{zz}}{\partial z} + \dfrac{\partial(i\,\mathfrak{i}_z)}{\partial l} = f_z \\[2mm]
\dfrac{\partial(i\,\mathfrak{s}_x)}{\partial x} + \dfrac{\partial(i\,\mathfrak{s}_y)}{\partial y} + \dfrac{\partial(i\,\mathfrak{s}_z)}{\partial z} + \dfrac{\partial(-\eta)}{\partial l} = iw.
\end{array}\right\}. \tag{42}$$

The time coordinate is chosen $l = it$, where the real time t is measured such that
the velocity of light comes out equal to 1.

[p. 1055]

$$
\begin{array}{cccc}
p_{xx} & p_{xy} & p_{xz} & i\mathfrak{t}_x \\
p_{yx} & p_{yy} & p_{yz} & i\mathfrak{t}_y \\
p_{zx} & p_{zy} & p_{zz} & i\mathfrak{t}_z \\
i\mathfrak{s}_x & i\mathfrak{s}_y & i\mathfrak{s}_z & -\eta
\end{array}
$$

is a symmetric tensor $(T_{\sigma v})$ of rank two (energy tensor);

$$
f_x,\, f_y,\, f_z,\, iw
$$

is a four-vector (K_σ). Of course, this is true for both only under linear orthogonal substitutions, which are the only ones admissible in the original theory of relativity. From a formal point of view, (42) says that (K_σ) is the divergence of the energy tensor $T_{\sigma v}$. Physically, the meanings are

p_{xx} etc., the "stress components"
$\mathfrak{t}$ the vector of momentum density
$\mathfrak{s}$ the vector of energy current
η the energy density
f the force vector per unit volume externally impressed upon the system
w the energy per unit volume and unit time supplied to the system.

The right-hand sides of equations (42) vanish if the system is a "complete" one.

Our task is now to find the generally-covariant equations that correspond to equations (42). It is clear that the generalized equations, too, are formally characterized by the fact that the divergence of a tensor of rank two is equated to a four-vector. However, with each such generalization one faces the difficulty that in generalized relativity theory—contrary to the original one—there are tensors of different character (covariant, contravariant, mixed, and furthermore the class of V-tensors); one therefore always has to make a choice. Yet this choice does not entail physical arbitrariness; it merely affects which variables are favored for representation.[4] The choice has to be made such that the equations are most comprehensive, and that the quantities used in them have the best descriptive physical

[p. 1056] meaning. It turns out that this aspect is best met when the tensor $T_{\sigma v}$ is represented by the mixed tensor $\mathfrak{T}^v_\sigma$, and the four-vector K_σ is represented by the covariant V-four-vector $\mathfrak{K}^v_\sigma$. The divergence is then to be formed according to (41b), whereupon one obtains as the generalization of (42) the generally-covariant equations

---

[4]This is connected to the fact that any tensor can be changed into one of another character by multiplying it with the fundamental tensor or with $\sqrt{-g}$, respectively.

$$\sum_\nu \frac{\partial \mathfrak{T}_\tau^\nu}{\partial x_\nu} = \frac{1}{2}\sum_{\mu\tau\nu} g^{\tau\mu}\frac{\partial g_{\mu\nu}}{\partial x_\sigma}\,\mathfrak{T}_\tau^\nu + \mathfrak{K}_\sigma. \tag{42a}$$

In keeping the above-listed names, we denote the components of $\mathfrak{T}_\sigma^\nu$ according to the scheme

[25]

	$\nu = 1$	$\nu = 2$	$\nu = 3$	$\nu = 4$
$\sigma = 1$	$-p_{xx}$	$-p_{xy}$	$-p_{xz}$	$-\mathfrak{i}_x$
$\sigma = 2$	$-p_{yx}$	$-p_{yy}$	$-p_{yz}$	$-\mathfrak{i}_y$
$\sigma = 3$	$-p_{zx}$	$-p_{zy}$	$-p_{zz}$	$-\mathfrak{i}_z$
$\sigma = 4$	$\mathfrak{s}_x$	$\mathfrak{s}_y$	$\mathfrak{s}_z$	η

$$\tag{43}$$

and the components of $\mathfrak{K}_\sigma$ according to the scheme

$\sigma = 1$	$-f_x$
$\sigma = 2$	$-f_y$
$\sigma = 3$	$-f_z$
$\sigma = 4$	w

$$\tag{44}$$

The purely covariant (or purely contravariant) tensor that is associated to $\mathfrak{T}_\sigma^\nu$ is symmetric. One understands easily that equations (42a) transform into equations (42) when the quantities $g_{\mu\nu}$ take the special values

$$\left.\begin{matrix} -1 & 0 & 0 & 0 \\ 0 & -1 & 0 & 0 \\ 0 & 0 & -1 & 0 \\ 0 & 0 & 0 & 1 \end{matrix}\right\}. \tag{45}$$

Discussion of (42a). First, we consider the special case that there is no gravitational field, i.e., that all the $g_{\mu\nu}$ are to be seen as constants. The first term on the right-hand side of (42a) must then vanish. Say that the system under consideration extends spatially, i.e., with respect to x_1, x_2, x_3, and is finite. The integral of a [p. 1057] quantity ϕ, when extended over the entire system, shall be denoted by $\overline{\phi}$. This type of integration over x_1, x_2, x_3 yields from (42a)

$$\frac{d\,\bar{\mathbf{i}}}{dx_4} = f$$

$$\frac{d\bar{\eta}}{dx_4} = w.$$

These are the theorems on the balances of momentum and energy in their conventional form. From them follows the constancy of momentum $\bar{\mathbf{i}}$ and energy $\bar{\eta}$ in time when no outer forces are present. The energy-momentum theorem expresses itself, in this case, as a conservation theorem proper that can be written in differential notation by the equation

$$\sum_{v} \frac{\partial \mathfrak{T}_{\sigma}^{v}}{\partial x_{v}} = 0 \qquad (42b)$$

when outer forces are absent ($\mathfrak{K}_{\sigma} = 0$).

There is no real conservation theorem for the (spatially finite) system under consideration when a gravitational field exists, i.e., when the $g_{\mu v}$ are not constant—even if the $\mathfrak{K}_{\sigma}$ vanish. The reason for this is that no equation of type (42b) prevails because the first term on the right-hand side of (42a) does not vanish. This corresponds to the physical fact that momentum and energy of a material system do change in a gravitational field in the course of time, because this field does transfer momentum and energy to the material system. The physical meaning of the first term on the right-hand side of (42a) is, therefore, analogous to that of the second term. The components of the first term, which we could call

$$- f_x^{(g)}, \; - f_y^{(g)}, \; - f_z^{(g)}, \; w^{(g)},$$

consequently express the negative momentum and energy resp. which is transferred (per unit volume and unit time), from the gravitational field to the material system.

However, for vanishing $\mathfrak{K}_{\sigma}$ one has to demand that for the material system and associated gravitational field combined, some theorems remain that express the constancy of total momentum and total energy of matter plus gravitational field. This leads to the statement that a complex of quantities t_{σ}^{v} must exist for the gravitational field such that the equations

$$\sum_{v} \frac{\partial (\mathfrak{T}_{\sigma}^{v} + t_{\sigma}^{v})}{\partial x_{v}} = 0 \qquad (42c)$$

[p. 1058] apply. This point can only be discussed in more detail when the differential laws of the gravitational field have been established.

One sees that for the action of a gravitational field upon material processes, the quantities

$$\Gamma^{\tau}_{v\sigma} = \frac{1}{2} \sum_{\mu} g^{\tau\mu} \frac{\partial g_{\mu v}}{\partial x_{\sigma}} \qquad (46) \qquad [27]$$

are of crucial importance, and for this reason we want to call them the "components of the gravitational field."

§10. Equations of Motion of Continuously Distributed Masses

Naturally measured quantities. It has already been emphasized that it is not possible in a generalized theory of relativity to select a coordinate system such that spatial and temporal coordinate differences are in a direct connection with the results obtained from measuring rods and clocks, as has been the case in the original theory of relativity. Such preferred choice of coordinates is only possible in the infinitesimally small by setting

$$ds^2 = \sum_{\mu v} g_{\mu v} dx_{\mu} dx_{v} = -d\xi_1^2 - d\xi_2^2 - d\xi_3^2 - d\xi_4^2. \qquad (46) \qquad \{11\}$$

The $d\xi$ are (see §2) just as measurable as the coordinates in the original theory of relativity, but they are not complete differentials. It is possible in the infinitesimally small to refer all quantities upon the coordinate system of the $d\xi$. When this is done, we call them "naturally measured" quantities. And we call the coordinate system the "normal system."

According to (17a) we have for infinitesimally small four-dimensional volumina

$$\sqrt{-g} \int dx_1 dx_2 dx_3 dx_4 = \int d\xi_1 d\xi_2 d\xi_3 d\xi_4. \qquad (47)$$

Let the volume under consideration consist of an infinitesimally short and infinitesimally thin four-dimensional thread. And let dv be the integral $\int dx_1 dx_2 dx_3$ extended over it. We now choose the $d\xi$ such that the $d\xi_4$-axis coincides with the axis of the thread, which effects $d\xi_4 = ds$ and the integral $\int d\xi_1 d\xi_2 d\xi_3$ to be the naturally measured volume dv_0 of the thread at rest. According to (47) we then find

$$\sqrt{-g}\, dv\, dx_4 = dv_0 ds \qquad (47a)$$

Unit of mass. The *comparison* of the masses of two mass points can be done by [p. 1059] the usual methods. We merely need a unit mass to measure masses. Let it be defined as the quantity of water that fits into a naturally measured unit volume 1 that is at rest. The mass of a material point is by definition an invariant under all transformations.

The scalar of density. By scalar density of continuously distributed matter we mean its mass per (co-moved) naturally measured unit volume. When it is permissible to ignore surface forces, then the scalar of density together with the components of velocity $\dfrac{dx_\mu}{ds}$ characterizes matter completely in the hydrodynamic sense.

The energy tensor of mass flow. Equations of motion. One can form a mixed *V*-tensor

$$\mathfrak{T}^\nu_\sigma = \rho_0 \sqrt{-g} \,\frac{dx_\nu}{ds} \sum_\mu g_{\sigma\mu} \frac{dx_\mu}{ds} \tag{48}$$

from the scalar ρ_0 and the contravariant four-vector $\left(\dfrac{dx_\mu}{ds}\right)$.

[28]   One can anticipate that $(\mathfrak{T}^\nu_\sigma)$ is the energy tensor of ponderable mass flow, and that equations (42a), in combination with (48), correspond to the Eulerian flow equations of incoherent masses, i.e., when surface forces can be ignored. We prove this by deriving from these equations those which we have previously given for the motion of a material point.

[29]   Let the extension of the masses under consideration in x_1, x_2, x_3 be infinitesimally small. We integrate (42a) in these variables over the entire "thread of flow" and use the abbreviation $dx_1 dx_2 dx_3 = dv$. We then get

$$\frac{d}{dx_4}\left\{\int \mathfrak{T}^4_\sigma dv\right\} = \sum_{\tau\nu}\left\{\Gamma^\tau_{\nu\sigma}\int \mathfrak{T}^\nu_\tau dv\right\} + \int \mathfrak{K}_\sigma dv. \tag{50}$$

Substituting for $\mathfrak{T}^\nu_\sigma$ the expression from (48) and considering that according to (47a)

$$dv = \frac{dv_0}{\sqrt{-g}}\frac{ds}{dx_4}, \tag{47b}$$

and furthermore that

$$m = \int \rho_0 dv_0, \tag{49}$$

[p. 1060] one arrives at the equation:

$$\frac{d}{dx_4}\left\{m\sum_\mu g_{\sigma\mu}\frac{dx_\mu}{ds}\right\} = \sum_{\nu\tau}\left\{\Gamma^\tau_{\nu\sigma}\frac{dx_\nu}{dx_4}m\sum_\mu g_{\tau\mu}\frac{dx_\mu}{ds}\right\} + \int \mathfrak{K}_\sigma dv. \tag{50a}$$

or, using as an abbreviation the covariant four-vector

$$\mathbf{I} = m\sum_\mu g_{\sigma\mu}\frac{dx_\mu}{ds} \tag{51}$$

one finally gets[5]

$$\frac{d\,\mathbf{I}_\sigma}{dx_4} = \sum_{v\tau} \Gamma^\tau_{v\sigma} \frac{dx_v}{dx_4}\,\mathbf{I}_\tau + \int \Re_\sigma\, dv. \tag{50b}$$

This is the equation of motion of a material point if the fourth coordinate ("time coordinate") is chosen as an independent variable. The components of $(\mathbf{I}_\sigma)$ have the physical meaning of being the negative components of momentum and energy, resp.; this follows from the scheme (43). In the special case of the original theory of relativity, i.e., when the $g_{\mu v}$ have the values given in (18), one gets

---

[5]At this point I might mention why, in my opinion, not equation (39) but rather equation (41) was used for the formulation of the energy-momentum theorem. According to (39), the energy tensor has to be seen as a contravariant V-tensor and the quantities $\begin{Bmatrix} \tau v \\ \mu \end{Bmatrix}$ as the components of the gravitational field. In this manner we would have been led, in §11, to the interpretation that the components of the *contravariant* four-vector $\left(\mathbf{I}^\sigma = m\dfrac{dx_\sigma}{ds} \right)$ represent the components of momentum and energy of the material point. We will show here in a very special case that such interpretation runs counter to our understanding of the essence of momentum.

Into a space without gravitational field, we introduce a coordinate system that deviates from a "normal system" only inasfar as the x_1-axis and the x_2-axis (judged from a normal system) enclose an angle ϕ different from $\dfrac{\pi}{2}$. Then we have

$$ds^2 = -dx_1^2 - dx_2^2 - 2dx_1 dx_2 \cos\phi - dx_3^2 + dx_4^2.$$

And in this case one gets, e.g., $-\mathbf{I}^2 = -m\,\dfrac{dx_2}{ds}$. This quantity vanishes if the point moves in the direction of the x_1-axis. On the other hand, it is clear that in the case under consideration an x_2-component of momentum actually exists, and it deviates from the x_1-component only by a factor of $\cos\phi$.

However, if one bases the momentum theorem on (41) and uses, therefore, according to (51) the *covariant* four-vector for the calculation of momentum and energy, then one

[30]   gets, in our case, $-\mathbf{I}_2 = -g_{12}m\,\dfrac{dx_1}{ds} = m\left(\dfrac{dx_1}{ds}\right)\cos\phi = (-\mathbf{I}_1)\cos\phi$, as it must be demanded.

$$
\left.\begin{array}{l}
-\mathbf{I}_1 = \dfrac{m\,\mathfrak{q}_x}{\sqrt{1 - q^2}} \\
\cdots\cdots\cdots \\
\mathbf{I}_4 = \dfrac{m}{\sqrt{1 - q^2}}
\end{array}\right\}
\tag{52}
$$

if $\mathfrak{q}$ represents the three-dimensional vector of velocity and q its magnitude. This is in agreement with the theory if we recall that we have—with the definition (18)—chosen the "light second" as our time unit.

[p. 1061]   Assume that $\mathfrak{K}_\sigma$ vanishes in (50b), that is, the outer forces vanish, but not those originating from the gravitational field. If then the equation is multiplied by $\dfrac{1}{m} \cdot \dfrac{dx_4}{ds}$, one obtains for the motion of a material point in a gravitational field, after a simple calculation, equation (23a), which is equivalent to (1). This confirms our anticipation that $\mathfrak{T}_\sigma^\nu$ in (48) is indeed the energy tensor of flowing matter.

The energy tensor of an ideal fluid. We now want to complete (48) such that we obtain the energy tensor of an ideal fluid while allowing for the surface forces (pressure) and the changes in energy due to changes in density.[6] Without difficulty, one can find the energy tensor at a point in the medium for that specific normal system whose $d\xi_4$-axis (at the point under consideration) coincides with the element of the four-dimensional line of flow.

Let ϕ be the (naturally measured) volume of such an amount of substance that, when brought to pressure 0, the volume ϕ_0 has mass 1. The naturally measured energy ε of this amount with volume ϕ is then—considering only adiabatic changes of state—

$$
1 - \int_{\phi_0}^{\phi} p\,d\phi,
$$

where p denotes the naturally measured pressure. According to (52), the energy of the unit mass at rest is 1 when pressure vanishes. The negatively taken integral is only a function of pressure, and we call it P. The energy unit per volume follows after multiplication with $\rho_0 = \dfrac{1}{\phi}$. Consequently, the energy density is

$$
\rho_0(1 + P).
$$

---

[6] In saying this, we restrict ourselves to adiabatic processes of flow in a fluid with a uniformly adiabatic equation of state.

With our special choice of coordinates, the tensor we are looking for is given by [p. 1062] the components

$$
\begin{matrix}
-p & 0 & 0 & 0 \\
0 & -p & 0 & 0 \\
0 & 0 & -p & 0 \\
0 & 0 & 0 & \rho_0(1 + P).
\end{matrix}
$$

With an arbitrarily chosen system of reference, this tensor obviously transforms into

$$
\mathfrak{T}_\sigma^\nu = -p\delta_\sigma^\nu\sqrt{-g} + \rho_0\sqrt{-g}\,(1 + \frac{p}{\rho_0} + P)\frac{dx_\nu}{ds}\sum_\mu g_{\sigma\mu}\frac{dx_\mu}{ds}. \tag{48a}
$$

[31]

After all, ρ_0, p, and P are scalars by definition. If we use the abbreviation

$$
\rho_0\sqrt{-g}\,(1 + \frac{p}{\rho_0} + P) = \rho^*,
$$

equations (42a) yield

$$
-\sqrt{-g}\,\frac{\partial p}{\partial x_\sigma} + \sum_{\mu\nu}\frac{\partial}{\partial x_\nu}\left(\rho^* g_{\sigma\mu}\frac{dx_\mu}{ds}\frac{dx_\nu}{ds}\right) = \frac{1}{2}\sum_{\mu\nu}\rho^*\frac{\partial g_{\mu\nu}}{\partial x_\sigma}\frac{dx_\mu}{ds}\frac{dx_\nu}{ds} + \mathfrak{K}_\sigma. \tag{53}
$$

These four equations determine the five unknown functions p and $\dfrac{dx_\nu}{ds}$ since the latter share the relation

$$
\sum_{\mu\nu} g_{\mu\nu}\frac{dx_\mu}{ds}\frac{dx_\nu}{ds} = 1
$$

and ρ is a known function of p, with a known adiabatic equation of state of our fluid. The $g_{\mu\nu}$ and $\mathfrak{K}_\sigma$ are to be considered as known. Equations (53) replace the Eulerian equations, inclusive of the continuity equation. This can easily be shown by specializing to the case of the original theory of relativity, provided one introduces the negligible terms resulting from this, such as velocities small compared to the speed of light and pressures so small that they have no perceptible influence upon inertia.

§11. The Electromagnetic Equations

[32]

The deliberations that will lead us to the generally-covariant equations of the electromagnetic processes are completely analogous to those we had to carry out when we integrated this subject matter into the original theory of relativity; we can, therefore, be brief here.

[p. 1063]   *Electromagnetic equations in vacuum.* Let $\mathfrak{F}^{\mu\nu}$ and $\mathfrak{F}^{\mu\nu*}$ be two dual and
[33]   contravariant V-six-vectors (see (24)). It then follows from (40) that the expressions

$$\sum_\nu \frac{\partial \mathfrak{F}^{\mu\nu}}{\partial x_\nu}, \; \sum_\nu \frac{\partial \mathfrak{F}^{\mu\nu*}}{\partial x_\nu}$$

are the components of contravariant V-four-vectors. One obtains the Maxwellian equations for a vacuum in a generally-covariant form if one sets these components equal to zero. It is indeed easily seen that these equations go into their Maxwellian form if one denotes the components of $\mathfrak{F}^{\mu\nu}$ and $\mathfrak{F}^{\mu\nu*}$ according to the scheme

$\mathfrak{F}^{23}$	$\mathfrak{F}^{31}$	$\mathfrak{F}^{12}$	$\mathfrak{F}^{14}$	$\mathfrak{F}^{24}$	$\mathfrak{F}^{34}$	$\mathfrak{F}^{23*}$	$\mathfrak{F}^{31*}$	$\mathfrak{F}^{12*}$	$\mathfrak{F}^{14*}$	$\mathfrak{F}^{24*}$	$\mathfrak{F}^{34*}$
$\mathfrak{h}_x$	$\mathfrak{h}_y$	$\mathfrak{h}_z$	$-\mathfrak{e}_x$	$-\mathfrak{e}_y$	$-\mathfrak{e}_z$	$-\mathfrak{e}_x^*$	$-\mathfrak{e}_y^*$	$-\mathfrak{e}_z^*$	$-\mathfrak{h}_x^*$	$-\mathfrak{h}_y^*$	$-\mathfrak{h}_z^*$

and observes, furthermore, that according to (24)

$$\mathfrak{h}^* = \mathfrak{h}$$
$$\mathfrak{e}^* = \mathfrak{e}$$

whenever the $g_{\mu\nu}$ take the special values given in (18).

Charge density, convection current. There is obviously an electrical charge density in the co-moving normal system. This density is a scalar by definition. Multiplying it with $\sqrt{-g}$ produces a V-scalar which we denote by $\rho_{(e)}$. Together with the contravariant four-vector $\dfrac{dx_\mu}{ds}$, it forms the contravariant V-four-vector of the convection current

$$\rho_{(e)} \frac{dx_\mu}{ds}.$$

The Lorentz equations in vacuum. When all interactions between matter and electromagnetic fields are reduced to the movement of electrical charges—as Lorentz does—one has to base everything upon the equations

$$\left. \begin{array}{l} \displaystyle\sum_\nu \frac{\partial \mathfrak{F}^{\mu\nu}}{\partial x_\nu} = \rho_e \frac{dx_\mu}{ds} \\[3mm] \displaystyle\sum_\nu \frac{\partial \mathfrak{F}^{\mu\nu*}}{\partial x_\nu} = 0 \end{array} \right\}. \tag{54}$$

They are the fundamental equations of Lorentz's theory of the electrons in general-
[p. 1064] covariant form. They explain by which laws the gravitational field acts upon the electromagnetic field.

Electromagnetic equations of moving bodies for the case considering only bodies with dielectric constant 1 and magnetic permeability 1. Let us consider the electric and magnetic polarization of bodies only insofar as they cause electric and magnetic charge densities; there shall be no electric or magnetic "polarization currents." Electric conduction currents along conductors, however, shall be considered. The general-covariant field equations for this case are found by taking into account (on the right-hand sides of the equations) the electric and magnetic convection currents as well as the electric conduction current along conductors.

Let ρ_e be the charge density of polarization electrons and conduction electrons combined, as has been previously defined. $\left(\rho_{(e)}\dfrac{dx_\mu}{ds}\right)$ is then the V-four-vector of the convection current, produced jointly by polarization electrons and conduction electrons.

Let ρ_m be the previously defined magnetic charge density that originates from the (rigid) magnetic polarization. $\left(\rho_{(m)}\dfrac{dx_\mu}{ds}\right)$ is then the V-four-vector of the magnetic convection current.

Another V-four-vector will correspond to the conduction current, and we shall denote by it $(\mathfrak{L}^\mu)$. It is determined by the fact that in the "normal system"

$$\mathfrak{L}^1 = -\lambda\mathfrak{F}^{14} \qquad \mathfrak{L}^2 = -\lambda\mathfrak{F}^{24} \qquad \mathfrak{L}^3 = -\lambda\mathfrak{F}^{34} \qquad \mathfrak{L}^4 = 0$$

and, on the other hand,

$$\frac{dx_1}{ds} = 0, \qquad \frac{dx_2}{ds} = 0, \qquad \frac{dx_3}{ds} = 0, \qquad \frac{dx_4}{ds} = 1. \qquad \{12\}$$

These conditions are met by setting

$$\mathfrak{L}^\mu = -\lambda\sum_{\alpha\beta} g_{\alpha\beta}\,\mathfrak{F}^{\mu\alpha}\,\frac{dx_\beta}{ds}. \tag{55}$$

The field equations then turn out to be

$$\left.\begin{aligned}
\sum_\nu \frac{\partial\mathfrak{F}^{\mu\nu}}{dx_\nu} &= \rho_{(e)}\frac{dx_\mu}{ds} + \mathfrak{L}^\mu \\
\sum_\nu \frac{\partial\mathfrak{F}^{\mu\nu*}}{dx_\nu} &= \rho_{(m)}\frac{dx_\mu}{ds}
\end{aligned}\right\}. \tag{56}$$

Field equations for isotropic, electrically and magnetically polarizable moving [p. 1065] *bodies.* We modify the case which we just considered, and now also take into account electric and magnetic polarization currents. It is assumed that in the co-moved normal system, the components of the field strengths are proportional to these polarizations.

We obtain the field equations for this case from (56) by adding on their right-hand sides the expressions for the V-four-vectors of the electric and magnetic polarization current, respectively. We represent the electric polarization by a contravariant V-four-vector ($\mathfrak{P}^{\mu}_{(e)}$) whose components in the co-moving normal system are determined by the equations

$$\mathfrak{P}^{1}_{(e)} = -\sigma_{(e)}\,\mathfrak{F}^{14}; \quad \mathfrak{P}^{2}_{(e)} = -\sigma_{(e)}\,\mathfrak{F}^{24}; \quad \mathfrak{P}^{3}_{(e)} = -\sigma_{(e)}\,\mathfrak{F}^{34}; \quad \mathfrak{P}^{4}_{(e)} = 0.$$

These conditions are met by the equations

$$\mathfrak{P}^{\mu}_{(e)} = -\sigma_{(e)}\sum_{\alpha\beta} g_{\alpha\beta}\,\mathfrak{F}^{\mu\alpha}\,\frac{dx_{\beta}}{ds}. \tag{57}$$

From this V-four-vector we form the V-six-vector

$$\mathfrak{P}^{\mu\nu}_{(e)} = \mathfrak{P}^{\mu}_{(e)}\,\frac{dx_{\nu}}{ds} - \mathfrak{P}^{\nu}_{(e)}\,\frac{dx_{\mu}}{ds}, \tag{58}$$

and from this the contravariant V-four-vector of the electric convection current

$$\sum_{\nu} \frac{\partial \mathfrak{P}^{\mu\nu}_{(e)}}{\partial x_{\nu}} \tag{59}$$

by forming the divergence according to (40). We note that the components of this vector in the normal system are

$$\frac{\partial(\sigma_{(e)}\mathfrak{e}_x)}{\partial t}; \quad \frac{\partial(\sigma_{(e)}\mathfrak{e}_y)}{\partial t}; \quad \frac{\partial(\sigma_{(e)}\mathfrak{e}_z)}{\partial t}; \quad -\left(\frac{\partial(\sigma_{(e)}\mathfrak{e}_x)}{\partial x} + \frac{\partial(\sigma_{(e)}\mathfrak{e}_y)}{\partial y} + \frac{\partial(\sigma_{(e)}\mathfrak{e}_z)}{\partial z}\right).$$

Therefore, if (59) is added on the right-hand side of the first equation (56), one obtains equations which, in the normal system, transform into the first Maxwellian equation system for bodies at rest. This justifies statements (57), (58), (59).

For the magnetic polarization we set, in analogy,

{13}

$$\mathfrak{P}^{\mu}_{(m)} = -\sigma_{(m)}\sum_{\alpha\beta} g_{\alpha\beta}\,\mathfrak{F}^{\mu\alpha*}\,\frac{dx_{\beta}}{ds}. \tag{57a}$$

$$\mathfrak{P}^{\mu\nu}_{(m)} = \mathfrak{P}^{\mu}_{(m)}\,\frac{dx_{\nu}}{ds} - \mathfrak{P}^{\nu}_{(m)}\,\frac{dx_{\mu}}{ds}, \tag{58a}$$

[p. 1066] which yields the components of the V-four-vector of the magnetic polarization current as

$$\sum_{\nu} \frac{\partial \mathfrak{P}^{\mu\nu}_{(m)}}{\partial x_{\nu}}. \tag{59}$$

One obtains in this manner the field equations

$$\left.\begin{array}{l} \displaystyle\sum_{\nu} \frac{\partial(\mathfrak{F}^{\mu\nu} - \mathfrak{P}^{\mu\nu}_{(e)})}{\partial x_{\nu}} = \rho_{(e)} \frac{dx_{\mu}}{ds} + \mathfrak{L}^{\mu} \\[4mm] \displaystyle\sum_{\nu} \frac{\partial(\mathfrak{F}^{\mu\nu*} - \mathfrak{P}^{\mu\nu}_{(m)})}{\partial x_{\nu}} = \rho_{(m)} \frac{dx_{\mu}}{ds}, \end{array}\right\} \qquad (60) \qquad \{14\}$$

where $\mathfrak{P}^{\mu\nu}_{(e)}$, $\mathfrak{P}^{\mu\nu}_{(m)}$, $\mathfrak{L}^{\mu}$ are connected to the six-vector of the field by the following relations

$$\left.\begin{array}{l} \mathfrak{P}^{\mu}_{(e)} = -\sigma_{(e)}\displaystyle\sum_{\alpha\beta} g_{\alpha\beta}\mathfrak{F}^{\mu\alpha}\frac{dx_{\beta}}{ds} \;\middle|\; \mathfrak{P}^{\mu}_{(m)} = -\sigma_{(m)}\displaystyle\sum_{\alpha\beta} g_{\alpha\beta}\mathfrak{F}^{\mu\alpha*}\frac{dx_{\beta}}{ds} \;\middle|\; \mathfrak{L}^{\mu} = -\lambda\displaystyle\sum_{\alpha\beta} g_{\alpha\beta}\mathfrak{F}^{\mu\alpha}\frac{dx_{\beta}}{ds} \\[4mm] \mathfrak{P}^{\mu\nu}_{(e)} = \mathfrak{P}^{\mu}_{(e)}\frac{dx_{\nu}}{ds} - \mathfrak{P}^{\nu}_{(e)}\frac{dx_{\mu}}{ds} \;\middle|\; \mathfrak{P}^{\mu\nu}_{(m)} = \mathfrak{P}^{\mu}_{(m)}\frac{dx_{\nu}}{ds} - \mathfrak{P}^{\nu}_{(m)}\frac{dx_{\mu}}{ds} \end{array}\right\} \quad (60a)$$

There are also no obstacles to carry out an energy-momentum summation along the lines of equation (42a); but the previous deliberations show with sufficient clarity how one has to proceed with the transformation of already known laws of nature into general-covariant ones.

D. The Differential Laws of the Gravitational Field

In the previous section we considered the coefficients $g_{\mu\nu}$ as given functions of the x_{ν}. And these coefficients are to be understood as the components of the gravitational potential. It remains to find the differential laws that are satisfied by these quantities. The epistemological satisfaction of the theory that has been developed up to here can be seen in the fact that this theory complies with the principle of relativity in the broadest meaning of the word. Seen under a formal aspect, this is based upon the feature that the equation systems are *general*, i.e., covariant under arbitrary substitutions of the x_{ν}.

The demand that the differential laws of the $g_{\mu\nu}$ must also be general-covariant appears therefore appropriate. However, we want to show that we have to restrict this demand if we want to satisfy the law of cause and effect. In fact, we shall prove that the laws that characterize the course of events in a gravitational field can impossibly be covariant in *all generality*.

§12. Proof of a Necessary Restriction in the Choice of Coordinates   [p. 1067]

[34]

We consider a finite part Σ of the continuum where some material process does not occur. The physical process in Σ is completely determined when the quantities $g_{\mu\nu}$ are given as functions of the x_ν relative to the coordinate system K which is used for the description. The totality of these functions is symbolically denoted by $G(x)$.

Let us introduce a new coordinate system K' which coincides with K outside of Σ, but deviates from K inside of Σ, such, however, that the $g'_{\mu\nu}$ relative to K' as well as the $g_{\mu\nu}$ (including their derivatives) are everywhere continuous. The totality of the $g'_{\mu\nu}$ is symbolically denoted by $G'(x')$. $G'(x')$ and $G(x)$ describe the same gravitational field. When we replace the coordinates x'_ν by the coordinates x_ν in the functions $g'_{\mu\nu}$, i.e., when we form $G'(x)$, then this $G'(x)$ also represents a gravitational field relative to K, which however, is not the same field as the factual (that is, the originally given) gravitational field.

If we assume the differential equations of the gravitational field to be everywhere covariant, then they are satisfied for $G'(x')$ relative to K' whenever they are satisfied for $G(x)$ relative to K. Therefore, they are also satisfied for $G'(x)$ relative to K. There are then two different solutions $G(x)$ and $G'(x)$ relative to K, even though the solutions coincide on the boundary of the domain Σ. In other words, *the course of events in this domain cannot be determined uniquely by general-covariant differential equations.*

Consequently, when we demand that the course of events in the gravitational field be a completely determined one, we are forced to restrict the choice of the coordinate system such as to make the introduction of a new coordinate system K' (as characterized above) impossible without a violation of these restrictions. The continuation of the coordinate system into the interior of the domain Σ cannot remain arbitrary.

§13. Covariance toward Linear Transformations. Adapted Coordinate Systems

Since we have seen that the coordinate system has to be subject to conditions, we must focus upon several kinds of specializations in the choice of coordinates. A very far-reaching specialization is obtained by admitting only linear transformations. Our theory would be deprived of its main support if we demand that the equations of

[p. 1068] physics be covariant merely toward *linear* transformations. A transformation to an accelerated or rotating system would no longer be an admissible transformation, and the physical equivalence of a "centrifugal field" and a gravitational field—emphasized in §1—would not be interpreted by the theory as to be, in essence, of like nature. On the other hand, it is advantageous to demand that linear transformations are also among the admissible transformations (as will be shown later). We have, therefore, to speak briefly about the modifications of the theory of covariants, set out in section

B, when only linear transformations instead of general ones are admitted. [35]

Covariant expressions relative to linear transformations. The algebraic properties of tensors that have been presented in §3 to §8 are not simplified if one limits oneself to linear transformations. In contrast, the rules for the formation of tensors by differentiation (§9) become significantly simpler.

It applies quite generally that

$$\frac{\partial}{\partial x'_{\rho}} = \sum_{\delta} \frac{\partial x_{\delta}}{\partial x'_{\rho}} \frac{\partial}{\partial x_{\delta}}.$$

Consequently one has, for example, for a covariant tensor of rank two, according to (5a), {15}

$$\frac{\partial A'_{\mu\nu}}{\partial x'_{\rho}} = \sum_{\alpha\beta\delta} \frac{\partial x_{\delta}}{\partial x'_{\rho}} \frac{\partial}{\partial x_{\delta}} \left(\frac{\partial x_{\alpha}}{\partial x'_{\mu}} \frac{\partial x_{\beta}}{\partial x'_{\nu}} A_{\alpha\beta} \right).$$

The derivatives $\dfrac{\partial x_{\alpha}}{\partial x'_{\mu}}$, etc., are independent of the x_{δ} under linear substitution,

and one has

$$\frac{\partial A'_{\mu\nu}}{\partial x'_{\rho}} = \sum_{\alpha\beta\delta} \frac{\partial x_{\alpha}}{\partial x'_{\mu}} \frac{\partial x_{\beta}}{\partial x'_{\nu}} \frac{\partial x_{\delta}}{\partial x'_{\rho}} \frac{\partial A_{\alpha\beta}}{\partial x_{\delta}}.$$

Therefore, $\left(\dfrac{\partial A_{\alpha\beta}}{\partial x_{\delta}} \right)$ is a covariant tensor of rank three.

It can be shown in all generality that differentiating the components of any tensor in its coordinates produces a tensor whose rank is raised by one, where the additional index carries covariant character. In other words, this is the operation of *extension* under restriction to linear transformations. And since extension combined with the algebraic operations is the very basis of forming covariants, we have command over the entire system of covariants relative to linear transformations. We now turn toward [p. 1069] deliberations that lead to a much limited choice of coordinates.

The transformation law of the integral J. Let H be a function of the $g^{\mu\nu}$ and {16}

their first derivatives $\dfrac{\partial g^{\mu\nu}}{\partial x_{\sigma}}$, where the latter are called $g_{\sigma}^{\mu\nu}$ for short. Now, J shall

be an integral extended over a finite part Σ of the continuum, thus

$$J = \int H\sqrt{-g}\,d\tau. \tag{61}$$

The coordinate system that is used first shall be K_1. We ask for the change ΔJ of J when we go from system K_1 to another one K_2, which is infinitesimally different from K_1. If $\Delta\phi$ denotes the increase—due to transformation—of an arbitrary quantity

ϕ at some point in the continuum, then one has, according to (17), first

$$\Delta(\sqrt{-g}\,d\tau) = 0 \tag{62}$$

and furthermore

$$\Delta H = \sum_{\mu\nu\sigma}\left(\frac{\partial H}{\partial g^{\mu\nu}}\Delta g^{\mu\nu} + \frac{\partial H}{\partial g_{\sigma}^{\mu\nu}}\Delta g_{\sigma}^{\mu\nu}\right). \tag{62a}$$

The $\Delta g^{\mu\nu}$ can be expressed by means of (8) in terms of the Δx_{μ} by taking the relations

$$\Delta g^{\mu\nu} = g^{\mu\nu\prime} - g^{\mu\nu}$$

$$\Delta x_{\mu} = x'_{\mu} - x_{\mu}$$

into account. One obtains

$$\Delta g^{\mu\nu} = \sum_{\alpha}\left(g^{\mu\alpha}\frac{\partial \Delta x_{\nu}}{\partial x_{\alpha}} + g^{\nu\alpha}\frac{\partial \Delta x_{\mu}}{\partial x_{\alpha}}\right) \tag{63}$$

$$\Delta g_{\sigma}^{\mu\nu} = \sum_{\alpha}\left\{\frac{\partial}{\partial x_{\sigma}}\left(g^{\mu\alpha}\frac{\partial \Delta x_{\nu}}{\partial x_{\alpha}} + g^{\nu\alpha}\frac{\partial \Delta x_{\mu}}{\partial x_{\alpha}}\right) - \frac{\partial g^{\mu\nu}}{\partial x_{\alpha}}\frac{\partial \Delta x_{\alpha}}{\partial x_{\sigma}}\right\}. \tag{63a}$$

The equations (62a), (63), (63a), present ΔH as linear homogeneous functions of the first and second derivatives of the Δx_{μ} in the coordinates.

We have, so far, not made any assumptions of how H shall depend upon the $g^{\mu\nu}$ and the $g_{\sigma}^{\mu\nu}$. Now we shall assume H to be invariant under linear transformations, i.e., ΔH shall vanish when $\dfrac{\partial^2 \Delta x_{\mu}}{\partial x_{\alpha}\partial x_{\sigma}}$ does. We obtain under this assumption

[36]

$$\frac{1}{2}\Delta H = \sum_{\mu\nu\tau\alpha} G^{\nu\alpha}\frac{\partial H}{\partial g_{\sigma}^{\mu\nu}}\frac{\partial^2 \Delta x_{\mu}}{\partial x_{\sigma}\partial x_{\alpha}}. \tag{64}$$

[p. 1070] By means of (64) and (62) one gets

$$\frac{1}{2}\Delta J = \int d\tau \sum_{\mu\nu\sigma\alpha} \frac{\partial H\sqrt{-g}}{\partial g_{\sigma}^{\mu\nu}}\frac{\partial^2 \Delta x_{\mu}}{\partial x_{\sigma}\partial x_{\alpha}}$$

and from this, by partial integration,

$$\frac{1}{2}\Delta J = \int d\tau \sum_{\mu}(\Delta x_{\mu}B_{\mu}) + F, \tag{65}$$

where we used the abbreviations

$$B_{\mu} = \sum_{\alpha\sigma\nu} \frac{\partial^2}{\partial x_{\sigma} \partial x_{\alpha}} \left(g^{\nu\alpha} \frac{\partial H\sqrt{-g}}{\partial g_{\sigma}^{\mu\nu}} \right) \tag{65a}$$

$$F = \int d\tau \sum_{\alpha\sigma\nu\mu} \frac{\partial}{\partial x_{\alpha}} \left[g^{\nu\alpha} \frac{\partial H\sqrt{-g}}{\partial g_{\sigma}^{\mu\nu}} \frac{\partial \Delta x_{\mu}}{\partial x_{\sigma}} - \frac{\partial}{\partial x_{\sigma}} \left(g^{\nu\sigma} \frac{\partial H\sqrt{-g}}{\partial g_{\alpha}^{\mu\nu}} \right) \Delta x_{\mu} \right]. \tag{65b} \quad \{17\}$$

F can be transformed into a surface integral. It vanishes when Δx_{μ} and $\dfrac{\partial \Delta x_{\mu}}{\partial x_{\sigma}}$ vanish at the boundary.

Adapted coordinate systems. We consider again our continuum and its domain   [37]
Σ, which we assume finite in all of its coordinates, and referred to the coordinate
system K. Starting from K we imagine successively introduced coordinate systems
K', K'' etc., all infinitely close to each other such that the Δx_{μ} and the $\dfrac{\partial \Delta x_{\mu}}{\partial \Delta x_{\alpha}}$ vanish
at the boundaries. We call all these systems "coordinate systems with coinciding
boundary coordinates." For every infinitesimal coordinate transformation between
neighboring coordinate systems of the totality K, K', K''... we have

$$F = 0,$$

so that instead of (65) we have the equation

$$\frac{1}{2}\Delta J = \sum_{\mu} \int d\tau \, \Delta x_{\mu} B_{\mu}. \tag{66} \quad [38]$$

Among all systems with coinciding boundary coordinates will be some for which J
attains an extremum as compared with the J-values of neighboring systems with
likewise coinciding boundary values. We call these coordinate systems "coordinate
systems adapted to the gravitational field." The equations

$$B_{\mu} = 0 \tag{67}$$

hold for these adapted systems according to (66) because the Δx_{μ} can be chosen
freely inside of Σ.

Inversely, (67) is a sufficient condition that the coordinate system is adapted to [p. 1071]
the gravitational field.

We avoid the difficulty mentioned in §13 by henceforth writing only differential
equations of the gravitational field that claim validity merely for adapted coordinate
systems. Under the limitation to adapted coordinate systems, it is indeed no longer
allowed to extend a coordinate system that is given for the exterior of Σ into the
interior of Σ in an arbitrary manner.

§14. The H-Tensor

Equation (65) leads us to a theorem that is of fundamental importance to the entire theory. If we vary the gravitational field of the $g_{\mu\nu}$ by an infinitely small amount, i.e., replace the $g_{\mu\nu}$ by $g^{\mu\nu} + \delta g^{\mu\nu}$, where the $\delta g^{\mu\nu}$ shall vanish in a zone of finite width adjacent to the boundary of Σ, then H becomes $H + \delta H$ and J becomes $J + \delta J$. We now claim that the equation

$$\Delta\{\delta J\} = 0 \tag{68}$$

always holds whichever way the $\delta g_{\mu\nu}$ might be chosen, provided the coordinate systems $(K_1$ and $K_2)$ are *adapted* coordinate systems relative to the unvaried gravitational field. This means that under the restriction to adapted coordinate systems, δJ is an invariant.

In order to prove this, we imagine the variations $\delta g^{\mu\nu}$ to be composed of two parts, and we therefore write

$$\delta g^{\mu\nu} = \delta_1 g^{\mu\nu} + \delta_2 g^{\mu\nu}, \tag{69}$$

where the two parts of the variation are chosen such that

a. The $\delta_1 g^{\mu\nu}$ are taken in a manner that the coordinate system K_1 is not only *adapted* to the (true) gravitational field of the $g^{\mu\nu}$ but also to the (varied) gravitational field of the $g^{\mu\nu} + \delta g^{\mu\nu}$. This means that not only the equations

$$B_\mu = 0$$

but also the equations

$$\delta_1 B_\mu = 0 \tag{70}$$

are valid. In other words, the $\delta_1 g^{\mu\nu}$ are not independent of each other; there are rather 4 differential equations between them.

b. The $\delta_2 g^{\mu\nu}$ are taken just as one would get them without changing the gravitational field, by mere variation of the coordinate system, specifically, by variation in that subdomain of Σ in which the $\delta g^{\mu\nu}$ differ from zero. Such a variation is determined by four mutually independent functions (variations of coordinates). Obviously, in general, $\delta_2 B_\mu \neq 0$.

The superposition of the two variations is therefore determined by

$$(10 - 4) + 4 = 10$$

mutually independent functions, and thus will be equivalent to an *arbitrary* variation of the $\delta g^{\mu\nu}$. Hence, the proof of our theorem will be completed when equation (68) is proven for the two partial variations.

Proof for the variation of δ_1: By δ_1-variation of (65) one obtains in a straightfor-ward manner

$$\frac{1}{2}\Delta(\delta_1 J) = \int d\tau \sum_{\mu} (\Delta x_{\mu} \delta_1 B_{\mu}) + \delta_1 F. \tag{65a}$$

Since the δ_1-variations of the $g^{\mu\nu}$ and all their derivatives vanish at the boundary of Σ, the quantity $\delta_1 F$ (which can be transformed into a surface integral) also vanishes according to (65b). After this and with (70), our equation (65a) turns into the relation

$$\Delta(\delta_1 J) = 0. \tag{68a}$$

Proof for the variation of δ_2: The variation $\delta_2 J$ is equivalent to an infinitesimal coordinate transformation under fixed coordinates of the boundary. Since the coordinate system is an adapted one relative to the unvaried gravitational field, it follows from the definition of adapted coordinate systems that

$$\delta_2 J = 0.$$

Next, we assume the variation of the gravitational field relative to the coordinate system K_1 to be chosen as a δ_2-variation; then we have

$$\delta_2(J_1) = 0.$$

If this variation is a δ_2-variation also relative to K_2—as we shall prove later—then we have an analogous equation relative to K_2, i.e.,

$$\delta_2(J_2) = 0.$$

The equation to be proven follows then by subtraction

$$\delta_2(\Delta J) = \Delta(\delta_2 J) = 0. \tag{68b}$$

We still have to show that the variation under consideration is a δ_2-variation also [p. 1073] relative to K_2. The unvaried tensors $g^{\mu\nu}$ relative to K_1 and K_2 be denoted symbolically by G_1 and G_2, respectively; similarly, the varied tensors $g^{\mu\nu}$ relative to K_1 and K_2 are G_1^* and G_2^*, respectively. The coordinate transformation T brings us from G_1 to G_2, also from G_1^* to G_2^* resp.; and the inverse transformation will be T^{-1}. Furthermore, the coordinate transformation t brings G_1 to G_1^*. Consequently, G_2^* is obtained from G_2 by the sequence of transformations

$$T^{-1} - t - T,$$

which is again a coordinate transformation. In this manner it is shown that the variation of the $g^{\mu\nu}$ which we have under consideration here is also a δ_2-variation relative to K_2.

The equation (68), which is to be proven, follows finally from (68a) and (68b).

From the proven theorem we deduce the existence of a complex of 10 components, which has tensorial character if we limit ourselves to adapted coordinate systems. According to (61) one has

$$dJ = \delta\left\{\int H\sqrt{-g}\,d\tau\right\}$$

$$= \int d\tau \sum_{\mu\nu\sigma} \left\{\frac{\partial(H\sqrt{-g})}{\partial g^{\mu\nu}}\delta g^{\mu\nu} + \frac{\partial(H\sqrt{-g})}{\partial g_\sigma^{\mu\nu}}\delta g_\sigma^{\mu\nu}\right\}$$

or, since $\delta g_\sigma^{\mu\nu} = \dfrac{\partial}{\partial x_\sigma}(\delta g^{\mu\nu})$, after partial integration and considering the vanishing

of the $\delta(g^{\mu\nu})$ at the boundary

$$\delta J = \int d\tau \sum_{\mu\nu} \delta g^{\mu\nu}\left\{\frac{\partial H\sqrt{-g}}{\partial g^{\mu\nu}} - \sum_\sigma \frac{\partial}{\partial x_\sigma}\left(\frac{\partial H\sqrt{-g}}{\partial g_\sigma^{\mu\nu}}\right)\right\}. \tag{71}$$

We have now proven that under limitation to adapted coordinate systems δJ is an invariant. Since the $\delta g^{\mu\nu}$ need differ from zero only in an infinitely small domain, and since $\sqrt{-g}\,d\tau$ is a scalar, the integral divided by $\sqrt{-g}$ is also an invariant, i.e., the quantity

$$\frac{1}{\sqrt{-g}}\sum \delta g^{\mu\nu}\,\mathfrak{E}_{\mu\nu}, \tag{72}$$

where

$$\mathfrak{E}_{\mu\nu} = \frac{\partial H\sqrt{-g}}{\partial g^{\mu\nu}} - \sum_\sigma \frac{\partial}{\partial x_\sigma}\left(\frac{\partial H\sqrt{-g}}{\partial g_\sigma^{\mu\nu}}\right). \tag{73}$$

Now, however, $\delta g^{\mu\nu}$ is a contravariant tensor just as $g^{\mu\nu}$ is, and the ratios of the $\delta g^{\mu\nu}$ can be chosen freely. From this follows that under limitation to adapted [p. 1074] coordinate systems and substitutions between them,

$$\frac{\mathfrak{E}_{\mu\nu}}{\sqrt{-g}}$$

is a covariant tensor and $\mathfrak{E}_{\mu\nu}$ itself is the corresponding covariant V-tensor and according to (73) a symmetric tensor.

§15. Derivation of the Field Equations

One may expect the tensor $\mathfrak{E}_{\mu\nu}$ to have a fundamental role in the field equations of gravitation that we want to find, and that those equations have to take the place that Poisson's equation has in the Newtonian theory. After the deliberations of §§13 and 14 we have to demand that the desired equations—as well as the tensor $\mathfrak{E}_{\mu\nu}$—are only covariant with respect to adapted coordinate systems. The equations we are looking for will have a strong correlation between the tensors $\mathfrak{E}_{\mu\nu}$ and $\mathfrak{T}_\sigma^\nu$ because we already saw, following after (42a), that the energy tensor $\mathfrak{T}_\sigma^\nu$ is decisive for the action of the gravitational field upon matter. It is therefore natural to assume the desired equations as

$$\mathfrak{E}_{\sigma\tau} = \kappa\,\mathfrak{T}_{\sigma\tau}. \tag{74}$$

κ is here a universal constant and $\mathfrak{T}_{\sigma\tau}$ is the symmetric covariant V-tensor, associated with the mixed energy tensor $\mathfrak{T}_\sigma^\nu$ by the relation

and
$$\left.\begin{aligned}
\mathfrak{T}_{\sigma\tau} &= \sum_\nu g_{\nu\tau}\,\mathfrak{T}_\sigma^\nu \\
\mathfrak{T}_\sigma^\nu &= \sum_\tau \mathfrak{T}_{\sigma\tau}\,g^{\nu\tau} \text{ resp.}
\end{aligned}\right\} . \tag{75}$$

The determination of the function H. The equations we are looking for are not yet completely given insofar as we have not yet determined the function H. Presently, we only know H to depend solely upon the $g^{\mu\nu}$ and the $g_\sigma^{\mu\nu}$, and to be a scalar under *linear* transformations.[7] A further condition that H must satisfy is found in the following manner.

The V-four-vector $(\mathfrak{R}_\sigma)$ of the force density vanishes in (42a) if $\mathfrak{T}_\sigma^\nu$ is the energy tensor of *all* the material processes in the domain under consideration. Equation (42a) then states that the divergence of the energy tensor $\mathfrak{T}_\sigma^\nu$ of the material processes vanishes; and the same applies—according to (74)—to the tensor $\mathfrak{E}_{\sigma\tau}$ or resp. for the mixed V-tensor $\mathfrak{E}_\sigma^\nu$ to be formed from it. Consequently, every gravitational field must [p. 1075] satisfy the relation (see (41b) and (34)):

$$\sum_{\nu\tau} \frac{\partial}{\partial x_\nu}(g^{\tau\nu}\,\mathfrak{E}_{\sigma\tau}) + \frac{1}{2}\sum_{\mu\nu}\frac{\partial g^{\mu\nu}}{\partial x_\sigma}\,\mathfrak{E}_{\mu\nu} = 0.$$

By means of (73) and (65a), this relation can be brought into the form

$$\sum_\nu \frac{\partial S_\sigma^\nu}{\partial x_\nu} - B_\sigma = 0, \tag{76}$$

where

$$S_\sigma^\nu = \sum_{\mu\tau}\left(g^{\nu\tau}\frac{\partial H\sqrt{-g}}{\partial g^{\sigma\tau}} + g_\mu^{\nu\tau}\frac{\partial H\sqrt{-g}}{\partial g_\mu^{\sigma\tau}} + \frac{1}{2}\delta_\sigma^\nu H\sqrt{-g} - \frac{1}{2}g_\sigma^{\mu\tau}\frac{\partial H\sqrt{-g}}{\partial g_\nu^{\mu\tau}}\right) \tag{76a}$$

---

[7]We would not have found the expression (65a) for B_μ without the latter limitation, which we introduced in §14. The consideration given in the following text would fail if we dropped this limitation. This fact is the justification for its introduction.

[40]   and $\delta_\sigma^\nu = 1$ or 0 depending upon $\sigma = \nu$ or $\sigma \neq \nu$.

The ten equations (74) can be used to determine the ten functions $g^{\mu\nu}$ if the $\mathfrak{T}_{\sigma\tau}$ are given. Furthermore, the $g^{\mu\nu}$ must also satisfy the four equations (67) because the coordinate system is to be an adapted one. We have, therefore, more equations than we have functions to be found. This is only possible if the equations are not all mutually independent of each other. It must be demanded that satisfying equations (74) implies that equations (67) are also satisfied. A glance at (76) and (76a) shows that this is achieved if S_σ^ν (a quantity which is a function of the $g^{\mu\nu}$ and the $g_\sigma^{\mu\nu}$ just as H is) vanishes identically for every combination of indices. H then has to be chosen in agreement with the conditions

$$S_\sigma^\nu \equiv 0. \tag{77}$$

Without being able to state a formal reason, I demand furthermore that H is an integral homogeneous function of the second degree in the $g_\sigma^{\mu\nu}$. In this case H is completely determined up to a constant factor. Since H shall be a scalar under linear transformations, it must[8] (considering what we just postulated) be a linear combination of the following five quantities:

[41]
{18}
$$\sum_{\mu\nu\sigma\tau} g_{\mu\nu} \frac{\partial g^{\mu\nu}}{\partial x_\sigma} \frac{\partial g^{\sigma\tau}}{\partial x_\tau}; \quad \sum_{\mu\nu\mu'\nu'\sigma\sigma'} g^{\sigma\sigma'} g_{\mu\nu} \frac{\partial g^{\mu\nu}}{\partial x_\sigma} g_{\mu'\nu'} \frac{\partial g^{\mu'\nu'}}{\partial x_{\sigma'}}; \quad \sum_{\sigma\sigma'\mu\nu} g_{\sigma\sigma'} \frac{\partial g^{\sigma\mu}}{\partial x_\mu} \frac{\partial g^{\sigma'\nu}}{\partial x_\nu};$$

$$\sum_{\mu\mu'\nu\nu'\sigma\sigma'} g_{\mu\mu'} g_{\nu\nu'} g^{\sigma\sigma'} \frac{\partial g^{\mu\nu}}{\partial x_\sigma} \frac{\partial g^{\mu'\nu'}}{\partial x_{\sigma'}}; \quad \sum_{\alpha\beta\sigma\tau} g_{\alpha\beta} \frac{\partial g^{\alpha\sigma}}{\partial x_\tau} \frac{\partial g^{\beta\tau}}{\partial x_\sigma}.$$

Conditions (77), finally, lead us to equate H—aside from a constant factor—to

[p. 1076]   the fourth one of these quantities. We therefore set[9] under consideration of (35) and

[42]   making free use of the constant,

$$H = \frac{1}{4} \sum_{\alpha\beta\tau\rho} g^{\alpha\beta} \frac{\partial g_{\tau\rho}}{\partial x_\alpha} \frac{\partial g^{\tau\rho}}{\partial x_\beta}. \tag{78}$$

We limit ourselves to show that this choice of H actually satisfies (77). Utilizing the relations

---

[8]The proof is simple but involved, and for this reason I delete it.

[9]Expressing H by the components $\Gamma_{\sigma\tau}^\nu$ of the gravitational field (see (46)), one obtains

$$H = -\sum_{\mu\rho\tau\tau'} g^{\tau\tau'} \Gamma_{\mu\tau}^\rho \Gamma_{\rho\tau'}^\mu.$$

$$dg = g \sum_{\sigma\tau} g^{\sigma\tau} dg_{\sigma\tau} = -g \sum_{\sigma\tau} g_{\sigma\tau} dg^{\sigma\tau}$$

$$dg_{\alpha\beta} = -\sum_{\mu\nu} g_{\alpha\mu} g_{\beta\nu} dg^{\mu\nu}$$

one obtains from (78)

[43]

$$
\left.
\begin{aligned}
\sum_{\tau} g^{\nu\tau} \frac{\partial H\sqrt{-g}}{\partial g^{\sigma\tau}} &= -\frac{1}{2} H\sqrt{-g}\,\delta_\sigma^\nu + \frac{1}{4}\sqrt{-g} \sum_{\mu\mu'\tau} g^{\nu\tau} \frac{\partial g_{\mu\mu'}}{\partial x_\sigma} \frac{\partial g^{\mu\mu'}}{\partial x_\tau} \\
&\quad - \frac{1}{2}\sqrt{-g} \sum_{\rho\rho'\kappa} g^{\rho\rho'} \frac{\partial g_{\sigma\kappa}}{\partial x_\rho} \frac{\partial g^{\nu\kappa}}{\partial x_{\rho'}} \\
\text{with} \quad \sum_{\mu\tau} g_\mu^{\nu\tau} \frac{\partial H\sqrt{-g}}{\partial g_\mu^{\sigma\tau}} &= \frac{1}{2}\sqrt{-g} \sum_{\rho\rho'\kappa} g^{\rho\rho'} \frac{\partial g_{\sigma\kappa}}{\partial x_\rho} \frac{\partial g^{\nu\kappa}}{\partial x_{\rho'}} \\
\text{and} \quad \frac{1}{2}\sum_{\mu\tau} g_\sigma^{\mu\tau} \frac{\partial H\sqrt{-g}}{\partial g_\nu^{\mu\tau}} &= \frac{1}{4}\sqrt{-g} \sum_{\mu\mu'\tau} g^{\nu\tau} \frac{\partial g_{\mu\mu'}}{\partial x_\sigma} \frac{\partial g^{\mu\mu'}}{\partial x_\tau}.
\end{aligned}
\right\}
$$

(79)

{19}

From these relations follows the claim made above.

Without using our physical knowledge of gravitation, we have arrived in a purely formal manner at quite distinct field equations. In order to get them into a more explicit notation, we multiply (74) by $g^{\nu\tau}$ and sum over the index τ. Considering (73), we thus get

$$\kappa \mathfrak{T}_\sigma^\nu = \sum_{\tau\alpha} g^{\nu\tau} \left(\frac{\partial H\sqrt{-g}}{\partial g^{\sigma\tau}} - \frac{\partial}{\partial x_\alpha} \left[\frac{\partial H\sqrt{-g}}{\partial g_\alpha^{\sigma\tau}} \right] \right)$$

(80)   [44]

or

$$-\sum_{\alpha\tau} \frac{\partial}{\partial x_\alpha} \left(g^{\nu\tau} \frac{\partial H\sqrt{-g}}{\partial g_\alpha^{\sigma\tau}} \right) = \kappa\mathfrak{T}_\sigma^\nu + \sum_{\alpha\tau} \left(-g^{\nu\tau} \frac{\partial H\sqrt{-g}}{\partial g^{\sigma\tau}} - g_\alpha^{\nu\tau} \frac{\partial H\sqrt{-g}}{\partial g_\alpha^{\sigma\tau}} \right).$$

(80a)   [45]

Since our coordinate system is an adapted one, the equation

$$\sum_{\alpha\tau\nu} \frac{\partial}{\partial x_\nu} \frac{\partial}{\partial x_\alpha} \left(g^{\nu\tau} \frac{\partial H\sqrt{-g}}{\partial g_\alpha^{\sigma\tau}} \right) = 0$$

[p. 1077]

also holds due to (67) and (65a) and, therefore, considering (80), also the equation

$$\sum_{\nu} \frac{\partial}{\partial x_\nu} \left\{ \mathfrak{T}_\sigma^\nu + \frac{1}{\kappa} \sum_{\alpha\tau} \left(-g^{\nu\tau} \frac{\partial H\sqrt{-g}}{\partial g^{\sigma\tau}} - g_\alpha^{\nu\tau} \frac{\partial H\sqrt{-g}}{\partial g_\alpha^{\sigma\tau}} \right) \right\} = 0.$$

(80b)   [46]

Utilizing (78), (79), and (46), we can replace equations (80a) and (80b) by the following ones:

$$\sum_{\alpha\beta} \frac{\partial}{\partial x^\alpha} (\sqrt{-g}\, g^{\alpha\beta} \Gamma^\nu_{\sigma\beta}) = -\kappa (\mathfrak{T}^\nu_\sigma + t^\nu_\sigma), \tag{81}$$

$$\sum_\nu \frac{\partial}{\partial x_\nu} (\mathfrak{T}^\nu_\sigma + t^\nu_\sigma) = 0, \tag{42c}$$

where

$$\Gamma^\nu_{\sigma\beta} = \frac{1}{2} \sum_\tau g^{\nu\tau} \frac{\partial g_{\sigma\tau}}{\partial x_\beta}, \tag{81a}$$

$$\left.\begin{aligned}
t^\nu_\sigma &= -\frac{\sqrt{-g}}{4\kappa} \sum_{\mu\mu'\rho\tau} \left(g^{\nu\tau} \frac{\partial g_{\mu\mu'}}{\partial x_\sigma} \frac{\partial g^{\mu\mu'}}{\partial x_\tau} - \frac{1}{2} \delta^\nu_\sigma g^{\rho\tau} \frac{\partial g_{\mu\mu'}}{\partial x_\rho} \frac{\partial g^{\mu\mu'}}{\partial x_\tau} \right) \\
&= \frac{\sqrt{-g}}{\kappa} \sum_{\mu\rho\tau\tau'} \left(g^{\nu\tau} \Gamma^\rho_{\mu\sigma} \Gamma^\mu_{\rho\tau} - \frac{1}{2} \delta^\nu_\sigma g^{\tau\tau'} \Gamma^\rho_{\mu\tau} \Gamma^\mu_{\rho\tau'} \right)
\end{aligned}\right\}. \tag{81b}$$

The equations (81) together with (81a) and (81b) are the differential equations of the gravitational field. Following the deliberations of §10, the equations (42c) represent the conservation laws of momentum and energy for matter and gravitational field combined. The t^ν_σ are those quantities, related to the gravitational field, which are in physical analogy to the components $\mathfrak{T}^\nu_\sigma$ of the energy tensor (V-tensor). It is to be emphasized that the t^ν_σ do not have tensorial covariance under arbitrary admissible transformations but only under *linear* transformations. Nevertheless, we call (t^ν_σ) the energy tensor of the gravitational field. A similar analogy applies to the components $\Gamma^\nu_{\sigma\beta}$ of the field strength of the gravitational field.

The system of equations (81) allows for a simple physical interpretation in spite of its complicated form. The left-hand side represents a kind of divergence of the gravitational field. As the right-hand side shows, this is caused by the components of the total energy tensor. A very important aspect of this is the result that the energy tensor of the gravitational field itself acts field-generatingly, just as does the energy tensor of matter.

[p. 1078]   *§16. Critical Remarks on the Foundation of the Theory*

It is the essence of the theory we derived here that the original theory of relativity holds in the infinitesimally small. This becomes obvious once we have shown that under a suitable choice of real-valued coordinates the quantities $g_{\mu\nu}$ assume the values

$$\begin{array}{rrrr} -1 & 0 & 0 & 0 \\ 0 & -1 & 0 & 0 \\ 0 & 0 & -1 & 0 \\ 0 & 0 & 0 & 1 \end{array}$$

at an arbitrarily given point. This is the case when the surface of second degree

$$\sum_{\mu\nu} g_{\mu\nu}\, \xi_\mu \xi_\nu = 1$$

has always (for every system of the $g_{\mu\nu}$ which occurs in our continuum) three imaginary-valued axes and one real-valued axis. If λ_1, λ_2, λ_3, λ_4 are the squares of the reciprocals of the semi-axes of the surface, the equation of fourth degree

$$|g_{\mu\nu} - \lambda\delta_\mu^\nu| = 0 = (\lambda_1 - \lambda)(\lambda_2 - \lambda)(\lambda_3 - \lambda)(\lambda_4 - \lambda)$$

is satisfied. Consequently,

$$\lambda_1\, \lambda_2\, \lambda_3\, \lambda_4 = g.$$

In order to prevent the $g^{\mu\nu}$ from reaching infinite values, one has to demand that g vanishes nowhere, since the $g^{\mu\nu}$ are equal to the minors of the $g_{\mu\nu}$-determinant divided by g. No λ can then ever become zero. Therefore, whenever $\lambda_1 < 0$, $\lambda_2 < 0$, $\lambda_3 < 0$, $\lambda_4 > 0$ for one point of the continuum, it is true for every point. Thus, the space-time character of our continuum in the neighborhood of all points remains the same as it was in the original theory of relativity. Mathematically this can be expressed by saying: among four mutually "orthogonal" line elements originating from one point, one element is always "time-like," the other three are "space-like."

However, this does not yet establish space-like or time-like relations *with the coordinate system* of the x_ν. In the original theory of relativity, every line element that deviates from zero *only* in dx_4 is time-like. The same statement cannot be claimed for our adapted coordinate systems. Considering sufficiently large parts of the universe, it is very well imaginable that no coordinate axis can be denoted as "time-axis," but that rather the line elements of *one* axis are in parts time-like, in [p. 1079] other parts space-like. The equivalence of the four dimensions of the world would then be not only a *formal* one but a *complete* one. For the time being one has to leave this important question an open one. [47]

An even deeper-reaching question of fundamental significance shall now be brought up—and I am not able to answer it. In the ordinary theory of relativity, every line that describes the movement of a material point, i.e., every line consisting only of time-like elements, is necessarily nonclosed, the reason being that such a line never contains elements for which dx_4 vanishes. An analogous statement cannot be claimed for the theory developed here. It is therefore a priori possible to imagine a point movement where the four-dimensional curve of the point is almost a closed one. In this case *one and the same* material point could exist in an arbitrarily small space-time domain in *several seemingly mutually independent representations*. This runs counter to my physical imagination in the most vivid manner. However, I am not able to demonstrate that the occurrence of such curves can be excluded from the theory that has been developed here.

[48]   Since I cannot help but see, after these confessions, a pitying smile creep over the face of the reader, I cannot suppress the following remark about the current opinion on the foundation of physics. Before Maxwell, the laws of nature with respect to their space dependence were in principle integral laws; this is to say that in elementary laws the distances between finitely distinct points did occur. Euclidean geometry is the basis for this description of nature. This geometry means originally only the essence of conclusions from geometric axioms; in this regard it has no physical content. But geometry becomes a physical science by adding the statement that two points of a "rigid" body shall have a distinct distance from each other that is independent of the position of the body. After this amendment, the theorems of this amended geometry are (in a physical sense) either factually true or not true. It is geometry in this extended sense which forms the basis of physics. Seen from this aspect, the theorems of geometry are to be looked at as integral laws of physics insofar as they deal with distance of points *at a finite range*.

Since MAXWELL, and by his work, physics has undergone a fundamental revision insofar as the demand gradually prevailed that distances of points at a finite range [p. 1080] should not occur in the elementary laws; i.e., theories of "action at a distance" are now replaced by theories of "local action." One forgot in this process that the EUCLIDEAN geometry too—as it is used in physics—consists of physical theorems that, from a physical aspect, are on an equal footing with the integral laws of NEWTONIAN mechanics of points. In my opinion, this is an inconsistent attitude of which we should free ourselves.

An attempt to free ourselves leads again, first, to the use of arbitrary parameters for the description of the four-dimensional continuum around us—instead of coordinates. Again, we arrive at the same considerations we have given in sections B and C of this paper—with the sole difference that a correlation of the $g_{\mu\nu}$ with the gravitational field is not postulated. But if we want to adhere to the demands of EUCLIDEAN geometry (in the sense stated above) we would have to replace the equations given in this section by others that derive from the following supposition: the coordinates x_ν can be chosen such that the $g_{\mu\nu}$ are independent of the x_ν. In this manner we are led to the demand that the components of the RIEMANN-CHRISTOFFEL tensor—as developed in §9—shall vanish. By this method, the theorems of EUCLIDEAN geometry would be reduced to differential laws. But by phrasing the situation in this manner, one realizes that a rigorous implementation of a "local action" theory is neither the most simple nor the next closest possibility that comes to mind.

E. Some Remarks on the Physical Content of the General Laws Developed

In the derivation of the laws I let myself be guided—inasfar as this was possible—by solely formal aspects. In order to leave the elaboration of the subject matter not too incomplete, we shall now also highlight the obtained results from their physical side. Not to be smothered by mathematical complications, we limit ourselves to the consideration of approximations.

§17. *The Establishment of Approximative Equations, Seen from Various Aspects*

One sees from the far-reaching usefulness of the equations of the original theory of relativity that in the space-time domain that is perceptible to us, the $g^{\mu\nu}$ can almost be treated like constants. Consequently, we put   [p. 1081]

$$\left.\begin{array}{l} g_{\mu\nu} = g_{\mu\nu 0} + h_{\mu\nu} \\ g^{\mu\nu} = g_0^{\mu\nu} + h^{\mu\nu} \end{array}\right\} \tag{82}$$

where the $g_{\mu\nu 0}$ and the $g_0^{\mu\nu}$ take the values

$$\left.\begin{array}{cccc} -1 & 0 & 0 & 0 \\ 0 & -1 & 0 & 0 \\ 0 & 0 & -1 & 0 \\ 0 & 0 & 0 & -1 \end{array}\right\}. \tag{82a}$$

The $h_{\mu\nu}$ and the $h^{\mu\nu}$ are treated as infinitesimally small quantities of first order; and neglecting infinitesimals of second order they obey the relations

$$h^{\mu\nu} = -h_{\mu\nu}.$$

The time coordinate is chosen as purely imaginary (as MINKOWSKI also does) and by this we accomplish that $(g_{44})_0 = g_0^{44} = -1$ and also the covariance of the systems of equations under linear *orthogonal* transformations. An imaginary choice for the time coordinate makes g_{14}, g_{24}, g_{34} and also $\sqrt{-g}$ imaginary. But the validity of the equations we have developed is retained, because one can go from a real-valued to an imaginary-valued time variable by means of a linear transformation. Fixing the values as was done in (82a) achieves the agreement of naturally measured lengths with coordinate length in the domain of consideration, up to infinitesimals of first order.

We now replace equations (81) and (81a) with others in which infinitesimally small quantities of second and higher order are neglected. t_σ^ν will then vanish and we obtain

$$\sum_\alpha \frac{\partial^2 h_{\sigma\nu}}{\partial x_\alpha^2} = i\kappa \mathfrak{T}_\sigma^\nu \tag{84}$$

$$\Gamma_{\sigma\beta}^\nu = -\frac{1}{2} \frac{\partial h_{\sigma\nu}}{\partial x_\beta}. \tag{84a}$$

We now introduce one further postulate of approximation by considering in $\mathfrak{T}_\sigma^\nu$ only those terms that correspond to ponderable matter, while terms from surface forces are ignored. Under these postulates, (48) represents the energy tensor. Since the $\mathfrak{T}_\sigma^\nu$ remain *finite* according to (48), one already obtains a far-reaching approximation if one neglects infinitesimals of first order in (48). In this manner one gets

[p. 1082]

$$\mathfrak{T}_\sigma^\nu = -i\rho_0 \frac{dx_\sigma}{ds_0} \frac{dx_\nu}{ds_0}. \tag{84b}$$

Substituting into (84) and writing the left-hand side $\Box h_{\sigma\nu}$, one obtains

$$\Box h_{\sigma\nu} = \kappa \rho_0 \frac{dx_\sigma}{dx_0} \frac{dx_\nu}{ds_0}. \tag{85}$$

x_1, x_2, x_3, are spatial coordinates in this equation and $x_4 = it$ is the (imaginary)

{20}   time coordinate, while $ds_0 = dt \sqrt{1 - \left(\dfrac{dx_1^2}{dt^2} + \dfrac{dx_2^2}{dt^2} + \dfrac{dx_3^2}{dt^2} \right)}$ is the element of MIN-

KOWSKI's "eigentime."

After we have replaced equations (81) with approximation equations, their similarity with the POISSON equation of the NEWTONIAN theory of gravitation hits the eye. We shall now replace the equations of a material point, (50b) and (51), by approximation equations. One obtains the coarsest approximation if one replaces (51) with

$$\mathbf{I}_\sigma = -m \frac{dx_\sigma}{ds_0}. \tag{86}$$

Introducing the three-dimensional velocity vector $\mathfrak{q}$ with magnitude q, this means the equations

$$\left.\begin{aligned}
-\mathbf{I}_1 &= \frac{m\ \mathfrak{q}_x}{\sqrt{1-q^2}} \\[6pt]
-\mathbf{I}_2 &= \frac{m\ \mathfrak{q}_y}{\sqrt{1-q^2}} \\[6pt]
-\mathbf{I}_3 &= \frac{m\ \mathfrak{q}_z}{\sqrt{1-q^2}} \\[6pt]
-\mathbf{I}_4 &= i\frac{m}{\sqrt{1-q^2}}
\end{aligned}\right\}. \tag{86a}$$

The choice of an imaginary time coordinate implies here that the energy is not represented by the quantity $\mathbf{I}_4$—as it is in (52)—but rather by $i\,\mathbf{I}_4$.

In the absence of external forces, one gets—due to (84a)—instead of (50b)

$$\frac{d(-\mathbf{I}_\sigma)}{dt} = -\frac{1}{2}\sum_{\nu\tau}\frac{\partial h_{\nu\tau}}{\partial x_\tau}\frac{dx_\nu}{dt}(-\mathbf{I}_\tau). \tag{87} \qquad [49]$$

Equations (85), (87), (86) replace the NEWTONIAN theory in first approximation.

NEWTON's theory as an approximation. We arrive at this approximation by [p. 1083] treating the velocity $\mathfrak{q}$ as infinitesimally small, and by retaining in the equations only those terms which contain the components of $\mathfrak{q}$ in the lowest power. In place of (85) one gets the equations

$$\left.\begin{aligned}
\Box h_{\sigma\nu} &= 0 \quad (\text{as long as not } \nu = \sigma = 4) \\
\Box h_{44} &= -\kappa\rho_o
\end{aligned}\right\} \tag{85a}$$

and in place of (87) one gets

$$\frac{d(m\ \mathfrak{q})}{dt} = \frac{m}{2}\,\mathrm{grad}\ h_{44}. \tag{87a}$$

From (85a) one concludes in this case (with suitable boundary conditions at infinity) that all $h_{\sigma\nu}$ vanish, except for h_{44}. From (87a) one concludes that $\left(\dfrac{-h_{44}}{2}\right)$ plays the role of the gravitational potential. Calling the latter quantity ϕ, one gets the equations

$$\left.\begin{aligned}
\Box\phi &= \frac{\kappa}{2}\,\rho_o \\[6pt]
\frac{d(m\ \mathfrak{q})}{dt} &= -m\,\mathrm{grad}\ \phi
\end{aligned}\right\}, \tag{88}$$

which agrees with NEWTON's theory, provided $\dfrac{\partial^2\phi}{\partial t^2}$ can be neglected compared to

[50] $$\frac{\partial^2 \phi}{\partial^2 x}.$$

The first one of the equations (88) is written in NEWTON's theory as

$$\frac{\partial^2 \phi}{\partial x^2} + \frac{\partial^2 \phi}{\partial y^2} + \frac{\partial^2 \phi}{\partial z^2} = 4\pi K \rho_o$$

and thus one gets

$$\frac{\kappa}{2} = 4\pi K.$$

Utilizing the second as the time unit, the constant K has the numerical value $6.7 \cdot 10^{-8}$; and if one chooses the light-second as the time unit, the value is $\frac{6.7 \cdot 10^{-8}}{(9 \cdot 10^{20})}$. Consequently, one gets

$$\kappa = 8\pi \frac{6.7 \cdot 10^{-8}}{9 \cdot 10^{20}} = 1.87 \cdot 10^{-27}. \tag{89}$$

For the naturally measured distance of neighboring space-time points, the NEWTONIAN approximation yields

$$ds^2 = \sum_{\mu\nu} g_{\mu\nu} dx_\mu dx_\nu = - dx^2 - dy^2 - dz^2 + (1 + 2\phi)dt^2.$$

[p. 1084] For a purely *spatial* distance one gets

$$- ds^2 = dx^2 + dy^2 + dz^2.$$

Coordinate lengths are here also equal to naturally measured lengths. The EUCLIDEAN geometry of distances is valid with the accuracy we considered here. For purely temporal distance we had

$$ds^2 = (1 + 2\phi)dt^2$$

or

$$ds = (1 + \phi)dt.$$

To the naturally measured duration ds belongs the time duration $\frac{ds}{(1 + \phi)}$. The clock rate is measured by $(1 + \phi)$ and, therefore, increases with the gravitational potential. One concludes from this that spectral lines of light, generated at the sun, show a redshift relative to corresponding spectral lines generated on earth, the shift amounting to

[51]

$$\frac{\Delta\lambda}{\lambda} = 2 \cdot 10^{-6}.$$

For light rays (ds = 0), one has

$$\frac{\sqrt{dx^2 + dy^2 + dz^2}}{dt} = 1 + \phi.$$

The speed of light is, therefore, independent in its direction but varies with the gravitational potential, and consequently one has a curved progression of light rays in a gravitational field.

Finally, we calculate momentum and energy of a material point in a NEWTONIAN field for which we do not use the equations (86a) but rather the rigorous equations (51). If we substitute into these for the $g_{\sigma\mu}$ the values

$$\begin{array}{cccc}
-1 & 0 & 0 & 0 \\
0 & -1 & 0 & 0 \\
0 & 0 & -1 & 0 \\
0 & 0 & 0 & -1 + h_{44}
\end{array}$$

and for the dx_ν the quantities

$$dx \quad dy \quad dz \quad idt,$$

furthermore limiting ourselves for h_{44} to quantities of first order, and replac-

ing $\left(\dfrac{-h_{44}}{2}\right)$ with ϕ, and neglecting terms of higher than second order in the velocities, one obtains

[p. 1085]

$$-\mathbf{I}_1 = m(1 - \phi)q_x$$

$$\cdots \cdots \cdots \cdots \cdots$$

$$i\,\mathbf{I}_4 = m\left[(1 + \phi) + \frac{1}{2}(1 - \phi)q^2\right].$$

Since $(-\mathbf{I}_1)$ is the X-component of the momentum and $(i\,\mathbf{I}_4)$ is the energy of the material point, one comes to the conclusion that the inertial mass increases with diminishing gravitational potential. This is very well in agreement with the spirit of the interpretation taken here. As there are no independent physical qualities of space in our theory, the inertia of mass is a consequence of the mutual action between each mass and all the other masses. This interaction must, therefore, increase when other masses are brought closer to the mass that is under consideration, i.e., if ϕ is decreased.

Additional notes by translator

{1}   There is no equation (5b).

{2}   Many English writers use "skew-symmetric" or "alternating" as synonyms for "anti-symmetric."

{3}   Many English writers use "vector product" or "cross product" as synonyms for "outer product."

{4}   Many English writers use "scalar product" or "dot product" as synonyms for "inner product."

{5}   "$\mu \pm \nu$" has been corrected to "$\mu \neq \nu$."

{6}   The indices "$\sigma\nu$" in the right-most term have been corrected to "$\mu\nu$."

{7}   Corrected to $\displaystyle\sum_{\alpha_1 s}$.

{8}   The index "α" in the right-most term has been corrected to "μ."

{9}   "$\partial(A_{\mu\nu}\sqrt{g}$" has been corrected to "$\partial(A_{\mu\nu}\sqrt{g})$."

{10} $\dfrac{\partial A_\tau}{\partial x_\mu}$ has been corrected to $\dfrac{\partial A_\mu}{\partial x_\tau}$ in fourth term on right-hand side.

{11}   The second use of (46) as an equation number in the original has no relation to the first use. This minor oversight in the German original is Einstein's.

{12}   "dx^3" has been corrected to "dx_3."

{13}   The summation has been placed over α and β.

{14}   "$dx_{(\mu)}$" has been corrected to "dx_μ."

{15}   (§5a) has been corrected to (5a).

{16}   Here and next line, "I" has been corrected to "J."

{17}   The summation has been placed over μ, α, σ, and ν.

{18}   The five quantities are:

$$\sum_{\mu\nu\sigma\tau} g_{\mu\nu}\frac{\partial g^{\mu\nu}}{\partial x_\sigma}\frac{\partial g^{\sigma\tau}}{\partial x_\tau} ; \quad \sum_{\mu\nu\mu'\nu'\sigma\sigma'} g^{\sigma\sigma'}g_{\mu\nu}\frac{\partial g^{\mu\nu}}{\partial x_\sigma}g_{\mu'\nu'}\frac{\partial g^{\mu'\nu'}}{\partial x_{\sigma'}} ; \quad \sum_{\sigma\sigma'\mu\nu} g_{\sigma\sigma'}\frac{\partial g^{\sigma\mu}}{\partial x_\mu}\frac{\partial g^{\sigma'\nu}}{\partial x_\nu} ;$$

$$\sum_{\mu\mu'\nu\nu'\sigma\sigma'} g_{\mu\mu'}g_{\nu\nu'}g^{\sigma\sigma'}\frac{\partial g^{\mu\nu}}{\partial x_\sigma}\frac{\partial g^{\mu'\nu'}}{\partial x_{\sigma'}} ; \quad \sum_{\alpha\beta\sigma\tau} g_{\alpha\beta}\frac{\partial g^{\alpha\sigma}}{\partial x_\tau}\frac{\partial g^{\beta\tau}}{\partial x_\sigma} .$$

{19}   The term "$g^{\mu\tau}$" on the right-hand side after $\sum$ has been corrected to "$g^{\nu\tau}$."

{20}   All three "dt" in the radicand have been corrected to "dt^2."

Doc. 10

Review of Alexander Brill, *The Principle of Relativity:*
An Introduction to the Theory

Not translated for this volume.

Doc. 11

Review of H. A. Lorentz, *The Principle of Relativity:*
Three Lectures . . .

Not translated for this volume.

Doc. 12

Expert Opinion on Legal Dispute between Anschütz & Co.
and Sperry Gyroscope Company

Not translated for this volume.

Doc. 13

Experimental Proof of Ampère's Molecular Currents
(with Wander J. de Haas)

Not translated for this volume.

Doc. 14

Experimental Proof of the Existence of
Ampère's Molecular Currents

(with Wander J. de Haas)

Doc. 14 is already in English. See *The Collected Papers of Albert Einstein,* vol. 6, pp. 172–189.

Doc. 15

Experimental Proof of Ampère's Molecular Currents

Not translated for this volume.

Doc. 16

Correction of My Joint Paper with J. W. de Haas: "Experimental Proof of Ampère's Molecular Currents"

Not translated for this volume.

Doc. 17

Comment on the Essay Submitted by Knapp: "The Shearing of the Light-Ether . . ."

Not translated for this volume.

Doc. 18

[p. 879]
[1]

Response to a Paper by M. von Laue: "A Theorem in Probability Calculus and Its Application to Radiation Theory"[1]

by A. Einstein

In the paper quoted, Laue brings the mathematical foundation of radiation statistics into a form which, as to precision and beauty, leaves nothing to be desired. However, as far as the application of these foundations to radiation theory is concerned, it seems to me that he has fallen victim to a critical error which requires urgent correction. Laue claims that the coefficients in a Fourier expansion, as they occur in the local oscillations of natural radiation, need not be statistically independent of each other. If this claim were justified, one would really have a very promising method to overcome the difficulties that are manifest in the theoretical "indigestibility" of all laws in which Planck's "h" plays an important role. This was precisely the reason that motivated me five years ago to investigate this question in more detail in a published paper, co-authored with L. Hopf.

[2]

The result of this not quite flawlessly executed paper is recognized by Laue as a correct conclusion from the basic assumptions made in it. What Laue denies is the permissibility of these basic assumptions, which can be phrased as follows:

If I obtain a completely disordered radiation (i.e., statistically mutually independent Fourier coefficients) by superimposing an infinite number of completely given identical radiations such that the total phases of this superimposed conglomerate are chosen at random, then natural radiation must a fortiori be statistically disordered.

This basic assumption appeared to me, then, as evident. The fact, however, that

[p. 880]

an experienced expert, such as Laue, does not share this opinion proves the opposite. I shall, therefore, in what follows give a proof that is free of said assumption and which—as I hope—will irrefutably demonstrate that our theory of undulation definitely demands the mutual statistical independence of the Fourier coefficients. However, before I enter into this proof, I want to show why the considerations in parts II and III of Laue's paper are, in my opinion, not a convincing proof.

Laue considers radiation from a large number of resonators that emit orthogonal to a layer (of thickness $c\tau$), over which they are irregularly distributed. In part II of his paper he assumes that all of these resonators oscillate simultaneously and under the same law; and in part III, that the oscillations of all resonators are governed by

---

Translator's note. Typographical errors occurring in the original have been corrected in this translation. See notes {1}—{3} and note [3].

the same statistical law which is assumed to be given. In both cases, the statistical independence of the Fourier coefficients in the development of the resulting radiation does not follow. From this one should not—in my opinion—deduce the admissibility of the hypothesis that this independence does not also exist in *natural radiation*. After all, it has not been shown that the degree of disorder from this irregular distribution of resonators in a layer of thickness $c\tau$ has to be the same as the one encountered in natural radiation.

The suspicion becomes more prominent when one sees in Laue's calculations that the degree of statistical dependence of two terms in the development of the resulting radiation characterized by indices p and p' is determined by a term, viz.,

$$\frac{\pi(p - p')\tau}{T},$$

i.e., *a quantity that depends upon the thickness of the layer*. But a statistical dependence of this kind in natural radiation—if such dependence could be found there—should have nothing to do with the method of generation of the radiation that is under consideration here.

In my opinion, therefore, none of the cases considered by Laue is equivalent to the disorder found in natural radiation, and his results do not allow one to conclude [p. 881] anything as far as natural radiation is concerned. Consequently, I uphold my previous claim and will try to support it with a new proof, while utilizing the theorems of probability upon which Laue has elaborated in his paper.

§1. Statistical Properties of Radiation Developed by Superposition of an Infinity of Mutually Independently Generated Radiations

Each one of the radiation components to be considered shall be represented in the time interval 0 to T by a Fourier expansion of the form

$$\sum_n a_n^{(\nu)} \cos 2\pi n \frac{t}{T} + b_n^{(\nu)} \sin 2\pi n \frac{t}{T}, \tag{1} \quad \{1\}$$

where the coefficients satisfy the probability law

$$dW = f^{(\nu)}(a_1^{(\nu)}\ldots a_z^{(\nu)}\ldots, b_1^{(\nu)}\ldots b_z^{(\nu)})da_1^{(\nu)}\ldots db_z^{(\nu)}\ldots, \tag{2}$$

which law can be a different one for each radiation component (ν). Furthermore, the law can be such that

$$
\left.
\begin{aligned}
\overline{a_n^{v}} &= \int a_n^{v} f^{(v)} da_1^{(v)} \dots db_z^{(v)} = 0 \\
\overline{b_n^{v}} &= \int b_n^{v} f^{(v)} da_1^{(v)} \dots db_z^{(v)} = 0
\end{aligned}
\right\}.
\tag{3}
$$

The resulting radiation is given in the time interval 0 to T by the expression

$$
\left.
\begin{aligned}
&\sum_n A_n \cos 2\pi n \frac{t}{T} + B_b \sin 2\pi n \frac{t}{T} \\
&= \sum_v \sum_n \left(a_n^{(v)} \cos 2\pi n \frac{t}{T} + b_n^{(v)} \sin 2\pi n \frac{t}{T} \right)
\end{aligned}
\right\}
\tag{4}
$$

and from this follows the validity of the relations

$$
\left.
\begin{aligned}
A_n &= \sum_v a_n^{(v)} \\
B_n &= \sum_v b_n^{(v)}
\end{aligned}
\right\}.
\tag{5}
$$

Which statistical law follows now for the Fourier coefficients $A_1 \dots B_z$?

[p. 882] Using considerations quite analogous to the ones in part I of Laue's paper, one finds the statistical law we are looking for to be as follows:

$$
dW = \text{const } e^{-\sum_{mn} (\alpha_{mn} A_m A_n + \beta_{mn} B_m B_n + 2\gamma_{mn} A_m B_n)} dA_1 \dots dB_z.
\tag{6}
$$

One can see from this relation that the superposition of infinitely many radiation components does not at all guarantee the statistical independence of the Fourier coefficients. The law (6), however, still allows us to reduce the question of statistical independence of the Fourier coefficients to a simpler one. The statistical independence is satisfied if and only if the exponent in the exponential function contains only the squares of the A_m and B_n, not, however, any products of these quantities, i.e.,

$$
\left.
\begin{aligned}
\alpha_{mn} &= \beta_{mn} = 0 \text{ for } m \neq n \\
\gamma_{\mu\nu} &= 0
\end{aligned}
\right\}
\tag{7}
$$

must be satisfied.

It is furthermore clear, because of (3) and (5), that under statistical independence the relations

$$
\left.
\begin{aligned}
\overline{A_m A_n} &= \overline{B_m B_n} = 0 \text{ for } m \neq n \\
\overline{A_m B_n} &= 0
\end{aligned}
\right\}
\tag{7a}
$$

must obtain. The number of conditions (7a) is equal to the number of conditions (7), and all conditions (7a) are mutually independent. It follows, therefore, that under the

validity of (6) the conditions (7a) are *sufficient* to guarantee the statistical independence of the Fourier coefficients.

In this manner we reach the following preliminary result: Since we have to
assume, for natural radiation, that its statistical properties are not changed by
superposition of incoherent partial radiations, one can say the equations (7a) are
sufficient conditions for natural radiation to assure the statistical independence of its
Fourier coefficients.

§2. Proof of the Statistical Independence of the Fourier Coefficients of Natural Radiation

Let $F(t)$ be a component of the radiation vector of stationary natural radiation, given
for an infinite time period. T is a large time span when compared to the oscillation   [p. 883]
period of the longest wavelength in the radiation. $F(t)$ is to be represented between t_0
and $t_0 + T$ by the Fourier series

$$\sum_n \left(A_n \cos 2\pi n \frac{t - t_0}{T} + B_n \sin 2\pi n \frac{t - t_0}{T} \right). \tag{4a}$$

The Fourier coefficients A_n, B_n, of $F(t)$ will obviously depend upon the choice
of time epoch t_0. By imagining the expansion to be carried out for very many and
arbitrarily selected t_0, we obtain statistical data for the derivation of the statistical
properties of the coefficients A_n, B_n that we must necessarily demand in natural
radiation.

In order to derive these properties, we first expand $F(t)$ into a Fourier series
between the times 0 and θ, where θ is very large compared to T. For this time
interval let there be

$$F(t) = \sum_\nu \alpha_\nu \cos \left(2\pi\nu \frac{t}{\theta} + \phi_n \right). \tag{8}$$

The coefficients A_n and B_n can be expressed with t_0 and the coefficients
α_ν and ϕ_ν from the expansion (8) if t_0 is chosen between $t = 0$ and $t = \theta - T$. One   {2}
next gets

$$
\begin{aligned}
A_n &= \frac{2}{T} \sum_\nu \left\{ \int_{t_0}^{t_0 + T} \alpha_\nu \cos \left(2\pi\nu \frac{t}{\theta} + \phi_\nu \right) \cos \left(2\pi n \frac{t - t_0}{T} \right) dt \right\} \\
B_n &= \frac{2}{T} \sum_\nu \left\{ \int_{t_0}^{t_0 + T} \alpha_\nu \cos \left(2\pi\nu \frac{t}{\theta} + \phi_\nu \right) \sin \left(2\pi n \frac{t - t_0}{T} \right) dt \right\}
\end{aligned}. \tag{9}
$$

After carrying out this integration and neglecting, in known manner, terms with the

[3]   factor $\dfrac{1}{\pi(\nu/\theta + n/T)}$ versus those with the factor $\dfrac{1}{\pi(\nu/\theta - n/T)}$, one obtains

$$\left.\begin{aligned}
A_n &= \sum_\nu \alpha_\nu \frac{\sin \pi\left(\nu \dfrac{T}{\theta} - n\right) \cos\left(\chi_{\nu n} + 2\pi\nu \dfrac{t_0}{\theta}\right)}{\pi\left(\nu \dfrac{T}{\theta} - n\right)} \\[3em]
B_n &= -\sum_\nu \alpha_\nu \frac{\sin \pi\left(\nu \dfrac{T}{\theta} - n\right) \cos\left(\chi_{\nu n} + 2\pi\nu \dfrac{t_0}{\theta}\right)}{\pi\left(\nu \dfrac{T}{\theta} - n\right)}
\end{aligned}\right\}
\tag{10}$$

[p. 884] where

$$\chi_{\nu n} = \pi\left(\nu \frac{T}{\theta} + n\right) + \phi_\nu .$$

The formulas (10) are only valid for values t_0 between $t_0 = 0$ and $t_0 = \theta - T$ because the expansion applies, due to (8), only to the time interval $0 - \theta$. However, we allow ourselves to apply formula (8) to the interval $0 - (\theta + T)$. In doing this, we replace the function $F(t)$ between the time values θ and $\theta + T$ with the values of $F(t)$ between 0 and T. This procedure will falsify our consideration of the mean values, but only by infinitesimal amounts because the time interval T is infinitesimal relative to θ. Progressing from this observation, we shall use the equations (10) such as if they would be valid in the entire interval $0 < t_0 < \theta$.

With the help of (10) we form the mean value $\overline{A_m A_n}$, i.e., the quantity

$$\overline{A_m A_n} = \frac{1}{\theta} \int_0^\theta A_m A_n \, dt.$$

The integral

$$\int_0^\theta \cos\left(\chi_{\mu m} + 2\pi\mu \frac{t_0}{\theta}\right) \cos\left(\chi_{\nu n} + 2\pi\nu \frac{t_0}{\theta}\right) dt_0$$

occurs during this process. But it vanishes with integers μ and ν when $\mu \neq \nu$, and for $\mu = \nu$ it has the value $\dfrac{\theta}{2}(-1)^{m - n}$. Considering this, the first equation (10) yields

$$\overline{A_m A_n} = \frac{(-1)^{m-n}}{2} \sum_\nu \alpha_\nu^2 \frac{\sin \pi\left(\nu \frac{T}{\theta} - m\right) \sin \pi\left(\nu \frac{T}{\theta} - n\right)}{\pi^2\left(\nu \frac{T}{\theta} - m\right)\left(\nu \frac{T}{\theta} - n\right)}$$

$$= \frac{1}{2} \sum_\nu \alpha_\nu^2 \frac{\sin^2 \pi\nu \frac{T}{\theta}}{\pi^2\left(\nu \frac{T}{\theta} - m\right)\left(\nu \frac{T}{\theta} - n\right)} . \tag{11}$$

It is a priori obvious that a statistical dependence between radiation components can only be expected between neighboring frequencies; m and n belong, therefore, into the same narrow spectral range, and the same holds for those values ν which contribute substantially to our sum.

The quotient on the right-hand side of (11) changes only slowly with ν because [p. 885] T/θ is a small quantity. Therefore, one can average relative to α_ν^2 over many [4] {3} sequential terms without noticeable error, and the mean value $\overline{\alpha_\nu^2}$ can be pulled as a constant before the sum since the summation extends only over a narrow spectral range. The summation over the quotients can then be transformed into an integral and one obtains:

$$\overline{A_m A_n} = \frac{1}{2} \alpha_\nu^2 \frac{\theta}{\pi T} \int \frac{\sin^2 x}{(x - m\pi)(x - n\pi)} dx. \tag{12}$$

Instead of taking the integral between the boundaries of the aforesaid spectral range, one can take it without noticeable error between $-\infty$ and $+\infty$.

The integral has the value π for $m = n$, but it always vanishes[1] when $m \neq n$ (m, n being integers). The vanishing of $\overline{A_m A_n}$ (for $m \neq n$) is herewith shown; proofs for the vanishing of $\overline{B_m B_n}$ (for $m \neq n$) and for $\overline{A_m B_n}$ can be carried out in analogy. From the vanishing of these mean values follows, according to §1, the

---

[1]The integral is equal to

$$\frac{1}{(m - n)\pi}\left\{\int_{-\infty}^{+\infty} \frac{\sin^2 x}{x - m\pi} dx - \int_{-\infty}^{+\infty} \frac{\sin^2 x}{x - n\pi} dx\right\}.$$

Each of the last two integrals is equal to

$$\int_{-\infty}^{+\infty} \frac{\sin^2 y}{y} dy = 0.$$

statistical independence of the Fourier coefficients, as has been claimed.

Note in Proof: Instead of averaging over many sequential terms in the sum (11), in order to get its value, one can also base it upon an infinity of mutually independent expansions (8) and average over these. This formation of the mean value, applied to (11), brings the correspondingly understood mean value $\overline{\alpha_\nu^2}$ in front of the summation symbol. The end result remains, of course, the same.

(Received on June 24, 1915)

Additional notes by translator

{1} "$\sum\limits_m$" has been corrected to "$\sum\limits_n$."

{2} "α_ν and ϕ_ν" have been corrected to "α_ν and ϕ_ν."

{3} "a_ν^2" has been corrected to "α_ν^2" and again one line further down, "$\overline{a_n^2}$" has been corrected to "$\overline{\alpha_\nu^2}$." Note [4] seems to contain misprints.

Doc. 19
Supplementary Expert Opinion

Not translated for this volume.

Doc. 20

My Opinion on the War[2]

[October 23—November 11, 1915][1]

The psychological roots of war are—in my opinion—biologically founded in the aggressive characteristics of the male creature. We "jewels of the creation" are not the only ones who can boast of this distinction; some animals outdo us on this point, e.g., the bull and the rooster. This aggressive tendency comes to the fore whenever individual males are placed side by side, and even more so when relatively close-knit societies have to deal with each other. Almost without fail they will end up in disputes that escalate into quarrels and murder unless special precautions are taken to prevent such occurrences. I will never forget <with> what <bloody and> honest hatred the schoolmates of my age felt for years against the first-graders of a school in a neighboring street. Innumerable fistfights occurred, resulting in many a hole in the heads of those little striplings. Who could doubt that vendetta and dueling spring from such feelings? I even believe that the *honor* [*Ehre*] that is so carefully groomed by us gains its major nourishment from such sources.

Understandably, the more modern organized states had to push these manifestations of primitive virile characteristics vigorously into the background. But wherever two nation states are next to each other and without a joint superpower above them, those feelings at times generate tensions in the moods [*Gemüt*]* that lead to catastrophes of war. In saying so, I consider so-called aims and causes of war as rather meaningless, because they are always found when passion needs them.

The more subtle intellects of all times have agreed that war is one of the worst enemies of human development, and that everything must be done to prevent it. Notwithstanding the unspeakably sad conditions of the present times, I have the conviction that it is possible, in the near future, to form a statelike organization in Europe that makes European wars impossible, just as now war between Bavaria and Württemberg is impossible in the German Reich. No friend of spiritual evolution should fail to stand up for this most important political aim of our time.[3]

One can ponder the question: Why does man in peacetime (when the social community suppresses almost every representation of rowdyism) not lose the capabilities and motivations that enable him to commit mass murder in war? The reasons seem to me as follows. When I look into the mind of the well-intentioned average citizen, I see a moderately illuminated cozy room. In one corner stands a

---

Translator's note. The psychologically multilayered and thus untranslatable German word "Gemüt" is here circumscribed with "moods."

well-tended shrine, the pride of the man of the house, and every visitor is loudly alerted to the presence of this shrine upon which is written in huge letters "patriotism." It is usually a taboo to open this shrine. Moreover, the master of the house barely knows, or does not know at all, that this shrine holds the moral requisites of bestial hatred and mass murder, which he dutifully takes out in case of war in order to use them. This type of shrine, dear reader, you will not find in my little room, and I would be happy if you would adopt the attitude that in that same corner of your little room a piano or a bookshelf would be a more fitting piece of furniture than the one you find only tolerable because you have been used to it since your early youth.

I have no intention of making a secret of my international sentiments. How close I feel for a human being or a human organization depends only upon how I judge their intentions and capabilities. The state to which I belong as a citizen does not play any role in my feelings [*Gemütsleben*], because I see it more like a matter of business, such as the relations with a life insurance company. (From what I have said above, there should be no doubt that I must strive to be a citizen of a country that presumably does not force me to take part in a war.)

But how can a powerless individual creature contribute to reaching this goal? Should perhaps everybody devote a considerable portion of his time and resources to politics? I really believe that the intellectually more mature people of Europe have sinned when they neglected to care for general political questions; yet, I do not see in political commitment the most important effectiveness of an individual in this matter. I rather believe that everybody should act individually in a sense that those feelings, which I have elaborately discussed before, are steered on a course that can no longer become a curse for the general public.

Every man who knows that he acted to his best knowledge and ability should feel honorable, without regard to the words and deeds of others. Words and actions of others or other groups cannot *violate* one's personal honor or the honor of one's group. Power-hunger and greed shall, as in the past, be treated as despicable vices; the same applies to hatred and quarrelsomeness. I do not suffer from an overestimation of the past, but in my opinion we have not made progress on this important point; we rather fell back. Every well-meaning person should work hard on himself and in his personal circle to improve in these respects. The heavy burdens which presently plague us in such a horrible way will then vanish too.

But why so many words when I can say it in one sentence, and in a sentence very appropriate for a Jew: Honor your Master Jesus Christ not only in words and songs, but rather foremost by your deeds.

Doc. 21

[p. 778]   Plenary Session of November 4, 1915

On the General Theory of Relativity

My efforts in recent years were directed toward basing a general theory of relativity,
[1]   also for nonuniform motion, upon the supposition of relativity. I believed indeed to
have found the only law of gravitation that complies with a reasonably formulated
postulate of general relativity; and I tried to demonstrate the truth of precisely this
solution in a paper[1] that appeared last year in the *Sitzungsberichte*.

Renewed criticism showed to me that this truth is absolutely impossible to show
in the manner suggested. That this seemed to be the case was based upon a
misjudgment. The postulate of relativity—*as far as I demanded it there*—is always
satisfied if the Hamiltonian principle is chosen as a basis. But in reality, it provides
no tool to establish the Hamiltonian function H of the gravitational field. Indeed,
equation (77) l.c. which limits the choice of H says only that H has to be an invariant
toward linear transformations, a demand that has nothing to do with the relativity of
accelerations. Furthermore, the choice determined by equation (78) l.c. does not
[3]   determine equation (77) in any way.

For these reasons I lost trust in the field equations I had derived, and instead
looked for a way to limit the possibilities in a natural manner. In this pursuit I arrived
at the demand of general covariance, a demand from which I parted, though with a
heavy heart, three years ago when I worked together with my friend GROSSMANN. As
a matter of fact, we were then quite close to that solution of the problem, which will
be given in the following.

Just as the special theory of relativity is based upon the postulate that all
equations have to be covariant relative to linear orthogonal transformations, so the
[p. 779]   theory developed here rests upon the postulate of the *covariance of all systems of
equations relative to transformations with the substitution determinant 1.*

Nobody who really grasped it can escape from its charm, because it signifies a
real triumph of the general differential calculus as founded by GAUSS, RIEMANN,
CHRISTOFFEL, RICCI, and LEVI-CIVITA.

---

[2]   [1]"Die formale Grundlage der Relativitätstheorie," *Sitzungsberichte* 41 (1914), pp.
1066–1077. Equations of this paper are quoted in the following with the additional note
"l.c." in order to keep them distinct from those in the present paper.

§1. Laws of Forming Covariants

I can be brief on the laws of forming covariants since I gave an elaborate description of the methods of absolute differential calculus in my paper of last year; thus we need only to investigate what will change in the theory of covariants if only substitutions of determinant 1 are permitted. The equation

$$d\tau' = \frac{\partial(x'_1 \ldots x'_4)}{\partial(x_1 \ldots x_4)x} d\tau,$$

which is valid for any substitutions, becomes, due to the premise in our theory, i.e.,

$$\frac{\partial(x'_1 \ldots x'_4)}{\partial(x_1 \ldots x_4)x} = 1 \tag{1}$$

now

$$d\tau' = d\tau \tag{2}$$

and the four-dimensional volume element $d\tau$ is therefore an invariant. Since furthermore (equation (17) l.c.) $\sqrt{-g}\,d\tau$ is an invariant toward arbitrary substitutions, it follows for the group that interests us now

$$\sqrt{-g'} = \sqrt{-g}. \tag{3}$$

The determinant of the $g_{\mu\nu}$ is therefore an invariant. Because of the scalar character of $\sqrt{-g}$ one can simplify the basic formulas of the formation of covariants, as compared to those of general covariance; which in short means, the factors $\sqrt{-g}$ and $1/\sqrt{-g}$ no longer occur in the basic formulas, and the distinction between tensors and V-tensors drops out. Specifically one gets:

1. The place of the tensors $G_{iklm} = \sqrt{-g}\,\delta_{iklm}$ and $G^{iklm} = \dfrac{1}{\sqrt{-g}}\,\delta_{iklm}$ (as in (19) [p. 780]

and (21a) l.c.) is now taken by the tensors

$$G_{iklm} = G^{iklm} = \delta_{iklm} \tag{4}$$

which are of a simpler structure.

2. The basic formulas (29) l.c. and (30) l.c. for the extension of tensors can, under our premise, not be replaced by simpler ones, but the equations that define divergence (representing a combination of the equations (30) l.c. and (31) l.c.) can be simplified. This can be written as

$$A^{\alpha_1 \ldots \alpha_l} = \sum_s \frac{\partial A^{\alpha_1 \ldots \alpha_l s}}{\partial x_s} + \sum_{s\tau}\left[\begin{Bmatrix} s\tau \\ \alpha_1 \end{Bmatrix}A^{\tau\alpha_2 \ldots \alpha_l s} + \ldots \begin{Bmatrix} s\tau \\ \alpha_l \end{Bmatrix}A^{\alpha_1 \cdot \alpha_{l-1}\tau s}\right] + \sum_{s\tau}\begin{Bmatrix} s\tau \\ s \end{Bmatrix}A^{\alpha_1 \ldots \alpha_l \tau}. \tag{5}$$

But according to (24) l.c. and (24a) l.c.

$$\sum_\tau \begin{Bmatrix} ST \\ S \end{Bmatrix} = \frac{1}{2} \sum_{\alpha s} g^{s\alpha} \left(\frac{\partial g_{s\alpha}}{\partial x_\tau} + \frac{\partial g_{\tau\alpha}}{\partial x_s} - \frac{\partial g_{s\tau}}{\partial x_\alpha} \right) = \frac{1}{2} \sum_s g^{s\alpha} \frac{\partial g_{s\alpha}}{\partial x_\tau} = \frac{\partial (lg\sqrt{-g})}{\partial x_\tau}. \quad (6)$$

And this quantity has the characteristics of a vector, due to (3). Consequently, the last term on the right-hand side of (5) is itself a contravariant tensor of rank l. We are therefore entitled to replace (5) by the simple definition of divergence, viz.,

$$A^{\alpha_1 \ldots \alpha_l} = \sum_s \frac{\partial A^{\alpha_1 \ldots \alpha_l s}}{\partial x_s} + \sum_{s\tau} \left[\begin{Bmatrix} ST \\ \alpha_1 \end{Bmatrix} A^{\tau \alpha_2 \ldots \alpha_l s} + \ldots \begin{Bmatrix} ST \\ \alpha_l \end{Bmatrix} A^{\alpha_1 \ldots \alpha_{l-1} \tau s} \right], \quad (5a)$$

and we shall do so throughout.

For example, the definition (37) l.c.

$$\Phi = \frac{1}{\sqrt{-g}} \sum_\mu \frac{\partial}{\partial x_\mu} (\sqrt{-g} A^\mu)$$

has to be replaced by the simpler definition

$$\Phi = \sum_\mu \frac{\partial A^\mu}{\partial x_\mu}, \quad (7)$$

and equation (40) l.c. for the divergence of the contravariant six-vector by the simpler

$$A^\mu = \sum_\nu \frac{\partial A^{\mu\nu}}{\partial x_\nu}. \quad (8)$$

In place of (41a) l.c. we have, due to our assumption,

$$A_\sigma = \sum_\nu \frac{\partial A_\sigma^\nu}{\partial x_\nu} - \frac{1}{2} \sum_{\mu\nu\tau} g^{\tau\mu} \frac{\partial g_{\mu\nu}}{\partial x_\sigma} A_\tau^\nu. \quad (9)$$

[p. 781]   A comparison with (41b) reveals that under our assumption the law of divergence is the same as that for the divergence of V-tensors in the general differential calculus. This remark applies to any divergence of tensors, as can be derived from (5) and (5a).

3. Our limitation to transformations of determinant 1 brings the farthest-reaching simplification for those covariants which are formed only from the $g_{\mu\nu}$ and their derivatives. It is shown in mathematics that these covariants can all be derived from the RIEMANN-CHRISTOFFEL tensor of rank four, which (in its covariant form) reads:

$$(ik, \ lm) = \frac{1}{2}\left(\frac{\partial^2 g_{im}}{\partial x_k \partial x_l} + \frac{\partial^2 g_{kl}}{\partial x_i \partial x_m} - \frac{\partial^2 g_{il}}{\partial x_k \partial x_m} - \frac{\partial^2 g_{mk}}{\partial x_l \partial x_i} \right)$$
$$+ \sum_{\rho\sigma} g^{\rho\sigma}\left(\begin{bmatrix} im \\ \rho \end{bmatrix}\begin{bmatrix} kl \\ \sigma \end{bmatrix} - \begin{bmatrix} il \\ \rho \end{bmatrix}\begin{bmatrix} km \\ \sigma \end{bmatrix} \right) \tag{10}$$

It is in the nature of gravitation that we are most interested in tensors of rank two, which can be formed by inner multiplication of this tensor of rank four with the $g_{\mu\nu}$. Due to the symmetry properties of the RIEMANNIAN tensor, apparent from (10), viz.,

$$(ik, \ lm) = (lm, \ ik)$$
$$(ik, \ lm) = -(ki, \ lm), \tag{11}$$

this multiplication can be formed only in *one* way; whereby one obtains the tensor

$$G_{im} = \sum_{kl} g^{kl}\,(ik, \ lm). \tag{12}$$

It is more advantageous for our purposes to derive this tensor from a different form of (10) which CHRISTOFFEL has given,[2] i.e.,

$$\{ik, \ lm\} = \sum_\rho g^{k\rho}\,(i\rho, \ lm) = \frac{\partial \left\{ \begin{matrix} il \\ k \end{matrix} \right\}}{\partial x_m} - \frac{\partial \left\{ \begin{matrix} im \\ k \end{matrix} \right\}}{\partial x_l} + \sum_\rho \left[\left\{ \begin{matrix} il \\ \rho \end{matrix} \right\}\left\{ \begin{matrix} \rho m \\ k \end{matrix} \right\} - \left\{ \begin{matrix} im \\ \rho \end{matrix} \right\}\left\{ \begin{matrix} \rho l \\ k \end{matrix} \right\} \right]. \tag{13}$$

When this tensor is multiplied (inner multiplication) with the tensor

$$\delta_k^l = \sum_\alpha g_{k\alpha}\,g^{\alpha l}$$

one obtains G_{im}, viz.,

$$G_{im} = \{il, \ lm\} = R_{im} + S_{im} \tag{13}$$

[p. 782]
{1}

$$R_{im} = -\frac{\partial \left\{ \begin{matrix} im \\ l \end{matrix} \right\}}{\partial x_l} + \sum_\rho \left\{ \begin{matrix} il \\ \rho \end{matrix} \right\}\left\{ \begin{matrix} \rho m \\ l \end{matrix} \right\} \tag{13a}$$

[4]

---

[2]A simple proof of the tensorial character of this expression can be found on page 1053 of my repeatedly quoted paper.

$$S_{im} = \frac{\partial \begin{Bmatrix} il \\ l \end{Bmatrix}}{\partial x_m} - \begin{Bmatrix} im \\ \rho \end{Bmatrix} \begin{Bmatrix} \rho l \\ l \end{Bmatrix}. \tag{13b}$$

Under the constraint to transformations with determinants 1, not only (G_{im}) is a tensor, but (R_{im}) and (S_{im}) also have tensorial character. It follows indeed from the fact that $\sqrt{-g}$ is a scalar, and because of (6), that $\begin{Bmatrix} il \\ l \end{Bmatrix}$ is a covariant four-vector. (S_{im}), however, is, due to (29) l.c., nothing other than the extension of this four-vector, which means it is also a tensor. From the tensorial character of (G_{im}) and (S_{im}) follows the same for (R_{im}), from (13). The tensor (R_{im}) is of utmost importance for the theory of gravitation.

§2. Notes on the Differential Laws of "Material" Processes

1. The energy-momentum theorem for matter (including electromagnetic processes in a vacuum.

According to the general considerations of the previous paragraph, equation (42a) l.c. has to be replaced by

$$\sum_{v} \frac{\partial T_\sigma^v}{\partial x_v} = \frac{1}{2} \sum_{\mu \tau v} g^{\tau \mu} \frac{\partial g_{\mu v}}{\partial x_\sigma} T_\tau^v + K_\sigma, \tag{14}$$

where T_σ^v is an ordinary tensor, K_σ an ordinary four-vector (not a V-tensor, V-vector, resp.). We have to attach a remark to this equation, because it is important for the following. The equations of conservation led me in the past to view the quantities

$$\frac{1}{2} \sum_{\mu} g^{\tau \mu} \frac{\partial g_{\mu v}}{\partial x_\sigma}.$$

as the natural expressions of the components of the gravitational field, even though the formulas of the absolute differential calculus seem to suggest the CHRISTOFFEL symbols $\begin{Bmatrix} v\sigma \\ \tau \end{Bmatrix}$ instead, as being the more natural quantities. The former view was a fateful prejudice. The preference for the CHRISTOFFEL symbols justifies itself especially because of the symmetry in their covariant indices (here v and σ) and, furthermore, because they occur in the fundamentally important equations of the geodesic line (23b) l.c.; and these latter are—from a physical point of view—the equations of motion of a material point in a gravitational field. Equation (14) cannot

[p. 783]

serve as a counterargument because the first term on its right-hand side can be brought into the form

$$\sum_{\nu\tau} \begin{Bmatrix} \sigma\nu \\ \tau \end{Bmatrix} T_\tau^\nu .$$

Therefore, from now on we shall call the quantities

$$\Gamma_{\mu\nu}^\sigma = -\begin{Bmatrix} \mu\nu \\ \sigma \end{Bmatrix} = -\sum_\alpha g^{\sigma\alpha}\begin{bmatrix} \mu\nu \\ \alpha \end{bmatrix} = -\frac{1}{2}\sum_\alpha g^{\sigma\alpha}\left(\frac{\partial g_{\mu\alpha}}{\partial x_\nu} + \frac{\partial g_{\nu\alpha}}{\partial x_\mu} - \frac{\partial g_{\mu\nu}}{\partial x_\alpha}\right) \quad (15)$$

the components of the gravitational field. K_ν vanishes when T_σ^ν denotes the energy tensor of all "material" processes, and the conservation theorem (14) takes the form

$$\sum_\alpha \frac{\partial T_\sigma^\alpha}{\partial x_\alpha} = -\sum_{\alpha\beta} \Gamma_{\sigma\beta}^\alpha T_\alpha^\beta . \qquad (14a)$$

We note that the equations of motion (23b) l.c. of a material point in a gravitational field take the form

$$\frac{d^2 x_\tau}{ds^2} = \sum_{\mu\nu} \Gamma_{\mu\nu}^\tau \frac{dx_\mu}{ds}\frac{dx_\nu}{ds}. \qquad (15) \qquad \{2\}$$

2. The considerations in paragraphs 10 and 11 of the quoted paper remain unchanged, except that the structures which were there called V-scalars and V-tensors are now ordinary scalars and tensors, respectively.

§3. The Field Equations of Gravitation

From what has been said, it seems appropriate to write the field equations of gravitation in the form

$$R_{\mu\nu} = -\kappa T_{\mu\nu} \qquad (16)$$

since we already know that these equations are covariant under any transformation of a determinant equal to 1. Indeed, these equations satisfy all conditions we can demand. Written out in more detail, and according to (13a) and (15), they are

$$\sum_\alpha \frac{\partial \Gamma_{\mu\nu}^\alpha}{\partial x_\alpha} + \sum_{\alpha\beta} \Gamma_{\mu\beta}^\alpha \Gamma_{\nu\alpha}^\beta = -\kappa T_{\mu\nu}. \qquad (16a)$$

We wish to show now that these field equations can be brought into the HAMILTO-   [p. 784]
NIAN form

$$\delta\left\{\int\left(\mathfrak{L} - \kappa\sum_{\mu\nu} g^{\mu\nu} T_{\mu\nu}\right) d\tau\right\}\Bigg|$$
$$\mathfrak{L} = \sum_{\sigma\tau\alpha\beta} g^{\sigma\tau}\Gamma^{\alpha}_{\sigma\beta}\Gamma^{\beta}_{\tau\alpha}$$

$$\tag{17}$$

where the $g^{\mu\nu}$ have to be varied while the $T_{\mu\nu}$ are to be treated as constants. The reason is that (17) is equivalent to the equations

$$\sum_{\alpha} \frac{\partial}{\partial x_{\alpha}}\left(\frac{\partial \mathfrak{L}}{\partial g^{\mu\nu}_{\alpha}}\right) - \frac{\partial \mathfrak{L}}{\partial g^{\mu\nu}} = -\kappa T_{\mu\nu}, \tag{18}$$

where $\mathfrak{L}$ has to be thought of as a function of the $g^{\mu\nu}$ and the $\dfrac{\partial g^{\mu\nu}}{\partial x_{\sigma}}(= g^{\mu\nu}_{\sigma})$. On the other hand, a lengthy but uncomplicated calculation yields the relations

$$\frac{\partial \mathfrak{L}}{\partial g^{\mu\nu}} = -\sum_{\alpha\beta} \Gamma^{\alpha}_{\mu\beta}\Gamma^{\beta}_{\nu\alpha} \tag{19}$$

$$\frac{\partial \mathfrak{L}}{\partial g^{\mu\nu}_{\alpha}} = \Gamma^{\alpha}_{\mu\nu}. \tag{19a}$$

These together with (18) provide the field equations (16a).

It can now also be easily shown that the principle of the conservation of energy and momentum is satisfied. Multiplying (18) by $g^{\mu\nu}_{\sigma}$ with summation over the indices μ and ν, one obtains after customary rearrangement

$$\sum_{\alpha\mu\nu} \frac{\partial}{\partial x_{\alpha}}\left(g^{\mu\nu}_{\sigma}\frac{\partial \mathfrak{L}}{\partial g^{\mu\nu}_{\alpha}}\right) - \frac{\partial \mathfrak{L}}{\partial x_{\sigma}} = -\kappa\sum_{\mu\nu} T_{\mu\nu}g^{\mu\nu}_{\sigma}.$$

According to (14), on the other hand, for the *total* energy tensor of matter one has

$$\sum_{\lambda} \frac{\partial T^{\lambda}_{\sigma}}{\partial x_{\lambda}} = -\frac{1}{2}\sum_{\mu\nu} \frac{\partial g^{\mu\nu}}{\partial x_{\sigma}} T_{\mu\nu}.$$

From the last two equations follows

$$\sum_{\lambda} \frac{\partial}{\partial x_{\lambda}}(T^{\lambda}_{\sigma} + t^{\lambda}_{\sigma}) = 0, \tag{20}$$

where

$$t^{\lambda}_{\sigma} = \frac{1}{2\kappa}\left(\mathfrak{L}\delta^{\lambda}_{\sigma} - \sum_{\mu\nu} g^{\mu\nu}_{\sigma}\frac{\partial \mathfrak{L}}{\partial g^{\mu\nu}_{\lambda}}\right) \tag{20a}$$

[p. 785] denotes the "energy tensor" of the gravitational field which, by the way, has tensorial character only under linear transformations. After a simple rearrangement, one gets from (20a) and (19a)

$$t_\sigma^\lambda = \frac{1}{2}\delta_\sigma^\lambda \sum_{\mu\nu\alpha\beta} g^{\mu\nu}\Gamma_{\mu\beta}^\alpha\Gamma_{\nu\alpha}^\beta - \sum_{\mu\nu\alpha} g^{\mu\nu}\Gamma_{\mu\sigma}^\alpha\Gamma_{\nu\alpha}^\lambda. \qquad (20b)$$

Finally, it is of interest to derive two scalar equations that result from the field equations. After multiplying (16a) by $g^{\mu\nu}$ with summation over μ and ν, we get after simple rearranging

$$\sum_{\alpha\beta} \frac{\partial^2 g^{\alpha\beta}}{\partial x_\alpha \partial x_\beta} - \sum_{\sigma\tau\alpha\beta} g^{\sigma\tau}\Gamma_{\sigma\beta}^\alpha\Gamma_{\tau\alpha}^\beta + \sum_{\alpha\beta} \frac{\partial}{\partial x_\alpha}\left(g^{\alpha\beta}\frac{\partial lg\sqrt{-g}}{\partial x_\beta}\right) = -\kappa\sum_\sigma T_\sigma^\sigma. \qquad (21)$$

On the other hand, multiplying (16a) by $g^{\nu\lambda}$ and summing over ν, we get

$$\sum_{\alpha\nu} \frac{\partial}{\partial x_\alpha}(g^{\nu\lambda}\Gamma_{\mu\nu}^\alpha) - \sum_{\alpha\beta\nu} g^{\nu\beta}\Gamma_{\nu\mu}^\alpha\Gamma_{\beta\alpha}^\lambda = -\kappa T_\mu^\tau,$$

or, also considering (20b),

$$\sum_{\alpha\nu} \frac{\partial}{\partial x_\alpha}(g^{\nu\lambda}\Gamma_{\mu\nu}^\alpha) - \frac{1}{2}\delta_\mu^\lambda \sum_{\mu\nu\alpha\beta} g^{\mu\nu}\Gamma_{\mu\beta}^\alpha\Gamma_{\nu\alpha}^\beta = -\kappa(T_\mu^\lambda + t_\mu^\lambda).$$

Taking (20) into account, and after simple rearranging, this yields

$$\frac{\partial}{\partial x_\mu}\left[\sum_{\alpha\beta} \frac{\partial^2 g^{\alpha\beta}}{\partial x_\alpha \partial x_\beta} - \sum_{\sigma\tau\alpha\beta} g^{\sigma\tau}\Gamma_{\sigma\beta}^\alpha\Gamma_{\tau\alpha}^\beta\right] = 0. \qquad (22)$$

However, we demand somewhat beyond that:

$$\sum_{\alpha\beta} \frac{\partial^2 g^{\alpha\beta}}{\partial x_\alpha \partial x_\beta} - \sum_{\sigma\tau\alpha\beta} g^{\sigma\tau}\Gamma_{\sigma\beta}^\alpha\Gamma_{\tau\alpha}^\beta = 0, \qquad (22a)$$

whereupon (21) becomes

$$\sum_{\alpha\beta} \frac{\partial}{\partial x_\alpha}\left(g^{\alpha\beta}\frac{\partial lg\sqrt{-g}}{\partial x_\beta}\right) = -\kappa\sum_\sigma T_\sigma^\sigma. \qquad (21a)$$

Equation (21a) shows the impossibility to choose the coordinate system such that $\sqrt{-g}$ equals 1, because the scalar of the energy tensor cannot be set to zero. [5]

Equation (22a) is a relation of the $g_{\mu\nu}$ alone; it would not be valid in a new coordinate system which would result from the original one by a forbidden transformation. The equation therefore shows how the coordinate system has to be adapted to the manifold.

§4. Some Remarks on the Physical Qualities of the Theory

A first approximation of the equations (22a) is

$$\sum_{\alpha\beta} \frac{\partial^2 g^{\alpha\beta}}{\partial x_\alpha \partial x_\beta} = 0.$$

This does not yet fix the coordinate system, because this would require 4 equations. We are therefore entitled to put for a first approximation arbitrarily

{3}
$$\sum_{\beta} \frac{\partial g^{\alpha\beta}}{\partial x_\beta} = 0. \tag{22}$$

For further simplification we want to introduce the imaginary time as a fourth variable. The field equations (16a) then take, as a first approximation, the form

$$\frac{1}{2}\sum_{\alpha} \frac{\partial^2 g_{\mu\nu}}{\partial x_\alpha^2} = \kappa T_{\mu\nu}, \tag{16b}$$

from which one sees immediately that it contains NEWTON's law as an approximation.—

[6] That the new theory complies with the relativity of motion follows from the fact that among the permissible transformations are those that correspond to a rotation of the new relative to the old system (with arbitrarily variable angular velocity), and also those where the origin of the new system performs an arbitrarily prescribed motion relative to that of the old one.

Indeed, the substitutions

$$\begin{aligned}
x' &= x \cos \tau + y \sin \tau \\
y' &= -x \sin \tau + y \cos \tau \\
z' &= z \\
t' &= t
\end{aligned}$$

and

$$\begin{aligned}
x' &= x - \tau_1 \\
y' &= y - \tau_2 \\
z' &= z - \tau_3 \\
t' &= t,
\end{aligned}$$

where τ and τ_1, τ_2, τ_3 respectively are arbitrary functions of t and substitutions with the determinant 1.

Additional notes by translator

{1}   The number (13) has already been assigned to an equation above.

{2}   The number (15) has already been assigned to an equation above.

{3}   The number (22) has already been assigned to an equation above.

Doc. 22

[p. 799]

On the General Theory of Relativity (Addendum)

by A. Einstein

In a recent investigation[1] I have shown how RIEMANN's theory of covariants in multidimensional manifolds can be utilized as a basis for a theory of the gravitational field. I now want to show here that an even more concise and logical structure of the theory can be achieved by introducing an admittedly bold additional hypothesis on the structure of matter.

The hypothesis whose justification we want to consider relates to the following topic. The energy tensor of "matter" T_μ^λ has a scalar $\sum_\mu T_\mu^\mu$, whose vanishing for the electromagnetic field is well known. In contrast, it seems to differ from zero for matter *proper*. Because, if we consider the most simple special case, that of an "incoherent" continuous fluid (with pressure neglected), then we are used to writing

$$T^{\mu\nu} = \sqrt{-g}\,\rho_0 \frac{dx_\mu}{ds} \frac{dx_\nu}{ds},$$

and we have

$$\sum_\mu T_\mu^\mu = \sum_{\mu\nu} g_{\mu\nu} T^{\mu\nu} = \rho_0 \sqrt{-g}.$$

The scalar of the energy tensor does not vanish in this approach.

One now has to remember that by our knowledge "matter" is not to be perceived as something primitively given or physically plain. There even are those, and not just a few, who hope to reduce matter to purely electrodynamic processes, which of course would have to be done in a theory more completed than MAXWELL's electrodynamics. Now let us just assume that in such completed electrodynamics the scalar of the energy tensor also would vanish! Would the result, shown above, prove that matter cannot be constructed in this theory? I think I can answer this question in the negative, because it might very well be that in "matter," to which the previous

[p. 800]

expression relates, gravitational fields do form an important constituent. In that case, $\sum T_\mu^\mu$ can appear positive for the entire structure while in reality only $\sum (T_\mu^\mu + t_\mu^\mu)$ is positive and $\sum T_\mu^\mu$ vanishes everywhere. *In the following we assume the conditions $\sum T_\mu^\mu = 0$ really to be generally true.*

---

[1] These *Sitzungsberichte*, p. 778.

Whoever does not categorically reject the possibility that gravitational fields could constitute an *essential* part of matter will find powerful support for this conception in the following.[2]

Derivation of the Field Equations

Our hypothesis allows us to take the last step that the idea of general relativity may consider as desirable. It allows us, namely, also to write the field equations of gravitation in a *general* covariant form. I have shown in the previous paper (equation (13)) that

$$G_{im} = \sum_l \{il, \, lm\} = R_{im} + S_{im} \tag{13}$$

is a covariant tensor. And we had

$$R_{im} = -\sum_l \frac{\partial \begin{Bmatrix} im \\ l \end{Bmatrix}}{\partial x_l} + \sum_{\rho l} \begin{Bmatrix} il \\ \rho \end{Bmatrix} \begin{Bmatrix} \rho m \\ l \end{Bmatrix} \tag{13a}$$

$$S_{im} = \sum_l \frac{\partial \begin{Bmatrix} il \\ l \end{Bmatrix}}{\partial x_m} - \sum_{\rho l} \begin{Bmatrix} im \\ \rho \end{Bmatrix} \begin{Bmatrix} \rho l \\ l \end{Bmatrix}. \tag{13b}$$

This tensor G_{im} is the only tensor available for the establishment of generally covariant equations of gravitation.

We have won generally covariant field equations if we agree that the field equations of gravitation should be

$$G_{\mu\nu} = -\kappa T_{\mu\nu}. \tag{16b}$$

These, together with the generally covariant laws, provided for by the absolute differential calculus, express the causal nexus for "material" processes in nature; and they express it in a form that emphasizes the fact that any special choice of coordinate system—which logically has nothing to do with nature's law anyway—is [p. 801] not used in the formulation of these laws.

---

[2]In writing this paper I was not yet aware that the hypothesis $\sum_\mu T_\mu^\mu = 0$ is, in principle, admissible.

Based upon this system one can—by retroactive choice of coordinates—return to those laws which I established in my recent paper, and without any actual change in these laws, because it is clear that we can introduce a new coordinate system such that relative to it

$$\sqrt{-g} = 1$$

holds everywhere. S_{im} then vanishes and one returns to the system of field equations

$$R_{\mu\nu} = -\kappa T_{\mu\nu} \tag{16}$$

of the recent paper. The formulas of absolute differential calculus degenerate exactly in the manner shown in said paper. And our choice of coordinates still allows only transformations of determinant 1.

The only difference in content between the field equations derived from general covariance and those of the recent paper is that the value of $\sqrt{-g}$ could not be prescribed in the latter. This value was rather determined by the equation

{1}
$$\sum_{\alpha\beta} \frac{\partial}{\partial x_\alpha}\left(g^{\alpha\beta} \frac{\partial lg\sqrt{-g}}{\partial x_\beta}\right) = -\kappa \sum_\sigma T_\sigma^\sigma. \tag{21a}$$

This equation shows that here $\sqrt{-g}$ can only be constant if the scalar of the energy tensor vanishes.

Under our present derivation $\sqrt{-g} = 1$ due to our arbitrary choice of coordinates. The vanishing of the scalar of the energy tensor of "matter" follows now from our field equations instead of from equation (21a). *The generally covariant field equations (16b), which form our starting point, do not lead to a contradiction only when the hypothesis, which we explained in the introduction, applies.* Then, however, we are also entitled to add to our previous field equation the limiting condition:

$$\sqrt{-g} = 1. \tag{21b}$$

Additional note by translator

{1} The "∂x_α" inside of the parentheses has been corrected to "∂x_β."

Doc. 23   [p. 420]

Comment on Our Paper:

"Experimental Proof of Ampère's Molecular Currents"[1]

by A. Einstein and W. L. de Haas

Our colleague BERLINER recently sent us two notes by S. J. BARNETT that were [2]
published in the July 30, 1915, and October, 1, 1915, issues of *Science*. From these [3]
it is plainly obvious that MAXWELL already had the idea to look for gyroscopic [4]
properties of magnets in order to test AMPÈRE's hypothesis. BARNETT writes: "The
experiment which I recently described in this journal can be viewed as a modification
of one undertaken a long time ago by MAXWELL, who seems to have been the first
one to think that a magnet must act like a gyroscope if the AMPÈRE currents are of
a truly material nature, as modern theory assumes." [5]

BARNETT began his experiments already six years ago, and now says that positive [6]
results have been achieved. He tried to demonstrate those magnetomotoric forces
which occur in an iron rod that is in rapid rotation. Experimentally, this problem is [7]
incomparably more difficult than the one undertaken by us, which is to verify the
angular momentum that occurs due to a change in magnetization. Mr. BARNETT's and
our own experiments complement each other in a gratifying manner.

---

[1]*Verh. d. D. Phys. Ges.* 17 (1915), p. 152.   [1]

Doc. 24

Explanation of the Perihelion Motion of Mercury
from the General Theory of Relativity

This translation is by Brian Doyle and is reprinted from *A Source Book in Astronomy and Astrophysics, 1900–1975*, edited by Kenneth R. Lang and Owen Gingerich. Copyright © 1979 by the President and Fellows of Harvard College. Reprinted by permission of Harvard University Press.

IN A WORK RECENTLY PUBLISHED in these reports, I set up the gravitational field equations that are covariant with respect to arbitrary transformations of determinant [1] 1. In a supplement I showed that these equations are generally covariant if the contraction of the energy tensor of "matter" vanishes, and I demonstrated that no important considerations oppose the introduction of this hypothesis, through which time and space are robbed of the last trace of objective [2] reality.[1]

In the present work I find an important confirmation of this most fundamental theory of relativity, showing that it explains qualitatively and quantitatively the secular rotation of the orbit of Mercury (in the sense of the orbital motion itself), which was discovered by Leverrier and which amounts to 45 [4] sec of arc per century.[2] Furthermore, I show that the theory has as a consequence a curvature of light rays due to gravitational fields twice as strong as was indicated in my earlier investigation.

THE GRAVITATIONAL FIELD

From my last two communications it follows that the gravitational field in a vacuum has to satisfy, upon properly choosing a reference frame, the equations

$$\sum_\alpha \frac{\partial \Gamma^\alpha_{\mu\nu}}{\partial x_\alpha} + \sum_{\alpha\beta} \Gamma^\alpha_{\mu\beta}\Gamma^\beta_{\nu\alpha} = 0, \tag{1}$$

where the $\Gamma^\alpha_{\mu\nu}$ are defined by the equations

$$\begin{aligned}
\Gamma^\alpha_{\mu\nu} &= -\begin{Bmatrix} \mu\nu \\ \alpha \end{Bmatrix} = -\sum_\beta g^{\alpha\beta}\begin{bmatrix} \mu\nu \\ \beta \end{bmatrix} \\
&= -\frac{1}{2}\sum_\beta g^{\alpha\beta}\left(\frac{\partial g_{\mu\beta}}{\partial x_\nu} + \frac{\partial g_{\nu\beta}}{\partial x_\mu} - \frac{\partial g_{\mu\nu}}{\partial x_\alpha}\right).
\end{aligned} \tag{2}$$

Let us make, moreover, the hypothesis established in the last communication, that the contraction of the energy tensor of "matter" always vanishes, so that, in addition, the determinantal condition is imposed:

$$|g_{\mu\nu}| = -1. \tag{3}$$

A point mass, the sun, is located at the origin of the coordinate system. The gravitational field this point mass produces can be calculated from these equations by means of successive approximations.

Nevertheless, we should consider that the $g_{\mu\nu}$ are still not completely determined mathematically by equations (1) and (3), because these equations are covariant with respect to arbitrary transformations of determinant 1. Yet we are justified in assuming that all these solutions can be reduced to one another by such transformations that they are distinguished (by the given boundary conditions) formally but not,

however, physically, from one another. Consequently, I am satisfied for the time being with deriving here a solution, without discussing the question whether the solution might be unique.

To proceed, let the $g_{\mu\nu}$ be given in the 0th approximation by the following scheme corresponding to the original theory of relativity:

$$\left.\begin{matrix}
-1 & 0 & 0 & 0 \\
0 & -1 & 0 & 0 \\
0 & 0 & -1 & 0 \\
0 & 0 & 0 & +1
\end{matrix}\right\}, \tag{4}$$

or, more briefly, [6]

$$\left.\begin{aligned}
g_{\rho\sigma} &= \delta_{\rho\sigma} \\
g_{\rho 4} &= g_{4\rho} = 0 \\
g_{44} &= 1
\end{aligned}\right\}. \tag{4a}$$

Here ρ and σ signify the indices 1, 2, 3; $\delta_{\rho\sigma}$ is equal to 1 or 0 if $\rho = \sigma$ or $\rho \neq \sigma$, respectively.

I assume in what follows that the $g_{\mu\nu}$ differ from the values given in equation (4a) only by quantities small compared to unity. I treat this deviation as a small quantity of first order, whereas functions of the nth degree in these deviations are treated as quantities of the nth order. Equations (1) and (3) together with equation (4a) enable us to calculate by successive approximations the gravitational field up to quantities of nth order exactly. The approximation given in equation (4a) forms the 0th approximation.

The solution has the following properties, which determine the coordinate system:

1. All components are independent of x_4.

2. The solution is spatially symmetric about the origin of the coordinate system, in the sense that we encounter the same solution again if we subject it to a linear orthogonal spatial transformation.

3. The equations $g_{\rho 4} = g_{4\rho} = 0$ are exactly valid for $\rho = 1, 2, 3$.

4. The $g_{\mu\nu}$ possess the values given in equation (4a) at infinity.

FIRST APPROXIMATION It is easy to verify that to quantities of first order the equations (1) and (3) are satisfied for the just-named four conditions by the assumed solution

$$\left.\begin{aligned}
g_{\rho\sigma} &= -\delta_{\rho\sigma} + \alpha\left(\frac{\partial^2 r}{\partial x_\rho \partial x_\sigma} - \frac{\delta_{\rho\sigma}}{r}\right) = -\delta_{\rho\sigma} - \alpha\frac{x_\rho x_\sigma}{r^3} \\
g_{44} &= 1 - \frac{\alpha}{r}
\end{aligned}\right\}. \tag{4b}$$

[7] The $g_{4\rho}$ ($g_{\rho4}$) are determined by condition 3, the r denotes the quantity $+\sqrt{x_1{}^2 + x_2{}^2 + x_3{}^2}$, and α is a constant determined by the mass of the sun.

That condition 3 is fulfilled to terms of first order we see immediately. More simply, the field equations (1) are also fulfilled in the first approximation. We need only to consider that upon neglect of quantities of second and higher order, the left side of equation (1) can be permuted successively through

$$\sum_\alpha \frac{\partial \Gamma^\alpha_{\mu\nu}}{\partial x_\alpha}$$

$$\sum_\alpha \frac{\partial}{\partial x_\alpha}\begin{bmatrix} \mu\nu \\ \alpha \end{bmatrix},$$

where α runs over only 1–3.

As we perceive from equation (4b), my theory implies that in the case of a resting mass, the components g_{11} up to g_{33} are in quantities of first order already different from 0. There-
[8] fore, as we shall see later, no disagreement with Newton's law arises in the first approximation. This theory, however, pro-duces an influence of the gravitational field on a light ray somewhat different from that given in my earlier work, because the velocity of light is determined by the equation

$$\sum g_{\mu\nu}\, dx_\mu\, dx_\nu = 0. \tag{5}$$

Upon the application of Huygen's principle, we find from equations (5) and (4b), after a simple calculation, that a light ray passing at a distance Δ suffers an angular deflection of magnitude $2\alpha/\Delta$, while the earlier calculation, which was not based upon the hypothesis $\sum T^\mu_\mu = 0$, had produced the value
[9] α/Δ. A light ray grazing the surface of the sun should experi-ence a deflection of 1.7 sec of arc instead of 0.85 sec of arc. In contrast to this difference, the result concerning the shift of the spectral lines by the gravitational potential, which was confirmed by Mr. Freundlich on the fixed stars (in order of magnitude), remains unaffected, because this result depends
[10] only on g_{44}.

Since we have obtained the $g_{\mu\nu}$ in the first approximation,
[11] we can also calculate the components $T^\alpha_{\mu\nu}$ of the gravitational field to the first approximation. From equations (2) and (4b) we have

[12]
$$\Gamma^\tau_{\rho\sigma} = -\alpha\left(\delta_{\rho\sigma}\frac{x_\tau}{r^2} - \frac{3}{2}\frac{x_\rho x_\sigma x_\tau}{r^5}\right), \tag{6a}$$

where ρ, σ, τ signify any one of the indices 1, 2, 3, and

$$\Gamma^\tau_{44} = \Gamma^4_{4\sigma} = -\frac{\alpha}{2}\frac{x_\sigma}{r^3}, \tag{6b}$$

where σ signifies the index 1, 2, or 3. Those components in which the index 4 appears once or three times vanish.

SECOND APPROXIMATION   It will subsequently be seen that we need to determine only three components Γ^τ_{44} exactly to quantities of the second order in order to be able to determine the orbits of the planets with the appropriate degree of accuracy. For this process, the last field equation, together with the general conditions we have imposed on our solution, suffices. The last field equation,

$$\sum_\sigma \frac{\partial \Gamma^\sigma_{44}}{\partial x_\sigma} + \sum_{\sigma\tau} \Gamma^\sigma_{4\tau}\Gamma^\tau_{4\sigma} = 0,$$

becomes upon consideration of equation (6b) and upon neglect of quantities of third and higher order

[13]
$$\sum_\sigma \frac{\partial \Gamma^\sigma_{44}}{\partial x_\sigma} = \frac{\alpha^2}{2r^4}.$$

From this we deduce, upon considering equation (6b) and the symmetry properties of our solution,

$$\Gamma^\sigma_{44} = -\frac{\alpha}{2}\frac{x_\sigma}{r^3}\left(1 - \frac{\alpha}{r}\right). \tag{6c}$$

THE MOTION OF THE PLANETS

The equation of motion of the point mass in the gravita-tional field yielded by the general theory of relativity reads

$$\frac{d^2 x_\nu}{ds^2} = \sum_{\sigma\tau} \Gamma^\nu_{\sigma\tau}\frac{dx_\sigma}{ds}\frac{dx_\tau}{ds}. \tag{7}$$

From this equation we first deduce that it contains the Newtonian equations of motion as a first approximation. Of course, if the motion of the planet takes place with a velocity less than the velocity of light, then dx_1, dx_2, dx_3 are smaller than dx_4. In consequence, we get a first approximation in which we consider on the right side only the term $\sigma = \tau = 4$. Upon considering equation (6b), we obtain

$$\left.\begin{aligned} \frac{d^2 x_\nu}{ds^2} &= \Gamma^\nu_{44} = -\frac{\alpha}{2}\frac{x_\nu}{r^3}\ (\nu = 1, 2, 3) \\ \frac{d^2 x_4}{ds^2} &= 0 \end{aligned}\right\} \tag{7a}$$

These equations show that we can set $s = x_4$ for the first approximation. Then the first three equations are exactly the Newtonian equations. If we introduce polar variables in the orbital plane, then, as is well known, the energy law and the law of areas yield the equations

$$\left.\begin{aligned} \frac{1}{2}u^2 + \Phi &= A \\ r^2\frac{d\phi}{ds} &= B \end{aligned}\right\}, \tag{8}$$

where A and B signify the constants of the energy law, and where

$$\left.\begin{aligned}\Phi &= -\frac{\alpha}{2r} \\[2mm] u^2 &= \frac{dr^2 + r^2\,d\phi^2}{ds^2}\end{aligned}\right\} \tag{8a}$$

is granted.

We have now to evaluate the equations to the next order. The last of the equations (7) yields, then, together with equation (6b),

$$\frac{d^2x_4}{ds^2} = 2\sum_\sigma \Gamma^4_{\sigma 4}\frac{dx_\sigma}{ds}\frac{dx_4}{ds} = -\frac{dg_{44}}{ds}\frac{dx_4}{ds},$$

or, correct to the first order,

$$\frac{dx_4}{ds} = 1 + \frac{\alpha}{r}. \tag{9}$$

We now turn to the first of the three equations (7). The right side yields:

a) for the index combination $\sigma = \tau = 4$

$$\Gamma^\nu_{44}\left(\frac{dx_4}{ds}\right)^2,$$

or considering equations (6c) and (9), correct to the second order,

$$-\frac{\alpha}{2}\frac{x_\nu}{r^3}\left(1 + \frac{\alpha}{r}\right);$$

b) for the index combination $\sigma \neq 4$, $\tau \neq 4$ (which alone still needs to be considered), upon considering the
[14] product $(dx_\sigma/ds)(dx_\tau/ds)$, using equation (8) to first order, correct to the second order,

$$-\frac{\alpha x_\nu}{r^3}\sum\left(\delta_{\sigma\tau} - \frac{3}{2}\frac{x_\sigma x_\tau}{r^2}\right)\frac{dx_\sigma}{ds}\frac{dx_\tau}{ds}.$$

The summation gives

$$-\frac{\alpha x_\nu}{r^3}\left(u^2 - \frac{3}{2}\left(\frac{dr}{ds}\right)^2\right).$$

Using this value we obtain for the equation of motion the form, correct to the second order,

$$\frac{d^2x_\nu}{ds^2} = -\frac{\alpha}{2}\frac{x_\nu}{r^3}\left(1 + \frac{\alpha}{r} + 2u^2 - 3\left(\frac{dr}{ds}\right)^2\right), \tag{7b}$$

which together with equation (9) determines the motion of the mass point. Moreover, it should be observed that equations (7b) and (9) for the case of circular motion give no deviation from Kepler's three laws.

From equation (7b) follows, above all, the exact validity of the equation

$$r^2\frac{d\phi}{ds} = B, \tag{10}$$

where B is a constant. The law of areas is therefore valid to second order if we use the "proper time" of the planet to measure time. In order to determine the secular rotation of the orbital ellipse from equation (7b), we substitute the terms of the first order in the parentheses most advantageously by means of equation (10) and the first of the equations (8), through which procedure the terms of second order on the right side are not altered. The parentheses take on the form

$$\left(1 - 2A + \frac{3B^2}{r^2}\right).$$

Finally, if we choose $s\sqrt{(1 - 2A)}$ as the time variable, and if we redesignate it as s, we have, with a somewhat different meaning of the constant B;

$$\frac{d^2x_\nu}{ds^2} = -\frac{\partial\Phi}{\partial x_\nu}, \qquad \Phi = -\frac{\alpha}{2}\left[1 + \frac{B^2}{r^2}\right]. \tag{7c} \qquad [15]$$

In order to determine the equation of the orbit, we now proceed exactly as in the Newtonian case. From equation (7c) we obtain first

$$\frac{dr^2 + r^2\,d\phi^2}{ds^2} = 2A - 2\Phi.$$

If we eliminate ds from this equation with the help of equation (10), we obtain

$$\left(\frac{dx}{d\phi}\right)^2 = \frac{2A}{B^2} + \frac{\alpha}{B^2}x - x^2 + \alpha x^3, \tag{11}$$

where we denote by x the quantity $1/r$. This equation differs from the corresponding one in Newtonian theory only in the last term on the right side.

The angle described by the radius vector between the perihelion and the aphelion is consequently given by the elliptic integral

$$\phi = \int_{\alpha_1}^{\alpha_2} \frac{dx}{\sqrt{\dfrac{2A}{B^2} + \dfrac{\alpha}{B^2}x - x^2 + \alpha x^3}},$$

where α_1 and α_2 signify the roots of the equation

$$\frac{2A}{B^2} + \frac{\alpha}{B^2} x - x^2 + \alpha x^3 = 0$$

and closely correspond to the neighboring roots of the equation that arises from this one by the omission of the last term.

Thus, it can be established with the precision demanded of us that

[16]
$$\phi = [1 + \alpha(\alpha_1 + \alpha_2)] \cdot \int_{\alpha_1}^{\alpha_2} \frac{dx}{\sqrt{-(x - \alpha_1)(x - \alpha_2)(1 - \alpha x)}},$$

or upon expansion of $(1 - \alpha x)^{-1/2}$,

$$\phi = [1 + \alpha(\alpha_1 + \alpha_2)] \cdot \int_{\alpha_1}^{\alpha_2} \frac{\left(1 + \frac{\alpha}{2} x\right) dx}{\sqrt{-(x - \alpha_1)(x - \alpha_2)}}.$$

The integration yields

$$\phi = \pi\left[1 + \frac{3}{4}\alpha(\alpha_1 + \alpha_2)\right],$$

or if we consider that α_1 and α_2 signify the reciprocal values of the maximum and minimum distances, respectively, from the sun,

$$\phi = \pi\left(1 + \frac{3}{2} \frac{\alpha}{a(1 - e^2)}\right). \tag{12}$$

Therefore, after a complete orbit, the perihelion advances by

$$\varepsilon = 3\pi \frac{\alpha}{a(1 - e^2)} \tag{13}$$

in the sense of the orbital motion, where the semimajor axis is denoted by a and the eccentricity by e. If we introduce the orbital period T (in seconds), we obtain

$$\varepsilon = 24\pi^3 \frac{a^2}{T^2 c^2(1 - e^2)}, \tag{14}$$

where c denotes the velocity of light in units of cm sec^{-1}. The calculation yields, for the planet Mercury, a perihelion advance of 43″ per century, while the astronomers assign 45″ ± 5″ per century as the unexplained difference between observations and the Newtonian theory. This theory therefore [17] agrees completely with the observations.

For Earth and Mars, the astronomers assign, respectively, forward motions of 11″ and 9″ per century, while our formula yields, respectively, 4″ and 1″ per century. Nevertheless, a small value seems to be proper to these assignments because of the small eccentricities of the orbits of these planets. A more certain confirmation of the perihelion motion will be made by determining the product of the motion with the eccentricity. [18] If we consider these quantities assigned by Newcomb,

	$e \dfrac{d\pi}{dt}$
Mercury	8.48″ ± 0.43
Venus	−0.05 ± 0.25
Earth	0.10 ± 0.13
Mars	0.75 ± 0.35,

for which I thank Dr. Freundlich, then we obtain the impression that the advance of the perihelion is, after all, demonstrated really only for Mercury. However, I prefer to relinquish a final decision to the astronomical specialists.

1. In a forthcoming communication it will be shown that [3] this hypothesis is unnecessary. It is because such a choice of reference frame is possible that the determinant $|g_{\mu\nu}|$ takes on the value −1. The following investigation is independent of this choice.

2. E. Freundlich recently wrote a noteworthy article on the impossibility of satisfactorily explaining the anomalies in the motion of Mercury on the basis of the Newtonian theory (*Astronomische Nachrichten 201*, 49 [1915]). [5]

Doc. 25

[p. 844]   Session of the physical-mathematical class on November 25, 1915

The Field Equations of Gravitation

by A. Einstein

In two recently published papers[1] I have shown how to obtain field equations of gravitation that comply with the postulate of general relativity, i.e., which in their general formulation are covariant under arbitrary substitutions of space-time variables.

Historically they evolved in the following sequence. First, I found equations that contain the NEWTONIAN theory as an approximation and are also covariant under arbitrary substitutions of determinant 1. Then I found that these equations are equivalent to generally-covariant ones if the scalar of the energy tensor of "matter" vanishes. The coordinate system could then be specialized by the simple rule that $\sqrt{-g}$ must equal 1, which leads to an immense simplification of the equations of the theory. It has to be mentioned, however, that this requires the introduction of the hypothesis that the scalar of the energy tensor of matter vanishes.

I now quite recently found that one can get away without this hypothesis about the energy tensor of matter merely by inserting it into the field equations in a slightly different way. The field equations for vacuum, onto which I based the explanation of the Mercury perihelion, remain unaffected by this modification. In order not to force the reader constantly to consult the previous publications, I repeat here the considerations in their entirety. [2]

One derives from the well-known RIEMANN-covariant of rank four the following covariant of rank two:

$$G_{im} = R_{im} + S_{im} \tag{1}$$

$$R_{im} = -\sum_{l} \frac{\partial \begin{Bmatrix} im \\ l \end{Bmatrix}}{\partial x_l} + \sum_{l\rho} \begin{Bmatrix} il \\ \rho \end{Bmatrix} \begin{Bmatrix} m\rho \\ l \end{Bmatrix} \tag{1a}$$

$$S_{im} = \sum_{l} \frac{\partial \begin{Bmatrix} il \\ l \end{Bmatrix}}{\partial x_m} - \sum_{l\rho} \begin{Bmatrix} im \\ \rho \end{Bmatrix} \begin{Bmatrix} \rho l \\ l \end{Bmatrix}. \tag{1b}$$

---

[1]*Sitzungsber.* 44, p. 778, and 46, p. 799 (1915). [1]

[p. 845]   The ten generally-covariant equations of the gravitational field in spaces where "matter" is absent are obtained by setting

$$G_{im} = 0. \tag{2}$$

These equations can be simplified by choosing the system of reference such that $\sqrt{-g} = 1$. S_{im} then vanishes because of (16), and one gets instead of (2)

$$R_{im} = \sum_{l} \frac{\partial \Gamma_{im}^{l}}{\partial x_{l}} + \sum_{\rho l} \Gamma_{i\rho}^{l} \Gamma_{ml}^{\rho} = 0 \tag{3}$$

$$\sqrt{-g} = 1. \tag{3a}$$

We have set here

$$\Gamma_{im}^{l} = -\begin{Bmatrix} im \\ l \end{Bmatrix}, \tag{4}$$

which quantities we call the "components" of the gravitational field.

When there is "matter" in the space under consideration, its energy tensor occurs on the right-hand sides of (2) and (3), respectively. We set

$$G_{im} = -\kappa \left(T_{im} - \frac{1}{2} g_{im} T \right), \tag{2a}$$

where

$$\sum_{\rho \sigma} g^{\rho \sigma} T_{\rho \sigma} = \sum_{\sigma} T_{\sigma}^{\sigma} = T. \tag{5}$$

T is the scalar of the energy tensor of "matter," and the right-hand side of (2a) is a tensor. If we specialize the coordinate system again in the familiar manner, we get in place of (2a) the equivalent equations

$$R_{im} = \sum_{l} \frac{\partial \Gamma_{im}^{l}}{\partial x_{l}} + \sum_{\rho l} \Gamma_{i\rho}^{l} \Gamma_{ml}^{\rho} = -\kappa \left(T_{im} - \frac{1}{2} g_{im} T \right) \tag{6}$$

$$\sqrt{-g} = 1. \tag{3a}$$

We assume, as usual, that the divergence of the energy tensor of matter vanishes when taken in the sense of the general differential calculus (energy-momentum theorem). Specializing the choice of coordinates according to (3a), this means basically that the T_{im} should satisfy the conditions

$$\sum_{\lambda} \frac{\partial T_{\sigma}^{\lambda}}{\partial x_{\lambda}} = -\frac{1}{2} \sum_{\mu \nu} \frac{\partial g^{\mu \nu}}{\partial x_{\sigma}} T_{\mu \nu} \tag{7}$$

or

$$\sum_{\lambda} \frac{\partial T_{\sigma}^{\lambda}}{\partial x_{\lambda}} = -\sum_{\mu\nu} \Gamma_{\sigma\nu}^{\mu} T_{\mu}^{\nu}. \tag{7a}$$

When one multiplies (6) by $\partial g^{im}/\partial x_{\sigma}$ and sums over i and m, one gets[2] because [p. 846] of (7) and because of the relation

$$\frac{1}{2}\sum_{im} g_{im} \frac{\partial g^{im}}{\partial x_{\sigma}} = - \frac{\partial lg\sqrt{-g}}{\partial x_{\sigma}} = 0$$

that follows from (3a), the conservation theorem of matter and gravitational field combined in the form

$$\sum_{\lambda} \frac{\partial}{\partial x_{\lambda}} (T_{\sigma}^{\lambda} + t_{\sigma}^{\lambda}) = 0, \tag{8}$$

where t_{σ}^{λ} (the "energy tensor" of the gravitational field) is given by

$$\kappa t_{\sigma}^{\lambda} = \frac{1}{2}\delta_{\sigma}^{\lambda} \sum_{\mu\nu\alpha\beta} g^{\mu\nu} \Gamma_{\mu\beta}^{\alpha}\Gamma_{\nu\alpha}^{\beta} - \sum_{\mu\nu\alpha} g^{\mu\nu} \Gamma_{\mu\sigma}^{\alpha} \Gamma_{\nu\alpha}^{\lambda}. \tag{8a}$$

The reasons that motivated me to introduce the second term on the right-hand sides of (2a) and (6) will only become transparent in what follows, but they are completely analogous to those just quoted (p. 785).

When we multiply (6) by g^{im} and sum over i and m, we obtain after a simple calculation

$$\sum_{\alpha\beta} \frac{\partial^2 g^{\alpha\beta}}{\partial x_{\alpha}\partial x_{\beta}} - \kappa(T + t) = 0, \tag{9}$$

where, corresponding to (5), we used the abbreviation

$$\sum_{\rho\sigma} g^{\rho\sigma} t_{\rho\sigma} = \sum_{\sigma} t_{\sigma}^{\sigma} = t. \tag{8b}$$

It should be noted that our additional term is such that the energy tensor of the gravitational field occurs in (9) on footing equal with the one of matter, which was not the case in equation (21) l.c.

Furthermore, one derives in place of equation (22) l.c. and in the same manner as there, with the help of the energy equation, the relations

---

[2]On the derivation see *Sitzungsber.* 44 (1915), pp. 784–785. For the following I ask   [3]
the reader also to consult, for a comparison, the deliberations given there on p. 785.

$$\frac{\partial}{\partial x_\mu}\left[\sum_{\alpha\beta} \frac{\partial^2 g^{\alpha\beta}}{\partial x_\alpha \partial x_\beta} - k(T + t)\right] = 0. \tag{10}$$

Our additional term insures that these equations carry no additional conditions when compared to (9); we thus need not make other hypotheses about the energy tensor of matter other than that it complies with the energy momentum theorem.

[p. 847]

    With this, we have finally completed the general theory of relativity as a logical structure. The postulate of relativity in its most general formulation (which makes space-time coordinates into physically meaningless parameters) leads with compelling necessity to a very specific theory of gravitation that also explains the movement of the perihelion of Mercury. However, the postulate of general relativity cannot reveal to us anything new and different about the essence of the various processes in nature than what the special theory of relativity taught us already. The opinions I recently voiced here in this regard have been in error. Every physical theory that complies with the special theory of relativity can, by means of the absolute differential calculus, be integrated into the system of general relativity theory—without the latter providing any criteria about the admissibility of such physical theory.

[4]

Doc. 26

On the Theory of Tetrode and Sackur for the Entropy Constant [p. 1]

[about February 15, 1916][1]

The following considerations—without offering anything substantially new—should facilitate a better understanding of the theory of Tetrode and Sackur.[2] In order to achieve this goal, I have removed all confusing paraphernalia. It is worthwhile to give some thought to this important theory because it contains Nernst's theorem as applied to crystallized solid bodies[3] and also Stern's formula of vapor pressure.[4]

§1. Entropy and Statistical Mechanics

The probability dW of the "micro"-state of a physical system (assumed to be connected to a heat reservoir of a relatively, infinitely large thermal capacity) is, according to Gibbs, given by the well-known law of the canonical distribution

$$dW = e^{\frac{\psi - E}{\Theta}} dq_1 \ldots dq_n dp_1 \ldots dp_n \qquad (1) \quad \{1\}$$

or in shorter form

$$dW = e^{\frac{\psi - E}{\Theta}} d\tau. \qquad (1a)$$

We assume preliminarily that the system can be viewed as a mechanical system as understood by classical mechanics. Θ is the absolute temperature, measured in suitable units; E the energy as a function of the $q_1 \ldots p_n$; ψ a quantity that is constant with respect to molecular movements, i.e., a quantity independent of $q_1 \ldots q_n$, $p_1 \ldots p_n$, which, as Gibbs has shown, is equal to the free energy of the system.

Considering that the total phase integral of (1) must equal unity, one has

$$+\psi = \bar{E} - \Theta S = -\Theta \lg \left\{ \int e^{-\frac{E}{\Theta}} d\tau \right\}, \qquad (2)$$

or for the entropy S the expression

Translator's note. The typographical errors in the original have been corrected here according to editorial notes [9], [10], [14], and [19].

$$S = \frac{\bar{E}}{\Theta} + \lg\left\{\int e^{-\frac{E}{\Theta}} d\tau\right\} \qquad (2a)$$

$\bar{E}$ denotes here the mean energy of the system under consideration. We make this formula the basis of all our subsequent considerations. We prefer it over formula (2) because—from a thermodynamic point of view—entropy is, in contrast to the free energy, a completely determined function of a ("macro") state, up to an additive constant.

The precise meaning of equation (2a) is as follows. The basic assumption (later to be modified) is that all molecular processes (including chemical ones) can be viewed as movements of the smallest particles governed by laws of classical mechanics. Macroscopically, i.e., phenomenologically in terms of the usual thermodynamic parameters (volume, pressure, etc.), the state of a system is determined when we know:

1)   the temperature Θ,

2)   the energy E as a function of the molecular variables $q_1...p_n$.

Every not purely thermal change of state in the system (e.g., a change in volume) is accounted for by a change of the function E. In order to express this, one views E as not only dependent upon $q_1...q_n$ but also dependent upon certain parameters a_λ; and every not purely thermal change of state in the system corresponds to a change in the values of the parameters a_λ. If S_1 and S_2 denote the entropy values of two states of the physical system considered, then—as statistical mechanics tells us—the transition of the system from the first into the second state involves a change in entropy, $S_2 - S_1$, which is equal to the change on the right-hand side of (2a).

From what has been said, one sees that equation (2a) remains valid if an additive constant is included on the right-hand side, or more precisely, a quantity that depends neither upon Θ nor on the other thermodynamic parameters of state a_λ. We shall use this feature in the following.

Equation (2a) produces the entropy difference of two states when the energy function, in the sense of molecular theory, is known. It therefore answers also in a very distinct manner the question of the constant of integration of entropy[5] whose

meaning has been especially clearly stressed by Nernst and answered by him through the known Nernst theorem (in a distinct manner).

Equation (2a) has the advantage over Nernst's theorem in the regard that its statement is not related to the zero point of absolute temperature, and therefore it can be tested without extrapolating experimental results down to absolute zero. But on the other hand, there seems to be, on first sight, a grave and almost devastating drawback in the use of this equation:

a) Equation (2a) presupposes that we have a complete system of molecular mechanics—which is not the case. Meanwhile, the equation can have only true meaning for us insofar as its results are largely independent of the peculiarities of the molecular picture chosen.[6]

b) Equation (2a) presupposes that molecular processes can be perceived as movements within the framework of classical mechanics. But we know from individual results, which are presently only loosely connected by quantum theory, that this postulate does not hold in nature.[7] It seems therefore <doubtful if, with this equation, one can successfully answer such general question> impossible to draw exact and valid conclusions from our equation (2a); and it is furthermore doubtful if the equation can successfully be used at all for the determination of the entropy constant.

We first address the consideration of the last-mentioned principal objection.

{2}

§2. Taking Quantum Theory into Account

While our present knowledge excludes an understanding of molecular processes within the laws of classical mechanics, we also know, on the other hand, that molecular mechanics allows, within far-reaching areas of macroscopic (thermody-namic) variables of state of a system, to give a representation as good as one can wish. Such "normal" areas of state are, however, separated by others in which molecular mechanics fails. For example, the thermal behavior of rarefied hydrogen gas above a certain temperature Θ_2 can be described by molecular mechanics with high accuracy, provided the molecule is viewed as consisting of two rigidly connected atoms; whereas below a certain temperature Θ_1 a molecular-kinetic interpretation is possible based upon the hypothesis that the molecule behaves kinetically like a single masspoint. Between Θ_1 and Θ_2, however, a molecular-mechanic interpretation is impossible.[8]

[p. 4]

This situation brings about that equation (2a) can be used for the calculation of an entropy difference $S_2 - S_1$ if the two states belong to the same "normal" area of state. Similarly, it is clear that (2a) cannot be used to calculate entropy differences if at least one state belongs to a non-normal area.

The question is now whether or not a bridge can be built between two "normal" areas that are separated by non-normal areas, i.e., if a rule can be found to calculate the entropy difference $S_2 - S_1$ in case S_1 and S_2 belong to two different normal areas. Of course, such connection can only be established by a hypothesis outside the domain of molecular mechanics.

The simplest hypothesis would be to use equation (2a) without any scruples, even in that case. But this hypothesis is to be rejected for the following reason. Since Θ has the dimension of energy, the entropy difference ΔS is a dimensionless quantity, i.e., independent of the choice of basic units. In contrast, the integral in the second term on the right-hand side of (2a) is not dimensionless; its dimension is rather

$$(ML^2T^{-1})^n,$$

where n denotes the degrees of freedom of the system.[9] This fact causes the difference $S_2 - S_1$, when calculated from (2a), to be dependent upon the choice of basic units if the degrees of freedom, n, in the regular area to which S_2 belongs, differ from the degrees of freedom in the other regular area to which S_1 belongs.

[p. 5]   This difficulty can be avoided if the right-hand side of (2a) is changed such that its value becomes independent of the choice of basic units. This can be effected by dividing the right-hand side integral by h^n, where h is a constant of dimension ML^2T^{-1}[10] which will reveal itself, by necessity, as numerically equal to Planck's well-known constant.

Due to its way of derivation, equation (2a) is only valid for processes where matter is neither added nor withdrawn from the system. Consequently, we do not change the content of the formula if we add to the right-hand side of (2a) an arbitrary constant which depends in any manner upon the total number of atoms N_α of the various kinds in the system.[11]

Mindful of the last two considerations, we replace equation (2a) with the equation

$$S = \frac{\bar{E}}{\Theta} + \lg \frac{\int e^{-\frac{E}{\Theta}} d\tau}{h^n \Pi(N_\alpha!)}, \qquad (2b)$$

where Π denotes the product symbol. This equation shall not only correctly represent the entropy differences within normal areas, but also the entropy differences of states

{3}   that belong to *different* normal (states[12]) {recte: areas}.[13]

First, by applying formula (2b) to the linear oscillator, which obviously is in a normal area of state at high temperatures as well as when near the point of absolute zero, we convince ourselves that we justifiably put the quantity h numerically equal

[p. 6]   to Planck's constant h. At high temperatures (Θ) the effective degrees of freedom n equal 1, and close to absolute zero they equal 0. The latter situation represents some difficulty because it is not immediately clear what the value of (2b) is for $n = 0$. We circumvent this with a little trick of the trade; we consider the total system, which is a combination of the first system to be investigated, and a second one, comprised of elementary structures of a different kind. Let the latter one be characterized by E', n', N'_α and also let the number of effective degrees of freedom, in both ranges of temperature that interest us, be greater than zero; furthermore, it should behave "normal" in both ranges. Then (2b) yields for the combination[14]

$$S_{tot} = S + S_1 = \frac{\overline{E} + \overline{E'}}{\Theta} + \lg \frac{\iint e^{-\frac{E + E'}{\Theta}} d\tau \, d\tau'}{h^{n + n'} \Pi(N_\alpha !) \Pi(N'_\alpha !)}.$$

$N_\alpha = 1$ if the nonprimed quantities refer to the one-dimensional oscillator. Integration over $d\tau$ becomes redundant and one must set $n = 0$ if the temperature is so low that its degree of freedom is no longer effective. Considering furthermore the expression for S', which results from (2b) after insertion of the primed quantities, it follows with $E = \text{const} = \overline{E}$ that

$$S_1 = 0$$

for the low temperatures that are of interest to us; they are denoted by the index "1." Considering this, (2b) now yields[15]

$$S_2 - S_1 = \frac{E_2}{\Theta_2} + \lg \frac{\Theta_2}{h\nu}. \qquad \text{[p. 7]}$$

The same result is obtained in a purely thermodynamic manner by basing by basing $\overline{E}$ on Planck's experimentally secured value for the specific heat, viz.,

$$\overline{E} = \frac{h\nu}{e^{\frac{h\nu}{\Theta}} - 1}.$$

One finds

$$S_2 - S_1 = \int_{\Theta_1}^{\Theta_2} \frac{d\overline{E}}{\Theta} = \left| \frac{\overline{E}}{\Theta} \right|_{\Theta_1}^{\Theta_2} + \int_{\Theta_1}^{\Theta_2} \frac{\overline{E} \, d\Theta}{\Theta^2},$$

which, after insertion of Planck's expression, takes the value given above for both, sufficiently high temperatures Θ_2 and sufficiently low Θ_1. This agreement, however, is tied to the need to equate the constant h in formula (2b) with Planck's constant h.—

The addition of the basically insignificant factor $\Pi(N_\alpha !)$ has the following meaning. In thermodynamics we are used to saying: if s is the entropy of one gram-mol of a substance, then βs is the entropy of β gram-mols. This statement is not free of arbitrariness, because there is no reversible process where we could produce β gram-mols out of *one* gram-mol. Therefore, we have a free choice in the integration constant of entropy for every value of β. It can easily be shown with the example of the single-atomic gas that we accommodate our formula (2b) to the free choice, which is customary in thermodynamics, by adding the factor $\Pi(N_\alpha !)$.

§3. Application of the Fundamental Formula (2b) to Some Typical Cases

[p. 8] It is undeniable that the fundamental equation (2b) is very likely correct; at least, it represents the simplest hypothesis one could wager. The question now, however, is whether the value of this equation is merely illusory, because of our gross ignorance of the details of the molecular picture that we used. We shall show that this is not at all the case, by treating two important and typical cases which are distinguished by the fact that they allow for an exact calculation of the right-hand side of (2b), without any knowledge about the action of the atomic forces.

The Crystallized Substance at Absolute Zero

Let the system under consideration be a chemically uniform crystal. Its thermal movement is such that the individual atoms under their interactions always remain in the neighborhood of certain fixed locations. The structure of the spatial lattice is such that the α-th kind of atoms have N_α fixed locations available. Phase space, therefore, splits into as many areas of equal valuation as there are possibilities of distribution for the atoms of the system under the structural law of the lattice. All possible distributions are obtained by permutating the atoms of the same kind among themselves. There are obviously

$$\Pi(N_\alpha!)$$

such possibilities. Since they are all of equal valuation, one obviously gets the integral in (2b) by forming the latter for *one* combination of atoms and then dividing {recte: multiplying} it by the given number of permutations.[16] Instead of (2b) one first gets

$$S = \frac{\bar{E}}{\Theta} + \lg\frac{\int e^{-\frac{E}{\Theta}} d\tau}{h^n} \,, \tag{3}$$

where n is the number of effective degrees of freedom, and the phase integral has to be formed such that only *one* specific distribution of atoms over the lattice has to be considered.

What is the value of S at absolute zero? For this case the formula fails at first. But by using the same trick we used above with the one-dimensional resonator, one sees that there too, one must set $n = 0$ and replace the phase integral by $e^{\frac{-E}{\Theta}}$. The right-hand side then vanishes, and one gets

$$S = 0. \tag{3a}$$

This is the expression of Nernst's theorem insofar as it applies at all according to this [p. 9] theory; it holds for crystallized, chemically uniform substances.

For mixed crystals the theory allows for deviations from Nernst's theorem.[17] For example, let the spatial lattice hold two kinds of atoms, α and β, which can replace each other arbitrarily. The number of permutations of equal valuation is then

$$(N_\alpha + N_\beta)!$$

For the entropy S at absolute zero, after an argument analogous to the one just given, results a value of

$$S = \lg \frac{(N_\alpha + N_\beta)!}{N_\alpha!\, N_\beta!}. \tag{3b}$$

The Ideal Gas with a Rigid Type of Molecule

We imagine a chemically homogeneous gas made up of N molecules. Let there be N_α atoms of the α-th kind in it; i.e., $\dfrac{N_\alpha}{N}$ per single molecule. First, we imagine the gas at such high temperature that we are in a normal area where we are allowed to assume all degrees of freedom of a molecule to be effective. We now apply the fundamental formula (2b) to this gas. Then one has to insert for n the threefold number of atoms in the system. The phase integral in the numerator of the second term splits again into separate partial areas of equal valuation, which correspond to the various possibilities to distribute the atoms among the molecules that are to be formed. Again, one can calculate the integral by extending it over just *one* of these partial areas, i.e., the integration is carried out as if the molecule compound were indestructible, and afterwards the integral is multiplied by the number of possible distributions Z.

I imagine a coordinate system rigidly connected with each one of the N molecules such that the fixed locations of corresponding atoms of these molecules have the same coordinates in all these coordinate systems; and we think of these coordinate systems as being numbered. It is clear that there are

$$\Pi(N_\alpha!) = \zeta$$

different ways to distribute our atoms in the prescribed manner among the coordinate [p. 10] systems. I now imagine all these ζ (possible) distributions to be realized separately, whereby I obtain for the gas ζ models that are only distinct by the way the molecules are distributed among the coordinate systems. In each one of the ζ models, I now allow all N systems to take all possible positions and orientations. How often do I have a realization of each distinct distribution of all the numbered atoms in the gas?

I now think of a distinct distribution of coordinate systems in the first one of the models. When I now exchange two of these systems (together with their atoms) with each other, I get a different distribution of atoms insofar as these atoms are imagined as being numbered. However, if I make a realization of this second distribution system in all models, I will find one model of this second distribution with exactly the same spatial distribution of atoms as I had in the first model of the first system distribution. Consequently, every distribution of atoms is realized $N!$ times in our "thought-experiment," corresponding precisely to the $N!$ permutations of the coordinate systems.

The number of realizations of every individual distribution of (numbered) atoms in space need not be exhausted by this. One rather finds more frequent repetitions of distributions if the molecule has symmetries. If, for example, there are p positions of the molecule's coordinate system such that fixed points of atoms coincide in pairs, then there are p^N configurations for p coordinate systems that allow for the same distribution of atoms. The same atomic distribution occurs, therefore, p^N times more frequently in our thought-experiment than if there were no symmetry properties of the molecule.

Consequently, our thought-experiment produces every configuration of atoms, in the total, $N!p^N$-fold. Our result would be too large by just this factor if we could multiply the phase integral (which corresponds to one single distribution of atoms over all systems) by $\zeta = \Pi(N_\alpha!)$, the latter being the individual distribution of atoms over N systems. The correct multiplier is therefore

$$\frac{\Pi(N_\alpha!)}{N!p^N}.$$

[p. 11]   Instead of (2b) we now get

$$S = \frac{\overline{E}}{\Theta} + \lg \frac{\int e^{\frac{-E}{\Theta}} d\tau}{h^n N! p^N},$$ (4)

where the phase integral is to be taken "under preservation of the molecule compound."

This formula does not allow for an immediate exact evaluation, because we have only rough clues about the laws of interaction of atoms within a molecule. But we can use it in those "normal" areas where the temperature is so low that the degrees of freedom, which correspond to the relative motion of atoms within the molecule, are "asleep." The phase integral is equal to the N-th power of the integral extended over the variables of state of *one* molecule

$$I = \int e^{\frac{-\eta}{\Theta}} d\tau,$$

where η is the energy function of a *single* molecule. We distinguish now between three cases, denoted by α, β, γ and characterized as follows:

α The rotational degrees of freedom are dormant.

β Two rotational degrees of freedom with equal moments of inertia are active (two-atomic gas).

γ All three degrees of freedom are active.

Case α. If m is the mass, x, y, z the coordinates of the center of gravity, and ξ, η, ζ the corresponding components of velocity, one has to put

$$\eta = \eta_0 + \frac{m}{2}(\xi^2 + \eta^2 + \zeta^2)$$

$$d\tau = m^3 dx\,dy\,dz\,d\xi\,d\eta\,d\zeta.$$

Integration yields

$$I_\alpha = e^{-\frac{\eta_0}{\Theta}} \cdot V(2\pi m\Theta)^{\frac{3}{2}} \qquad (5\alpha)$$

For the other two cases we merely list the results of the simple calculation[18]

$$I_\beta = e^{-\frac{\eta_0}{\Theta}} \cdot V(2\pi m\Theta)^{\frac{3}{2}} \cdot 8\pi^2 I\Theta \qquad (5\beta) \quad \text{[p. 12]}$$

$$I_\gamma = e^{-\frac{\eta_0}{\Theta}} V(2\pi m\Theta)^{\frac{3}{2}} \cdot 2^{\frac{9}{2}}\pi^{\frac{7}{2}}(I_1 I_2 I_3)^{\frac{1}{2}}\Theta^{\frac{3}{2}}. \qquad (5\gamma)$$

For $\dfrac{\overline{E}}{\Theta}$ one has to put in the three cases $N\left(\dfrac{\eta_0}{\Theta} + \dfrac{3}{2}\right); N\left(\dfrac{\eta_0}{\Theta} + \dfrac{5}{2}\right); N\left(\dfrac{\eta_0}{\Theta} + 3\right),$

respectively,[19] and for n one has $3N, 5N, 6N$, respectively. The values for p can often be found from the chemical formula. For example, for HCl $p = 1$, for H_2 $p = 2$, for CH_4 $p = 12$.

The problem is herewith completely solved for gases with a rigid molecule. The importance of the formulas for the calculation of chemical equilibria are sufficiently known; so I don't have to go into details. Here, we shall only apply them to the transition from "single-atomic" to two-atomic H_2, where specific heat per mol (measured in the customary temperature units) increases during the transition from

$\dfrac{3}{2}R$ to $\dfrac{5}{2}R$.

For the transition under consideration, formula (4) in combination with (5α) and (5β) yields

$$S_2 - S_1 = \frac{\bar{E}}{\Theta_2} + \lg\frac{8\pi^2 I \Theta_2}{h^2} \, ,$$

where S is the entropy and E the energy of the rotational movement alone. Thermodynamically one gets

$$S_2 - S_1 = \int_{\Theta_1}^{\Theta_2} \frac{d\bar{E}}{\Theta} = \int \frac{d\bar{E}}{d\Theta} d\lg\Theta = \left| \lg\Theta\frac{d\bar{E}}{d\Theta} \right|_{\Theta_1}^{\Theta_2} - \int_{\Theta_1}^{\Theta_2} \lg\Theta dc \quad ,$$

where $c = \dfrac{d\bar{E}}{d\Theta}$ is the rotation-related specific heat of the molecule, which increases between the two states from 0 to 1. The last formula, therefore, yields

$$S_2 - S_1 = \lg\Theta_2 - \lg\bar{E},$$

where $\overline{\Theta}$ is a mean value of the conversion temperature of the specific heat, defined by

$$\lg\overline{\Theta} = \int_{c=0}^{c=1} \lg\Theta dc \tag{6}$$

[p. 13] a value that can be taken from experience. Equating the two expressions for $S_2 - S_1$ one obtains

$$\overline{\Theta} = \frac{h^2}{8\pi^2 I} \tag{7}$$

This equation yields the moment of inertia of the hydrogen molecule.

———————————

Additional notes by translator

{1} "p_n" has been corrected to "dp_n."
{2} The text between angular braces is so marked in the original in order to indicate that it is superseded by the text outside.
{3} The corrective substitution of German words by editorial notes has been translated and inserted in braces {} with the notation "recte."

Translations of editorial notes

[5]   Insofar as this question can be posed at all clearly. Not the absolute values of
      the entropy constants are thermodynamically defined, but always only the
      differences of the entropy constants of two domains of state of one and the
      same material system are defined.

[13] In this formula, n is the number of degrees of freedom of the system, which
      has to be assumed in order to represent the actual thermal behavior of the
      system in the normal area that has to be considered. <The place of the phase

      integral for a "dormant" degree of freedom has to be taken by the value $e^{-\frac{E}{\Theta}}$;
      this is especially to be considered in cases where the system has only
      "dormant" degrees of freedom (at absolute zero). Because it is clear,>.

Doc. 27

[p. 184]   Plenary session of February 3, 1916

A New Formal Interpretation of
Maxwell's Field Equations of Electrodynamics

by A. Einstein

The current covariance-theoretical interpretation of the electrodynamic equations originates with MINKOWSKI. It can be characterized as follows. The components of the electrodynamic field form a six-vector (antisymmetric tensor of rank two). There is a second six-vector associated to the first one (and is dual to it) whose components have in the special case of the original theory of relativity the same values as the first one, but are distinct in the way the components are associated with the four coordinate axes. The two systems of MAXWELLian equations are obtained by setting the divergence of the first one equal to zero, and the divergence of the other one

[1]   equal to the four-vector of the electric current.

The introduction of the dual six-vector makes its covariance-theoretical representation relatively involved and confusing. Especially the derivations of the conservation theorems of momentum and energy are complicated, particularly in the case of the general theory of relativity, because it also considers the influence of the gravitational field upon the electromagnetic field. The following formulation avoids the concept of the dual six-vector and thus achieves a considerable simplification in the system. Next, we will immediately treat the case in the general theory of relativity.[1]

§1. The Field Equations

Let ϕ_ν be the components of a covariant four-vector, the four-vector of the

[p. 185] electromagnetic potential. We form from it the components $F_{\rho\sigma}$ of the covariant six-vector of the electromagnetic field according to the system of equations

---

[2]   [1]My paper "Die formale Grundlage der allgemeinen Relativitätstheorie" (these *Sitzungsberichte* 41 [1914], p. 1030) will in the following be assumed as known; the codicil "l.c." in the following text always refers to this paper.

$$F_{\rho\sigma} = \frac{\partial \phi_\rho}{\partial x_\sigma} - \frac{\partial \phi_\sigma}{\partial x_\rho}. \tag{1}$$

It follows from (28a) l.c. that $F_{\rho\sigma}$ is really a covariant tensor. It follows from (1) that the system of equations

$$\frac{\partial F_{\rho\sigma}}{\partial x_\tau} + \frac{\partial F_{\sigma\tau}}{\partial x_\rho} + \frac{\partial F_{\tau\rho}}{\partial x_\sigma} = 0 \tag{2}$$

is satisfied, and it also represents the most natural form of the second system of MAXWELL's equations (FARADAY's law of induction). First, one recognizes that (2) is a general covariant system of equations, because it evolves as a consequence of the general covariant system (1). Furthermore, that the left-hand side of (2) is a covariant tensor of rank three is proven by a threefold application of (29) l.c. upon $F_{\rho\sigma}$, $F_{\sigma\tau}$, $F_{\tau\rho}$, and by the extension in the indices τ, ρ, σ resp., and then adding the three resulting expressions together, utilizing of course the antisymmetric character of $F_{\rho\sigma}$. This tensor of rank three is antisymmetric, because the antisymmetric character of $F_{\rho\sigma}$ enforces that the left-hand side of (2) suffers a change in sign without change of value whenever two indices are permuted. The system (2) can, therefore, be completely replaced by the four equations

$$\left.\begin{aligned}
\frac{\partial F_{23}}{\partial x_4} + \frac{\partial F_{34}}{\partial x_2} + \frac{\partial F_{42}}{\partial x_3} &= 0 \\[2mm]
\frac{\partial F_{34}}{\partial x_1} + \frac{\partial F_{41}}{\partial x_3} + \frac{\partial F_{13}}{\partial x_4} &= 0 \\[2mm]
\frac{\partial F_{41}}{\partial x_2} + \frac{\partial F_{12}}{\partial x_4} + \frac{\partial F_{24}}{\partial x_1} &= 0 \\[2mm]
\frac{\partial F_{12}}{\partial x_3} + \frac{\partial F_{23}}{\partial x_1} + \frac{\partial F_{31}}{\partial x_2} &= 0
\end{aligned}\right\}; \tag{2a}$$

they are generated when the indices τ, ρ, σ are sequentially given the values 2, 3, 4, then resp. 3 4 1, then 4 1 2, then 1 2 3.

In the generally familiar special case where there is no gravitational field, one has to equate

$$\left.\begin{aligned}
F_{23} &= \mathfrak{h}_x & F_{14} &= \mathfrak{e}_x \\
F_{31} &= \mathfrak{h}_y & F_{24} &= \mathfrak{e}_y \\
F_{12} &= \mathfrak{h}_z & F_{34} &= \mathfrak{e}_z
\end{aligned}\right\}. \tag{3}$$

The equations (2a) then yield the field equations   [p. 186]

$$\left.\begin{array}{r} \dfrac{\partial \mathfrak{h}}{\partial t} + \operatorname{curl} \mathfrak{e} = 0 \\[2mm] \operatorname{div} \mathfrak{h} = 0 \end{array}\right\}. \tag{2b}$$

These last equations can be retained in the case of the general theory of relativity if one stays with the defining equations (3), i.e., when the six-vector ($\mathfrak{e}$, $\mathfrak{h}$) is treated as a covariant six-vector.

Regarding the first system of MAXWELL's equations, we stay with the generalization of MINKOWSKI's scheme, which has been elaborated upon in §11 of the repeatedly quoted paper. We introduce the covariant V-six-vector

{1}

$$\mathfrak{F}^{\mu\nu} = \sqrt{-g} \sum_{\alpha\beta} g^{\mu\alpha} g^{\nu\beta} F_{\alpha\beta} \tag{4}$$

and demand that the divergence of this contravariant six-vector equals the contravariant V-four-vector $\mathfrak{J}^{\mu}$ of the electric current density in vacuum, viz.,

$$\sum_{\nu} \frac{\partial \mathfrak{F}^{\mu\nu}}{\partial x_{\nu}} = \mathfrak{J}^{\mu}. \tag{5}$$

This system of equations is truly equivalent to the first system of MAXWELL, as can be seen by calculating $\mathfrak{F}^{\mu\nu}$ from (4) for the case of special relativity where the $g_{\mu\nu}$ have the values

$$\begin{array}{cccc} -1 & 0 & 0 & 0 \\ 0 & -1 & 0 & 0 \\ 0 & 0 & -1 & 0 \\ 0 & 0 & 0 & +1. \end{array}$$

(3) and (4) yield for this special case

$$\left.\begin{array}{ll} \mathfrak{F}^{23} = \mathfrak{h}_x & \mathfrak{F}^{14} = -\mathfrak{e}_x \\ \mathfrak{F}^{31} = \mathfrak{h}_y & \mathfrak{F}^{24} = -\mathfrak{e}_y \\ \mathfrak{F}^{12} = \mathfrak{h}_z & \mathfrak{F}^{34} = -\mathfrak{e}_z \end{array}\right\} \tag{6}$$

By setting in addition

$$\mathfrak{J}^{1} = \mathfrak{i}_x, \quad \mathfrak{J}^{2} = \mathfrak{i}_y, \quad \mathfrak{J}^{3} = \mathfrak{i}_z, \quad \mathfrak{J}^{4} = \rho \tag{7}$$

(5) takes the familiar form

$$\left.\begin{array}{r} \operatorname{curl} \mathfrak{h} - \dfrac{\partial \mathfrak{e}}{\partial t} = \mathfrak{i} \\[2mm] \operatorname{div} \mathfrak{e} = \rho \end{array}\right\}. \tag{5b}$$

[p. 187] Equations of the form (5b) also apply in the general theory of relativity, but the (three-dimensional) vectors $\mathfrak{e}$ and $\mathfrak{h}$ are no longer the same as in (2b). One rather would have to introduce two new vectors $\mathfrak{e}'$ and $\mathfrak{h}'$ which are in a pretty complex relationship with $\mathfrak{e}$ and $\mathfrak{h}$, which is determined by equation (4).

We note in summarizing that the new generalization of MAXWELL's system is completely given by equations (2), (4), and (5). It deviates from the previous one only in form, but not in content.

§2. Ponderomotive Force and the Energy-Momentum Theory[2]

By inner multiplication of the covariant six-vector $F_{\sigma\mu}$ of the electromagnetic field with the V-four-vector $\mathfrak{J}^{\mu}$ of the electric current density, we construct the covariant V-four-vector

$$\mathfrak{K}_{\sigma} = \sum_{\mu} F_{\sigma\mu}\mathfrak{J}^{\mu}. \tag{8}$$

Its components are, according to (3), in conventional three-dimensional notation

$$\mathfrak{K}_1 = \rho\mathfrak{e}_x + [\mathfrak{i}, \mathfrak{h}]_x$$
$$\mathfrak{K}_2 = \rho\mathfrak{e}_y + [\mathfrak{i}, \mathfrak{h}]_y$$
$$\mathfrak{K}_3 = \rho\mathfrak{e}_z + [\mathfrak{i}, \mathfrak{h}]_z$$
$$\mathfrak{K}_4 = -(\mathfrak{i}, \mathfrak{e}).$$

$\mathfrak{K}_{\sigma}$ is therefore for the electromagnetic field precisely that V-vector that was introduced in equation (42a) l.c. as the four-vector of force density. $\mathfrak{K}_1$, $\mathfrak{K}_2$, $\mathfrak{K}_3$ are the negative components of momentum per units of volume and time; $\mathfrak{K}_4$ is the energy per units of volume and time transmitted to the field.

In order to obtain the components of $\mathfrak{T}_{\sigma}^{\nu}$ of the energy tensor of the electromagnetic field, we merely have to form—from equation (7) and the field equations—the equivalent of equation (42a) l.c. One first gets from (7) and (5)

$$\mathfrak{K}_{\sigma} = \sum_{\mu\nu} F_{\sigma\mu}\frac{\partial\mathfrak{F}^{\mu\nu}}{\partial x_{\nu}} = \sum_{\mu\nu}\frac{\partial}{\partial x_{\nu}}(F_{\sigma\mu}\mathfrak{F}^{\mu\nu}) - \sum_{\mu\nu}\mathfrak{F}^{\mu\nu}\frac{\partial F_{\sigma\mu}}{\partial x_{\nu}}. \tag{2}$$

The second term on the right-hand side can be rewritten with (2) as   [p. 188]

$$\sum_{\mu\nu}\mathfrak{F}^{\mu\nu}\frac{\partial F_{\sigma\mu}}{\partial x_{\nu}} = -\frac{1}{2}\sum_{\mu\nu}\mathfrak{F}^{\mu\nu}\frac{\partial F_{\mu\nu}}{\partial x_{\sigma}} = -\frac{1}{2}\sum_{\mu\nu\alpha\beta}\sqrt{-g}\,g^{\mu\alpha}g^{\nu\beta}F_{\alpha\beta}\frac{\partial F_{\mu\nu}}{\partial x_{\sigma}} \tag{3}$$

and for reasons of symmetry, the last expression can also be written as

---

[2]We owe a different treatment of the same topic to H. A. LORENTZ (*Koninkl. Akad. van*   [3]
Wetensch. 23 [1915], p. 1085).

{4}
$$- \frac{1}{4} \sum_{\mu\nu\alpha\beta} \left[\sqrt{-g}\, g^{\mu\alpha} g^{\nu\beta} F_{\alpha\beta} \frac{\partial F_{\mu\nu}}{\partial x_\sigma} + \sqrt{-g}\, g^{\mu\alpha} g^{\nu\beta} \frac{\partial F_{\alpha\beta}}{\partial x_\sigma} F_{\mu\nu} \right].$$

This, however, can now be written as

{4}
$$- \frac{1}{4} \frac{\partial}{\partial x_\sigma} \left(\sum_{\mu\nu\alpha\beta} \sqrt{-g}\, g^{\mu\alpha} g^{\nu\beta} F_{\alpha\beta} F_{\mu\nu} \right) + \frac{1}{4} \sum_{\mu\nu\alpha\beta} F_{\alpha\beta} F_{\mu\nu} \frac{\partial}{\partial x_\sigma} (\sqrt{-g}\, g^{\mu\alpha} g^{\nu\beta}).$$

The first one of these terms is, in shorter notation,

$$- \frac{1}{4} \frac{\partial}{\partial x_\sigma} (\sum_{\mu\nu} \mathfrak{F}^{\mu\nu} F_{\mu\nu}),$$

and the second one is, after differentiation and some rearranging,

[4] {5}
$$- \frac{1}{2} \sum_{\mu\nu\rho\tau} \mathfrak{F}^{\mu\tau} F_{\mu\nu} g^{\nu\rho} \frac{\partial g_{\rho\tau}}{\partial x_\sigma} + \frac{1}{8} \sum_{\alpha\beta\rho\tau} \mathfrak{F}^{\alpha\beta} F_{\alpha\beta} g^{\rho\tau} \frac{\partial g_{\rho\tau}}{\partial x_\sigma}.$$

Finally, taking all four of the calculated terms together, one obtains the relation

{6}
$$\sum_\nu \frac{\partial \mathfrak{T}_\sigma^\nu}{\partial x_\nu} - \frac{1}{2} \sum_{\mu\nu\tau} g^{\tau\mu} \frac{\partial g_{\mu\nu}}{\partial x_\sigma} \mathfrak{T}_\tau^\nu = \mathfrak{K}_\sigma, \tag{8a}$$

where

$$\mathfrak{T}_\sigma^\nu = \sum_{\alpha\beta} \left(- \mathfrak{F}^{\nu\alpha} F_{\sigma\alpha} + \frac{1}{4} \mathfrak{F}^{\alpha\beta} F_{\alpha\beta} \delta_\sigma^\nu \right). \tag{9}$$

{7}
δ_σ^ν is the mixed tensor whose components are 1 or 0 depending upon $\sigma = \nu$ or $\sigma \neq \nu$, respectively. A comparison of equation (8a) with (42a) l.c. shows that (8a) is the energy-momentum equation of the electromagnetic field, where the components of the energy tensor are given by (9). With the help of (3) and (6) it is easily seen that the energy tensor of the electromagnetic field found here coincides with the one in the previous theory. However, the newly found form is more comprehensive than the one found under the previous treatment of this topic.

Additional notes by translator

{1}   The repeated typographical error "$\sqrt{\ }$" in the German original, instead of a correct "V," is henceforth corrected in the translation without further note.

{2}   All summations imply that μ, ν are summation indices and have been added here.

{3} The first two summations are over μ, ν, the third one over μ, ν, α, β, and have been added here.

{4} The summations extend over the indices μ, ν, α, β, and have been added here.

{5} The summation indices in the first sum are μ, ν, ρ, τ; in the second sum they are α, β, ρ, τ. They have been added here.

{6} The second sum extends over the three indices μ, ν, τ, which have been added here.

{7} "$\sigma = n$" has been corrected to "$\sigma \neq \nu$."

Doc. 28

[p. 173]

A Simple Experiment to Demonstrate
Ampère's Molecular Currents

by A. Einstein

(Presented in the session of February 25, 1915)

The following describes a simple experiment which can serve to demonstrate Ampère's molecular currents in lectures. It is a variation of an experiment which I did together with Mr. DE HAAS.[1]

We want to demonstrate the (apparent) angular momentum that a tiny iron rod exhibits under a reversion of its magnetization, so that the orbits of the electrons change their orientation. A major difficulty for a simple demonstration of this effect lies in the fact that the purely magnetic forces, exerted onto the rod by the magnetizing field, are very large compared to the forces to be measured. In order to minimize these difficulties, the rod is not *permanently* exposed to the magnetic field, but only during times so short (about 1/1000 of a second) that the field just suffices to reverse the *remanent* magnetization of the rod.

[p. 174]   The arrangement that was ultimately used can be seen (in a vertical axial cut-away) in the sketch (fig. 1) below. In the middle of the tiny iron rod *S* (diameter 1.4

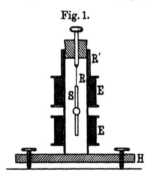

Fig. 1.

millimeters, length about 10 centimeters) that is to be investigated is a small mirror, and the rod is suspended by a thread of quartz of several centimeters in length and about 10 μ in diameter. At the top, this thread of quartz is glued to a copper spindle with a turning knob, the latter bored through the center of a cork that has friction. The cork is pressed into a wooded extension *R'* of a pressure-coil pipe *R*; the latter is mounted on a board *H* with balancing screws. Around the pipe *R* are

[2]

two (e.g., serially connected) coils Σ which together have about 4000 windings, and there is space left between the coils for a glass window (not shown in the sketch) through which a light-pointer can be (objectively) observed when the rod executes oscillations.

In series with the two coils is a capacitor of 2 microfarads. The open circuit is connected to a 120-volt direct current supply via a commutator of 500 to 1000 ohms

---

[1]   [1]*Verh. d. D. Phys. Ges.* 17 (1915), p. 152. In the meantime BARNETT [*Phys. Rev.* (2) 6 (1915), p. 171], succeeded in also showing the reverse of the effect that is described above.

ohmic resistance (in order to avoid electrical oscillations). The turning of the commutator charges the capacitor in reverse, during which time a field of short duration is induced in the coil Σ, which in turn reverses the remanent magnetism of the rod S. Further required is an arrangement to compensate for the geomagnetic field, e.g., a horizontally movable magnetic rod at about the height of the mirror.

The experiment is conducted in the following manner. First, one has to compensate for the earth's magnetic field to a degree that the resting position of the tiny rod is exactly the same in both directions of magnetization. (The reason why this is not initially the case is caused by the fact that the poles of the rod are never completely on its axis of rotation, whereupon the field of the earth produces, in general, an angular momentum that has to be removed with the utmost care.)

Next, one brings a small magnet temporarily close in order to induce torsional oscillations that can be easily followed with the eyes since the light-pointer is set up [p. 175] with oscillation periods of 1 to 2 seconds. Now one can begin to reverse the commutator whenever the light-pointer passes through its resting position, which means the tiny rod receives an angular momentum, by the effect to be demonstrated, whenever its angular velocity is at a maximum. One achieves in this manner an easily noticeable amplification, resp. attenuation of the oscillations. It is also easily demonstrated that sign and order of magnitude of the effect comply with theory.

Attention must also be drawn to the following points. It is important that the point of suspension of the tiny rod is as precise as possible in line with the major axis of inertia (smallest momentum of inertia). If this is not the case with good approximation, the horizontal deviations of the location of the apparatus will give cause to initiate torsional oscillations of the tiny rod; but if the centering is a good one, the experiment can be done well without special suspension of the whole apparatus (to avoid vibrations of it); e.g., one can place it on a wall console.

The poorer the centering of the tiny rod, the more sensitive is the apparatus to variations in the earth's field, and the more difficult it is also to compensate the earth's field sufficiently.

There were originally considerable difficulties with the centric suspension of the tiny rod, but Herr JAEGER kindly assisted me in overcoming them. Finally, the [3] following almost droll method achieved the goal. The tiny rod is fastened

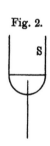
Fig. 2.

vertically (and loosely!) to a stand such that the end that is to be suspended points downward. The copper spindle with the threat of quartz is suitably supported, also upside down, vertically underneath the cork. Hereby, the height is to be carefully adjusted such that the quartz thread (held by the moistened fingers and directed straight up) just does not touch the plane endpoint of the tiny rod (see sketch in fig. 2). The end of S is now heated with the very small flame of a gas burner (made from a tapered glass pipe) until a droplet of colophony sticks to the tiny rod [p. 176] when brought from underneath with a finger. The droplet melts and curves under the action of capillary forces into a very symmetric bulge. If one now brings the thread of quartz into contact with the colophony, the capillary forces will suck it into the droplet as deeply as possible, thus centering it automatically. After cooling, the suspension is finished. After this, one can use wax to fasten the little mirror (3 mm^2 and made from a microscope slide) onto the tiny rod.

It should also be pointed out that the effect to be demonstrated cannot be confused with the purely magnetically induced angular momenta of the tiny rod, which will be present due to the combination of nonvertical positioning and asymmetry of the coil's fields with the asymmetries in the tiny rod. The reason is that every commutation changes both the field and the magnetic poles in sign, relative to the previous state, so that the resulting angular momentum has the same sign in both cases and thus cannot cause a systematic amplification or attenuation of the oscillations. By the way, if such angular momenta become noticeable, they can easily be eliminated by adjusting of the balancing screws.

Finally one more point which, originally, gave me some headaches. The effect demanded by the theory is large enough to be demonstrated when the tiny rod is stimulated from its resting position by just one single commutation. The experiment, however, always produced in this case only a rather fast vibration of the tiny rod, without a noticeable torsional oscillation. A simple analysis shows that this originates with the eccentric suspension of the tiny rod. The reason being that, about (almost) the main axis of inertia, a torsional oscillation is initiated whose directional force does not come from the torsion of the thread, but rather from the weight of the tiny rod.[2]

[p. 177]   Here I want to express my sincerest thanks to my colleagues JAEGER and
[4]   ORLICH, the first-named for his kind assistance, the other for producing the special coil, mentioned above. The experiment was first conducted with the kind permission
[5]   of Herr WARBURG at the Reichsanstalt.

---

[2]This kind of oscillation can be easily observed on an elongated, not too slender body when it is in asymmetric suspension, e.g., like a larger pair of scissors hanging by a thread from one of its grip holes. If this hole is given a slight push at the point of suspension, and perpendicular to the plane of the edges, the motion mentioned above sets in.

Doc. 29

Ernst Mach

[p. 101]

by A. Einstein

Recently Ernst Mach left us—a man who in our times had the greatest influence upon   [1]
the epistemological orientation of natural scientists, a man of rare independence of
judgment. In him, the direct joy of seeing and understanding—Spinoza's *amor dei
intellectualis*—was so prevalent that he could look into the world with the curious
eyes of a child, even in his old age, and thus be perfectly happy to enjoy himself by
understanding interwoven connections.

But why does a properly gifted natural scientist care for epistemology at all? Is
there no more important work in his field? That's what I hear many a colleague say,
and I sense that even more feel that way. I cannot share this conviction. When I think
of the most able students I met in my teaching—and I mean those who distinguished
themselves not just by mere skills but by their independence in judgment—then I
must state that they all had a lively interest in epistemology. They liked to start
discussions about the final goals and methods of the sciences, and showed without
doubt, by the hardheadedness with which they defended their views, that the topic
was important to them. And this is not surprising.

If I turn toward a science not for external reasons such as earning an income, or
for ambition, and also not—or at least not exclusively—for the mere sportive joy and
the fun of brain-acrobatics, then I must ask myself the question: what is the final goal
that the science I am devoted to will and can reach? To what extent are its general
results "true"? What is essential and what is based only on accident in its development?

In order to appreciate Mach's merits, one must not ask the question: what has
Mach found in all these general questions that no man before him saw or thought of?
Again and again, the truth in these things must be chiseled out by energetic
personalities, and must always be commensurate with the needs of the times in which
the sculptor works. If this truth is not always created anew, we will finally lose it
altogether. Hence it is difficult, albeit also not so essential, to answer the question: [p. 102]
what did Mach teach us that would, in principle, be new when compared to Bacon
and Hume? Or, as far as the general epistemological point of view toward individual
sciences is concerned, what distinguishes him basically from Stuart Mill, Kirchhoff,
Hertz, Helmholtz? It is a fact that Mach has had tremendous impact upon our
generation of natural scientists, in particular with his historical-critical writings where
he follows the evolution of individual sciences with so much love, where he probes
practically the most remote brain cells of researchers who broke new paths in their

[2]   fields. I even believe that those who consider themselves to be opponents of Mach barely know how many of his views they absorbed, in a manner of speaking, with their mother's milk.

Science is, according to Mach, nothing but the comparison and orderly arrangement of factually given contents of our consciousness, in accord with certain gradually acquired points of view and methods. Therefore, physics and psychology differ from each other not so much in the subject matter, but rather only in the points of view of the arrangement and connection of the various topics. It appears that Mach sought as his most important task to demonstrate how this order evolved in the particular sciences he knew. As a result of this activity of orderly arrangement, one obtains the abstract concepts and the laws (rules) of their connection. Both are chosen such that together they form a scheme of order into which the facts can be ranged securely and comprehensively. From what has been said, concepts make sense only insofar as they can be pointed out in things; also the points of view according to which concepts are associated with things (analysis of concepts).

The significance of personalities like Mach lies by no means only in the fact that they satisfy the philosophical needs of their times, an endeavor which the hard-nosed specialist may dismiss as a luxury. Concepts that have proven useful in ordering things can easily attain an authority over us such that we forget their wordly origin and take them as immutably given. They are then rather rubber-stamped as a "sine-qua-non of thinking" and an "a priori given," etc. Such errors make the road of scientific progress often impassable for long times. Therefore, it is not at all idle play when we are trained to analyze the entrenched concepts, and point out the circumstances that promoted their justification and usefulness and how they evolved from the experience at hand. This breaks their all too powerful authority. They are removed when they cannot properly legitimize themselves; they are corrected when their association with given things was too sloppy; they are replaced by others when a new system can be established that, for various reasons, we prefer.

The specialized scientist who concentrates more on details views such analysis often as superfluous, as far-out, and, occasionally, as out-right ridiculous. But the situation changes when one of the traditionally used concepts is to be replaced by a more precise one, because the development of the science in question demands it. Then those who did not cleanly use their own concepts raise vigorous protest and complain of a revolutionary endangerment of their most sacred treasures. Mixed into this outcry are the voices of those philosophers who believe they cannot dispense with the disputed concepts because they have incorporated them into their treasure chest of "absolutes" and "a prioris," or, in short, because they have already proclaimed their fundamental immutability.

The reader can already guess that I am touching on certain concepts of the philosophy of space and time, and also mechanics, which have undergone a modification by the theory of relativity. Nobody can deny that epistemologists paved the road for progress; and for myself, I know at least that Hume and Mach have helped me a lot, both directly and indirectly. I ask the reader to look at Mach's opus, [3] "Die Mechanik in ihrer Entwicklung," especially the deliberations under 6. and 7. of chapter 2 ("Newton's opinions on time, space and movement" and "Comprehensive critique of Newton's list"). There one finds superbly explained thoughts which, up to now, have by no means become commonplace among physicists. These portions are especially attractive because they are tied to Newton with verbatim quotations. Here are just some of these nuggets:

Newton: "The absolute, true and mathematical time flows, by itself and by its nature, uniformly and without relation to any extraneous object. It is also denoted as duration."

"Relative, apparent, and ordinary time can be felt as an extraneous, either accurate or changing, measure of duration, and is commonly used as hour, day, [p. 103] month, year, instead of true time."

Mach: ". . . when a thing *A* changes in the course of time, it means only that the circumstances of a thing *A* depend upon the circumstances of another thing *B*. The oscillations of a pendulum occur in time if the excursion depends upon its positions relative to Earth. Since we do not need to consider, during the observation of a pendulum, its dependence upon the position relative to Earth—we could compare it to any other thing—(. . .) it is easy to get the opinion that all these things are irrelevant. . . . We are not able to measure the change of things against time. Time is rather an abstraction which we reach through the change of things, because we are not dependent upon a distinct measure, since all things hang mutually together."

Newton: "Absolute space remains, by its nature and without relation to extraneous objects, always the same and immovable."

"Relative space is a measure or a movable part of the former and, through our senses, is denoted by its position relative to other bodies, and is usually taken as immovable space."

This is followed by the corresponding definitions of "absolute motion" and "relative motion." Then follows:

"The active causes by which absolute and relative motion differ are the centrifugal forces, pointing away from the axis of motion. These forces do not exist if the circular motion is only relative; but they are smaller or larger depending upon the ratio of (absolute) motion."

Next follows a description of the well-known experiment with the pail, intended to illustrate the foregoing reasoning.

It is interesting to look at Mach's critique of this position, and I quote from it a few particularly characteristic portions. "When we see a body *K* change its direction and velocity only because of the influence of another body *K'*, then we cannot even reach this conclusion unless there are other bodies *A, B, C,* . . . relative to which we judge the motion of the body *K*. We do in fact know a relation of the body *K* toward *A, B, C.* . . . If we now suddenly want to ignore *A, B, C,* . . . and instead talk about the behavior of the body *K* in absolute space, we are liable to make a double error. First, we cannot know how *K* would behave in the absence of *A, B, C,* . . .; but then we would also have no means to judge the behavior of the body *K* in order to test our statement—which, consequently, has no meaning in natural science."

"The movement of a body *K* can always and only be judged relative to other bodies *A, B, C,* . . . . As there is always a sufficient number of bodies available, which are either relatively fixed against each other or only slowly changing in their relative positions, we are not dependent upon a specific one and can occasionally ignore one or the other. In this manner, the opinion arises that these bodies are completely irrelevant."

"Newton's experiment with the rotating water container teaches us that the relative rotation of water toward the walls of the vessel do not stir up noticeable centrifugal forces, but that they are rather caused by the relative rotation toward the mass of the earth and the other heavenly bodies. Nobody can tell the result of the experiment if the walls of the vessel would become thicker and more massive until, finally, they reach several miles in thickness. . . ."

[4]   These quotations show that Mach clearly recognized the weak points of classical mechanics, and thus came close to demand a general theory of relativity—and this almost half a century ago! It is not improbable that Mach would have hit on relativity theory when in his time—when he was in fresh and youthful spirit—physicists would have been stirred by the question of the meaning of the constancy of the speed of light. In the absence of this stimulation, which flows from Maxwell-Lorentzian electrodynamics, even Mach's critical urge did not suffice to raise a feeling for the

[5]   need of a definition of simultaneity for spatially distant events.

The contemplations on Newton's experiment with the pail demonstrate how close his mind was to the demands of relativity in a wider sense (relativity of accelerations). Admittedly, the vivid consciousness was missing that the equivalence of inertial and gravitational mass elicits a postulate of relativity in a wider sense, because we are not in a position to decide by experiments if the falling of a body relative to a coordinate system is caused by the presence of a gravitational field or by a state of acceleration of the coordinate system.

[p. 104] In his intellectual development, Mach was not a philosopher who chose natural science as an object of speculation, but rather an ardent researcher of nature with many interests who obviously found pleasure in researching detailed questions off the

[6]   trodden path of general interest. Testimony to this are the almost uncountable individual investigations from the fields of physics and empirical psychology which

he published partly by himself, partly together with students. Among his experimental investigations in physics, those of the acoustic waves generated by bullets are best known. While the basic idea was not new, the investigations revealed an extraordinary experimental talent. He succeeded in photographing the density distribution of air around a supersonic bullet, and thus he shed new light upon a group of acoustic processes about which nothing was known before him. Everyone who likes the subjects of physics will enjoy his popular lecture about it.   [7]

All of Mach's philosophical studies sprang from his desire to find a point of view from which the various branches of science, to which he dedicated his lifelong labor, can be seen as an integrated endeavor. He comprehended all science as a striving for order among individual elementary experiences which he called "sensations" (*Empfindungen*). This choice of word is probably the reason why those who are less familiar with his works often mistook the sober and careful thinker for a philosophical idealist and solipsist.

When reading Mach's works, one really feels the cozy ease the author must have felt when he effortlessly wrote his concise and striking sentences. But not only the enjoyment of good style and mere intellectual joy make the reading of his books so attractive; it is also the kind, the humanitarian and hopeful attitude that gleams between the lines when he talks about human things in general. These sentiments and convictions also made him immune to another disease of our time, from which few are spared today—national fanaticism. In his popular essay "On phenomena of Flying Projectiles" he could not refrain, in the last paragraph, from expressing his hope of an understanding among nations.   [8]

(Received March 14, 1916)

Doc. 30

[p. 769]

The Foundation of the General Theory of Relativity

by A. Einstein

[This first page was missing in the existing translation.]

The theory which is presented in the following pages conceivably constitutes the farthest-reaching generalization of a theory which, today, is generally called the "theory of relativity"; I will call the latter one—in order to distinguish it from the first named—the "special theory of relativity," which I assume to be known. The generalization of the theory of relativity has been facilitated considerably by Minkowski, a mathematician who was the first one to recognize the formal equivalence of space coordinates and the time coordinate, and utilized this in the construction of the theory. The mathematical tools that are necessary for general relativity were readily available in the "absolute differential calculus," which is based upon the research on non-Euclidean manifolds by Gauss, Riemann, and Christoffel, and which has been systematized by Ricci and Levi-Civita and has already been applied to problems of theoretical physics. In section B of the present paper I developed all the necessary mathematical tools—which cannot be assumed to be known to every physicist—and I tried to do it in as simple and transparent a manner as possible, so that a special study of the mathematical literature is not required for the understanding of the present paper. Finally, I want to acknowledge gratefully my friend, the mathematician Grossmann, whose help not only saved me the effort of studying the pertinent mathematical literature, but who also helped me in my search for the field equations of gravitation.

[1]
[2]
[3]
[4]
[5]
[6]

[The balance of this translation is reprinted from H. A. Lorentz et al., *The Principle of Relativity*, trans. W. Perrett and G. B. Jeffery (Methuen, 1923; Dover rpt., 1952).]

THE FOUNDATION OF THE GENERAL THEORY OF RELATIVITY

By A. EINSTEIN

A. FUNDAMENTAL CONSIDERATIONS ON THE POSTULATE OF RELATIVITY

§ 1. Observations on the Special Theory of Relativity

THE special theory of relativity is based on the following postulate, which is also satisfied by the mechanics of Galileo and Newton.

If a system of co-ordinates K is chosen so that, in relation to it, physical laws hold good in their simplest form, the *same* laws also hold good in relation to any other system of co-ordinates K' moving in uniform translation relatively to K. This postulate we call the " special principle of relativity." The word " special " is meant to intimate that the principle is restricted to the case when K' has a motion of uniform translation relatively to K, but that the equivalence of K' and K does not extend to the case of non-uniform motion of K' relatively to K.

Thus the special theory of relativity does not depart from classical mechanics through the postulate of relativity, but through the postulate of the constancy of the velocity of light *in vacuo*, from which, in combination with the special principle of relativity, there follow, in the well-known way, the relativity of simultaneity, the Lorentzian transformation, and the related laws for the behaviour of moving bodies and clocks.

The modification to which the special theory of relativity has subjected the theory of space and time is indeed far-reaching, but one important point has remained unaffected.

For the laws of geometry, even according to the special theory of relativity, are to be interpreted directly as laws relating to the possible relative positions of solid bodies at rest; and, in a more general way, the laws of kinematics are to be interpreted as laws which describe the relations of measuring bodies and clocks. To two selected material points of a stationary rigid body there always corresponds a distance of quite definite length, which is independent of the locality and orientation of the body, and is also independent of the time. To two selected positions of the hands of a clock at rest relatively to the privileged system of reference there always corresponds an interval of time of a definite length, which is independent of place and time. We shall soon see that the general theory of relativity cannot adhere to this simple physical interpretation of space and time.

§ 2. The Need for an Extension of the Postulate of Relativity

In classical mechanics, and no less in the special theory of relativity, there is an inherent epistemological defect which was, perhaps for the first time, clearly pointed out by Ernst Mach. We will elucidate it by the following example :—Two fluid bodies of the same size and nature hover freely in space at so great a distance from each other and from all other masses that only those gravitational forces need be taken into account which arise from the interaction of different parts of the same body. Let the distance between the two bodies be invariable, and in neither of the bodies let there be any relative movements of the parts with respect to one another. But let either mass, as judged by an observer at rest relatively to the other mass, rotate with constant angular velocity about the line joining the masses. This is a verifiable relative motion of the two bodies. Now let us imagine that each of the bodies has been surveyed by means of measuring instruments at rest relatively to itself, and let the surface of S_1 prove to be a sphere, and that of S_2 an ellipsoid of revolution. Thereupon we put the question—What is the reason for this difference in the two bodies ? No answer can

[7]

be admitted as epistemologically satisfactory,* unless the reason given is an *observable fact of experience*. The law of causality has not the significance of a statement as to the world of experience, except when *observable facts* ultimately appear as causes and effects.

Newtonian mechanics does not give a satisfactory answer to this question. It pronounces as follows :—The laws of mechanics apply to the space R_1, in respect to which the body S_1 is at rest, but not to the space R_2, in respect to which the body S_2 is at rest. But the privileged space R_1 of Galileo, thus introduced, is a merely *factitious* cause, and not a thing that can be observed. It is therefore clear that Newton's mechanics does not really satisfy the requirement of causality in the case under consideration, but only apparently does so, since it makes the factitious cause R_1 responsible for the observable difference in the bodies S_1 and S_2.

The only satisfactory answer must be that the physical system consisting of S_1 and S_2 reveals within itself no imaginable cause to which the differing behaviour of S_1 and S_2 can be referred. The cause must therefore lie *outside* this system. We have to take it that the general laws of motion, which in particular determine the shapes of S_1 and S_2, must be such that the mechanical behaviour of S_1 and S_2 is partly conditioned, in quite essential respects, by distant masses which we have not included in the system under consideration. These distant masses and their motions relative to S_1 and S_2 must then be regarded as the seat of the causes (which must be susceptible to observation) of the different behaviour of our two bodies S_1 and S_2. They take over the rôle of the factitious cause R_1. Of all imaginable spaces R_1, R_2, etc., in any kind of motion relatively to one another, there is none which we may look upon as privileged *a priori* without reviving the above-mentioned epistemological objection. *The laws of physics must be of such a nature that they apply to systems of reference in any kind of motion.* Along this road we arrive at an extension of the postulate of relativity.

In addition to this weighty argument from the theory of

---

* Of course an answer may be satisfactory from the point of view of epistemology, and yet be unsound physically, if it is in conflict with other experiences.

knowledge, there is a well-known physical fact which favours an extension of the theory of relativity. Let K be a Galilean system of reference, i.e. a system relatively to which (at least in the four-dimensional region under consideration) a mass, sufficiently distant from other masses, is moving with uniform motion in a straight line. Let K' be a second system of reference which is moving relatively to K in *uniformly accelerated* translation. Then, relatively to K', a mass sufficiently distant from other masses would have an accelerated motion such that its acceleration and direction of acceleration are independent of the material composition and physical state of the mass.

Does this permit an observer at rest relatively to K' to infer that he is on a " really " accelerated system of reference? The answer is in the negative; for the above-mentioned relation of freely movable masses to K' may be interpreted equally well in the following way. The system of reference K' is unaccelerated, but the space-time territory in question is under the sway of a gravitational field, which generates the accelerated motion of the bodies relatively to K'.

[8]

This view is made possible for us by the teaching of experience as to the existence of a field of force, namely, the gravitational field, which possesses the remarkable property of imparting the same acceleration to all bodies.* The mechanical behaviour of bodies relatively to K' is the same as presents itself to experience in the case of systems which we are wont to regard as " stationary " or as " privileged." Therefore, from the physical standpoint, the assumption readily suggests itself that the systems K and K' may both with equal right be looked upon as " stationary," that is to say, they have an equal title as systems of reference for the physical description of phenomena.

It will be seen from these reflexions that in pursuing the general theory of relativity we shall be led to a theory of gravitation, since we are able to "produce" a gravitational field merely by changing the system of co-ordinates. It will also be obvious that the principle of the constancy of the velocity of light *in vacuo* must be modified, since we easily

[9]

* Eötvös has proved experimentally that the gravitational field has this property in great accuracy.

recognize that the path of a ray of light with respect to K′ must in general be curvilinear, if with respect to K light is propagated in a straight line with a definite constant velocity.

§ 3. The Space-Time Continuum.   Requirement of General Co-Variance for the Equations Expressing General Laws of Nature

In classical mechanics, as well as in the special theory of relativity, the co-ordinates of space and time have a direct physical meaning. To say that a point-event has the X_1 co-ordinate x_1 means that the projection of the point-event on the axis of X_1, determined by rigid rods and in accordance with the. rules of Euclidean geometry, is obtained by measuring off a given rod (the unit of length) x_1 times from the origin of co-ordinates along the axis of X_1. To say that a point-event has the X_4 co-ordinate $x_4 = t$, means that a standard clock, made to measure time in a definite unit period, and which is stationary relatively to the system of co-ordinates and practically coincident in space with the point-event,* will have measured off $x_4 = t$ periods at the occurrence of the event.

This view of space and time has always been in the minds of physicists, even if, as a rule, they have been unconscious of it. This is clear from the part which these concepts play in physical measurements ; it must also have underlain the reader's reflexions on the preceding paragraph (§ 2) for him to connect any meaning with what he there read. But we shall now show that we must put it aside and replace it by a more general view, in order to be able to carry through the postulate of general relativity, if the special theory of relativity applies to the special case of the absence of a gravitational field.

In a space which is free of gravitational fields we introduce a Galilean system of reference K (x, y, z, t), and also a system of co-ordinates K′ (x', y', z', t') in uniform rotation relatively to K.   Let the origins of both systems, as well as their axes

---

* We assume the possibility of verifying " simultaneity " for events immediately proximate in space, or—to speak more precisely—for immediate proximity or coincidence in space-time, without giving a definition of this fundamental concept.

of Z, permanently coincide. We shall show that for a space-time measurement in the system K′ the above definition of the physical meaning of lengths and times cannot be maintained. For reasons of symmetry it is clear that a circle around the origin in the X, Y plane of K may at the same time be regarded as a circle in the X′, Y′ plane of K′. We suppose that the circumference and diameter of this circle have been measured with a unit measure infinitely small compared with the radius, and that we have the quotient of the two results. If this experiment were performed with a measuring-rod at rest relatively to the Galilean system K, the quotient would be π. With a measuring-rod at rest relatively to K′, the quotient would be greater than π. This is readily understood if we envisage the whole process of measuring from the "stationary" system K, and take into consideration that the measuring-rod applied to the periphery undergoes a Lorentzian contraction, while the one applied along the radius does not. Hence Euclidean geometry does not apply to K′. The notion of co-ordinates defined above, which presupposes the validity of Euclidean geometry, therefore breaks down in relation to the system K′. So, too, we are unable to introduce a time corresponding to physical requirements in K′, indicated by clocks at rest relatively to K′. To convince ourselves of this impossibility, let us imagine two clocks of identical constitution placed, one at the origin of co-ordinates, and the other at the circumference of the circle, and both envisaged from the "stationary" system K. By a familiar result of the special theory of relativity, the clock at the circumference—judged from K—goes more slowly than the other, because the former is in motion and the latter at rest. An observer at the common origin of co-ordinates, capable of observing the clock at the circumference by means of light, would therefore see it lagging behind the clock beside him. As he will not make up his mind to let the velocity of light along the path in question depend explicitly on the time, he will interpret his observations as showing that the clock at the circumference "really" goes more slowly than the clock at the origin. So he will be obliged to define time in such a way that the rate of a clock depends upon where the clock may be.

[10]

We therefore reach this result:—In the general theory of relativity, space and time cannot be defined in such a way that differences of the spatial co-ordinates can be directly measured by the unit measuring-rod, or differences in the time co-ordinate by a standard clock.

The method hitherto employed for laying co-ordinates into the space-time continuum in a definite manner thus breaks down, and there seems to be no other way which would allow us to adapt systems of co-ordinates to the four-dimensional universe so that we might expect from their application a particularly simple formulation of the laws of nature. So there is nothing for it but to regard all imaginable systems of co-ordinates, on principle, as equally suitable for the description of nature. This comes to requiring that:—

The general laws of nature are to be expressed by equations which hold good for all systems of co-ordinates, that is, are co-variant with respect to any substitutions whatever (generally co-variant).

It is clear that a physical theory which satisfies this postulate will also be suitable for the general postulate of relativity. For the sum of *all* substitutions in any case includes those which correspond to all relative motions of three-dimensional systems of co-ordinates. That this requirement of general co-variance, which takes away from space and time the last remnant of physical objectivity, is a natural one, will be seen from the following reflexion. All our space-time verifications invariably amount to a determination of space-time coincidences. If, for example, events consisted merely in the motion of material points, then ultimately nothing would be observable but the meetings of two or more of these points. Moreover, the results of our measurings are nothing but verifications of such meetings of the material points of our measuring instruments with other material points, coincidences between the hands of a clock and points on the clock dial, and observed point-events happening at the same place at the same time.

The introduction of a system of reference serves no other purpose than to facilitate the description of the totality of such coincidences. We allot to the universe four space-time variables x_1, x_2, x_3, x_4 in such a way that for every point-event

[11]

there is a corresponding system of values of the variables $x_1 \ldots x_4$. To two coincident point-events there corresponds one system of values of the variables $x_1 \ldots x_4$, i.e. coincidence is characterized by the identity of the co-ordinates. If, in place of the variables $x_1 \ldots x_4$, we introduce functions of them, x'_1, x'_2, x'_3, x'_4, as a new system of co-ordinates, so that the systems of values are made to correspond to one another without ambiguity, the equality of all four co-ordinates in the new system will also serve as an expression for the space-time coincidence of the two point-events. As all our physical experience can be ultimately reduced to such coincidences, there is no immediate reason for preferring certain systems of co-ordinates to others, that is to say, we arrive at the requirement of general co-variance.

§ 4. The Relation of the Four Co-ordinates to Measurement in Space and Time

It is not my purpose in this discussion to represent the general theory of relativity as a system that is as simple and logical as possible, and with the minimum number of axioms; but my main object is to develop this theory in such a way that the reader will feel that the path we have entered upon is psychologically the natural one, and that the underlying assumptions will seem to have the highest possible degree of security. With this aim in view let it now be granted that :—

For infinitely small four-dimensional regions the theory of relativity in the restricted sense is appropriate, if the co-ordinates are suitably chosen.

For this purpose we must choose the acceleration of the infinitely small (" local ") system of co-ordinates so that no gravitational field occurs; this is possible for an infinitely small region. Let X_1, X_2, X_3, be the co-ordinates of space, and X_4 the appertaining co-ordinate of time measured in the appropriate unit.* If a rigid rod is imagined to be given as the unit measure, the co-ordinates, with a given orientation of the system of co-ordinates, have a direct physical meaning

---

* The unit of time is to be chosen so that the velocity of light *in vacuo* as measured in the " local " system of co-ordinates is to be equal to unity.

in the sense of the special theory of relativity. By the special theory of relativity the expression

$$ds^2 = - dX_1^2 - dX_2^2 - dX_3^2 + dX_4^2 \quad . \quad . \quad (1)$$

then has a value which is independent of the orientation of the local system of co-ordinates, and is ascertainable by measurements of space and time. The magnitude of the linear element pertaining to points of the four-dimensional continuum in infinite proximity, we call ds. If the ds belonging to the element $dX_1 \ldots dX_4$ is positive, we follow Minkowski in calling it time-like ; if it is negative, we call it space-like.

To the " linear element " in question, or to the two infinitely ·proximate point-events, there will also correspond definite differentials $dx_1 \ldots dx_4$ of the four-dimensional co-ordinates of any chosen system of reference. If this system, as well as the " local " system, is given for the region under consideration, the dX_ν will allow themselves to be represented here by definite linear homogeneous expressions of the dx_σ :—

$$dX_\nu = \sum_\sigma a_{\nu\sigma} dx_\sigma \quad . \qquad . \qquad . \quad (2)$$

Inserting these expressions in (1), we obtain

$$ds^2 = \sum_{\tau\sigma} g_{\sigma\tau} dx_\sigma dx_\tau, . \qquad . \qquad . \quad (3)$$

where the $g_{\sigma\tau}$ will be functions of the x_σ. These can no longer be dependent on the orientation and the state of motion of the " local " system of co-ordinates, for ds^2 is a quantity ascertainable by rod-clock measurement of point-events infinitely proximate in space-time, and defined independently of any particular choice of co-ordinates. The $g_{\sigma\tau}$ are to be chosen here so that $g_{\sigma\tau} = g_{\tau\sigma}$; the summation is to extend over all values of σ and τ, so that the sum consists of 4×4 terms, of which twelve are equal in pairs.

The case of the ordinary theory of relativity arises out of the case here considered, if it is possible, by reason of the particular relations of the $g_{\sigma\tau}$ in a finite region, to choose the system of reference in the finite region in such a way that the $g_{\sigma\tau}$ assume the constant values

$$\left.\begin{array}{cccc} -1 & 0 & 0 & 0 \\ 0 & -1 & 0 & 0 \\ 0 & 0 & -1 & 0 \\ 0 & 0 & 0 & +1 \end{array}\right\} \quad . \quad . \quad . \quad (4)$$

We shall find hereafter that the choice of such co-ordinates is, in general, not possible for a finite region.

From the considerations of § 2 and § 3 it follows that the quantities $g_{\tau\sigma}$ are to be regarded from the physical standpoint as the quantities which describe the gravitational field in relation to the chosen system of reference. For, if we now assume the special theory of relativity to apply to a certain four-dimensional region with the co-ordinates properly chosen, then the $g_{\sigma\tau}$ have the values given in (4). A free material point then moves, relatively to this system, with uniform motion in a straight line. Then if we introduce new space-time co-ordinates x_1, x_2, x_3, x_4, by means of any substitution we choose, the $g^{\sigma\tau}$ in this new system will no longer be constants, but functions of space and time. At the same time the motion of the free material point will present itself in the new co-ordinates as a curvilinear non-uniform motion, and the law of this motion will be independent of the nature of the moving particle. We shall therefore interpret this motion as a motion under the influence of a gravitational field. We thus find the occurrence of a gravitational field connected with a space-time variability of the g_{σ} . So, too, in the general case, when we are no longer able by a suitable choice of co-ordinates to apply the special theory of relativity to a finite region, we shall hold fast to the view that the $g_{\sigma\tau}$ describe the gravitational field.

Thus, according to the general theory of relativity, gravitation occupies an exceptional position with regard to other forces, particularly the electromagnetic forces, since the ten functions representing the gravitational field at the same time define the metrical properties of the space measured.

B. MATHEMATICAL AIDS TO THE FORMULATION OF GENERALLY COVARIANT EQUATIONS

Having seen in the foregoing that the general postulate of relativity leads to the requirement that the equations of physics shall be covariant in the face of any substitution of the co-ordinates $x_1 \ldots x_4$, we have to consider how such generally covariant equations can be found. We now turn to this purely mathematical task, and we shall find that in its solution a fundamental rôle is played by the invariant ds given in equation (3), which, borrowing from Gauss's theory of surfaces, we have called the "linear element."

The fundamental idea of this general theory of covariants is the following :—Let certain things ("tensors") be defined with respect to any system of co-ordinates by a number of functions of the co-ordinates, called the "components" of the tensor. There are then certain rules by which these components can be calculated for a new system of co-ordinates, if they are known for the original system of co-ordinates, and if the transformation connecting the two systems is known. The things hereafter called tensors are further characterized by the fact that the equations of transformation for their components are linear and homogeneous. Accordingly, all the components in the new system vanish, if they all vanish in the original system. If, therefore, a law of nature is expressed by equating all the components of a tensor to zero, it is generally covariant. By examining the laws of the formation of tensors, we acquire the means of formulating generally covariant laws.

§ 5. Contravariant and Covariant Four-vectors

Contravariant Four-vectors.—The linear element is defined by the four "components" dx_ν, for which the law of transformation is expressed by the equation

$$dx'_\sigma = \sum_\nu \frac{\partial x'_\sigma}{\partial x_\nu} dx_\nu \quad . \qquad . \qquad . \qquad . \quad (5)$$

The dx'_σ are expressed as linear and homogeneous functions of the dx_ν. Hence we may look upon these co-ordinate differentials as the components of a "tensor" of the particular kind which we call a contravariant four-vector. Any thing which is defined relatively to the system of co-ordinates by four quantities A^ν, and which is transformed by the same law

$$A'^\sigma = \sum_\nu \frac{\partial x'_\sigma}{\partial x_\nu} A^\nu, \quad . \qquad . \qquad . \qquad . \quad (5a)$$

we also call a contravariant four-vector. From (5a) it follows at once that the sums $A^\sigma \pm B^\sigma$ are also components of a four-vector, if A^σ and B^σ are such. Corresponding relations hold for all "tensors" subsequently to be introduced. (Rule for the addition and subtraction of tensors.)

Covariant Four-vectors.—We call four quantities A_ν the components of a covariant four-vector, if for any arbitrary choice of the contravariant four-vector B^ν

$$\sum_\nu A_\nu B^\nu = \text{Invariant} \qquad . \qquad . \qquad . \quad (6)$$

The law of transformation of a covariant four-vector follows from this definition. For if we replace B^ν on the right-hand side of the equation

$$\sum_\sigma A'_\sigma B'^\sigma = \sum_\nu A_\nu B^\nu$$

by the expression resulting from the inversion of (5a),

$$\sum_\sigma \frac{\partial x_\nu}{\partial x'_\sigma} B'^\sigma,$$

we obtain

$$\sum_\sigma B'^\sigma \sum_\nu \frac{\partial x_\nu}{\partial x'_\sigma} A_\nu = \sum_\sigma B'^\sigma A'_\sigma.$$

Since this equation is true for arbitrary values of the B'^σ, it follows that the law of transformation is

$$A'_\sigma = \sum_\nu \frac{\partial x_\nu}{\partial x'_\sigma} A_\nu \qquad . \qquad . \qquad . \quad (7)$$

Note on a Simplified Way of Writing the Expressions.— A glance at the equations of this paragraph shows that there is always a summation with respect to the indices which occur twice under a sign of summation (e.g. the index ν in (5)), and only with respect to indices which occur twice. It is therefore possible, without loss of clearness, to omit the sign of summation. In its place we introduce the convention:— If an index occurs twice in one term of an expression, it is always to be summed unless the contrary is expressly stated.

[12]

The difference between covariant and contravariant four-vectors lies in the law of transformation ((7) or (5) respectively). Both forms are tensors in the sense of the general remark above. Therein lies their importance. Following Ricci and

Levi-Civita, we denote the contravariant character by placing
the index above, the covariant by placing it below.

§ 6. Tensors of the Second and Higher Ranks

Contravariant Tensors.—If we form all the sixteen pro-
ducts $A^{\mu\nu}$ of the components A^{μ} and B^{ν} of two contravariant
four-vectors

$$A^{\mu\nu} = A^{\mu}B^{\nu} \quad . \quad . \quad . \quad . \quad (8)$$

then by (8) and (5a) $A^{\mu\nu}$ satisfies the law of transformation

$$A'^{\sigma\tau} = \frac{\partial x'_{\sigma}}{\partial x_{\mu}} \frac{\partial x'_{\tau}}{\partial x_{\nu}} A^{\mu\nu} \quad . \quad . \quad . \quad (9)$$

We call a thing which is described relatively to any system
of reference by sixteen quantities, satisfying the law of transformation (9), a contravariant tensor of the second rank.   Not
every such tensor allows itself to be formed in accordance
with (8) from two four-vectors, but it is easily shown that
any given sixteen $A^{\mu\nu}$ can be represented as the sums of the
$A^{\mu}B^{\nu}$ of four appropriately selected pairs of four-vectors.
Hence we can prove nearly all the laws which apply to the
tensor of the second rank defined by (9) in the simplest
manner by demonstrating them for the special tensors of the
type (8).

Contravariant Tensors of Any Rank.—It is clear that, on
the lines of (8) and (9), contravariant tensors of the third and
higher ranks may also be defined with 4^3 components, and so
on.   In the same way it follows from (8) and (9) that the
contravariant four-vector may be taken in this sense as a
contravariant tensor of the first rank.

Covariant Tensors.—On the other hand, if we take the
sixteen products $A_{\mu\nu}$ of two covariant four-vectors A_{μ} and B_{ν},

$$A_{\mu\nu} = A_{\mu}B_{\nu}, \quad . \quad . \quad . \quad (10)$$

the law of transformation for these is

$$A'_{\sigma\tau} = \frac{\partial x_{\mu}}{\partial x'_{\sigma}} \frac{\partial x_{\nu}}{\partial x'_{\tau}} A_{\mu\nu} \quad . \quad . \quad . \quad (11)$$

This law of transformation defines the covariant tensor of
the second rank.   All our previous remarks on contravariant
tensors apply equally to covariant tensors.

NOTE.—It is convenient to treat the scalar (or invariant) both as a contravariant and a covariant tensor of zero rank.

Mixed Tensors.—We may also define a tensor of the second rank of the type

$$A_\mu^\nu = A_\mu B^\nu \qquad . \qquad . \qquad . \qquad . \quad (12)$$

which is covariant with respect to the index μ, and contravariant with respect to the index ν. Its law of transformation is

$$A_\sigma'^\tau = \frac{\partial x'_\tau}{\partial x_\nu}\frac{\partial x_\mu}{\partial x'_\sigma}A_\mu^\nu \qquad . \qquad . \qquad . \quad (13)$$

Naturally there are mixed tensors with any number of indices of covariant character, and any number of indices of contravariant character. Covariant and contravariant tensors may be looked upon as special cases of mixed tensors.

Symmetrical Tensors.—A contravariant, or a covariant tensor, of the second or higher rank is said to be symmetrical if two components, which are obtained the one from the other by the interchange of two indices, are equal. The tensor $A^{\mu\nu}$, or the tensor $A_{\mu\nu}$, is thus symmetrical if for any combination of the indices μ, ν,

$$A^{\mu\nu} = A^{\nu\mu}, \qquad . \qquad . \qquad . \quad (14)$$

or respectively,

$$A_{\mu\nu} = A_{\nu\mu}. \qquad . \qquad . \qquad . \quad (14a)$$

It has to be proved that the symmetry thus defined is a property which is independent of the system of reference. It follows in fact from (9), when (14) is taken into consideration, that

$$A'^{\sigma\tau} = \frac{\partial x'_\sigma}{\partial x_\mu}\frac{\partial x'_\tau}{\partial x_\nu}A^{\mu\nu} = \frac{\partial x'_\sigma}{\partial x_\mu}\frac{\partial x'_\tau}{\partial x_\nu}A^{\nu\mu} = \frac{\partial x'_\sigma}{\partial x_\nu}\frac{\partial x'_\tau}{\partial x_\mu}A^{\mu\nu} = A'^{\tau\sigma}.$$

The last equation but one depends upon the interchange of the summation indices μ and ν, i.e. merely on a change of notation.

Antisymmetrical Tensors.—A contravariant or a covariant tensor of the second, third, or fourth rank is said to be antisymmetrical if two components, which are obtained the one from the other by the interchange of two indices, are equal and of opposite sign. The tensor $A^{\mu\nu}$, or the tensor $A_{\mu\nu}$, is therefore antisymmetrical, if always

$$A^{\mu\nu} = - A^{\nu\mu}, \quad . \qquad . \qquad . \qquad . \quad (15)$$

or respectively,

$$A_{\mu\nu} = - A_{\nu\mu} \quad . \qquad . \qquad . \qquad . \quad (15a)$$

Of the sixteen components $A^{\mu\nu}$, the four components $A^{\mu\mu}$ vanish; the rest are equal and of opposite sign in pairs, so that there are only six components numerically different (a six-vector). Similarly we see that the antisymmetrical tensor of the third rank $A^{\mu\nu\sigma}$ has only four numerically different components, while the antisymmetrical tensor $A^{\mu\nu\sigma\tau}$ has only one. There are no antisymmetrical tensors of higher rank than the fourth in a continuum of four dimensions.

§ 7. Multiplication of Tensors

Outer Multiplication of Tensors.—We obtain from the components of a tensor of rank n and of a tensor of rank m the components of a tensor of rank $n + m$ by multiplying each component of the one tensor by each component of the other. Thus, for example, the tensors T arise out of the tensors A and B of different kinds,

$$\begin{aligned}
T_{\mu\nu\sigma} &= A_{\mu\nu}B_{\sigma}, \\
T^{\mu\nu\sigma\tau} &= A^{\mu\nu}B^{\sigma\tau}, \\
T^{\sigma\tau}_{\mu\nu} &= A_{\mu\nu}B^{\sigma\nu}.
\end{aligned}$$

The proof of the tensor character of T is given directly by the representations (8), (10), (12), or by the laws of transformation (9), (11), (13). The equations (8), (10), (12) are themselves examples of outer multiplication of tensors of the first rank.

" Contraction " of a Mixed Tensor.—From any mixed tensor we may form a tensor whose rank is less by two, by equating an index of covariant with one of contravariant character, and summing with respect to this index (" contraction "). Thus, for example, from the mixed tensor of the fourth rank $A^{\sigma\tau}_{\mu\nu}$, we obtain the mixed tensor of the second rank,

$$A^{\tau}_{\nu} = A^{\mu\tau}_{\mu\nu} \quad (= \sum_{\mu} A^{\mu\tau}_{\mu\nu}),$$

and from this, by a second contraction, the tensor of zero rank,

$$A = A^{\nu}_{\nu} = A^{\mu\nu}_{\mu\nu}.$$

The proof that the result of contraction really possesses the tensor character is given either by the representation of a tensor according to the generalization of (12) in combination with (6), or by the generalization of (13).

Inner and Mixed Multiplication of Tensors.—These consist in a combination of outer multiplication with contraction.

Examples.—From the covariant tensor of the second rank $A_{\mu\nu}$ and the contravariant tensor of the first rank B^{σ} we form by outer multiplication the mixed tensor

$$D^{\sigma}_{\mu\nu} = A_{\mu\nu}B^{\sigma}.$$

On contraction with respect to the indices ν and σ, we obtain the covariant four-vector

$$D_{\mu} = D^{\nu}_{\mu\nu} = A_{\mu\nu}B^{\nu}.$$

This we call the inner product of the tensors $A_{\mu\nu}$ and B^{σ}. Analogously we form from the tensors $A_{\mu\nu}$ and $B^{\sigma\tau}$, by outer multiplication and double contraction, the inner product $A_{\mu\nu}B^{\mu\nu}$. By outer multiplication and one contraction, we obtain from $A_{\mu\nu}$ and $B^{\sigma\tau}$ the mixed tensor of the second rank $D^{\tau}_{\mu} = A_{\mu\nu}B^{\nu\tau}$. This operation may be aptly characterized as a mixed one, being "outer" with respect to the indices μ and τ, and "inner" with respect to the indices ν and σ.

We now prove a proposition which is often useful as evidence of tensor character. From what has just been explained, $A_{\mu\nu}B^{\mu\nu}$ is a scalar if $A_{\mu\nu}$ and $B^{\sigma\tau}$ are tensors. But we may also make the following assertion: If $A_{\mu\nu}B^{\mu\nu}$ is a scalar *for any choice of the tensor* $B^{\mu\nu}$, then $A_{\mu\nu}$ has tensor character. For, by hypothesis, for any substitution,

$$A'_{\sigma\tau}B'^{\sigma\tau} = A_{\mu\nu}B^{\mu\nu}.$$

But by an inversion of (9)

$$B^{\mu\nu} = \frac{\partial x_{\mu}}{\partial x'_{\sigma}}\frac{\partial x_{\nu}}{\partial x'_{\tau}}B'^{\sigma\tau}.$$

This, inserted in the above equation, gives

$$\left(A'_{\sigma\tau} - \frac{\partial x_{\mu}}{\partial x'_{\sigma}}\frac{\partial x_{\nu}}{\partial x'_{\tau}}A_{\mu\nu}\right)B'^{\sigma\tau} = 0.$$

This can only be satisfied for arbitrary values of $B'^{\sigma\tau}$ if the bracket vanishes. The result then follows by equation (11). This rule applies correspondingly to tensors of any rank and character, and the proof is analogous in all cases.

The rule may also be demonstrated in this form: If B^μ and C^ν are any vectors, and if, for all values of these, the inner product $A_{\mu\nu}B^\mu C^\nu$ is a scalar, then $A_{\mu\nu}$ is a covariant tensor. This latter proposition also holds good even if only the more special assertion is correct, that with any choice of the four-vector B^μ the inner product $A_{\mu\nu}B^\mu B^\nu$ is a scalar, if in addition it is known that $A_{\mu\nu}$ satisfies the condition of symmetry $A_{\mu\nu} = A_{\nu\mu}$. For by the method given above we prove the tensor character of $(A_{\mu\nu} + A_{\nu\mu})$, and from this the tensor character of $A_{\mu\nu}$ follows on account of symmetry. This also can be easily generalized to the case of covariant and contravariant tensors of any rank.

Finally, there follows from what has been proved, this law, which may also be generalized for any tensors: If for any choice of the four-vector B^ν the quantities $A_{\mu\nu}B^\nu$ form a tensor of the first rank, then $A_{\mu\nu}$ is a tensor of the second rank. For, if C^μ is any four-vector, then on account of the tensor character of $A_{\mu\nu}B^\nu$, the inner product $A_{\mu\nu}B^\nu C^\mu$ is a scalar for any choice of the two four-vectors B^ν and C^μ. From which the proposition follows.

§ 8. Some Aspects of the Fundamental Tensor $g_{\mu\nu}$

The Covariant Fundamental Tensor.—In the invariant expression for the square of the linear element,

$$ds^2 = g_{\mu\nu}dx_\mu dx_\nu,$$

the part played by the dx_μ is that of a contravariant vector which may be chosen at will. Since further, $g_{\mu\nu} = g_{\nu\mu}$, it follows from the considerations of the preceding paragraph that $g_{\mu\nu}$ is a covariant tensor of the second rank. We call it the "fundamental tensor." In what follows we deduce some properties of this tensor which, it is true, apply to any tensor of the second rank. But as the fundamental tensor plays a special part in our theory, which has its physical basis in the peculiar effects of gravitation, it so happens that the relations to be developed are of importance to us only in the case of the fundamental tensor.

The Contravariant Fundamental Tensor.—If in the determinant formed by the elements $g_{\mu\nu}$, we take the co-factor of each of the $g_{\mu\nu}$ and divide it by the determinant $g = |\,g_{\mu\nu}\,|$, we obtain certain quantities $g^{\mu\nu}(\,= g^{\nu\mu})$ which, as we shall demonstrate, form a contravariant tensor.

By a known property of determinants

$$g_{\mu\sigma}g^{\nu\sigma} = \delta_{\mu}^{\nu} \quad . \qquad . \qquad . \qquad . \quad (16)$$

where the symbol δ_{μ}^{ν} denotes 1 or 0, according as $\mu = \nu$ or $\mu \neq \nu$.

Instead of the above expression for ds^2 we may thus write

$$g_{\mu\sigma}\delta_{\nu}^{\sigma}dx_{\mu}dx_{\nu}$$

or, by (16)

$$g_{\mu\sigma}g_{\nu\tau}g^{\sigma\tau}dx_{\mu}dx_{\nu}.$$

But, by the multiplication rules of the preceding paragraphs, the quantities

$$d\xi_{\sigma} = g_{\mu\sigma}dx_{\mu}$$

form a covariant four-vector, and in fact an arbitrary vector, since the dx_{μ} are arbitrary.   By introducing this into our expression we obtain

$$ds^2 = g^{\sigma\tau}d\xi_{\sigma}d\xi_{\tau}.$$

Since this, with the arbitrary choice of the vector $d\xi_{\sigma}$, is a scalar, and $g^{\sigma\tau}$ by its definition is symmetrical in the indices σ and τ, it follows from the results of the preceding paragraph that $g^{\sigma\tau}$ is a contravariant tensor.

It further follows from (16) that δ_{μ}^{ν} is also a tensor, which we may call the mixed fundamental tensor.

The Determinant of the Fundamental Tensor.—By the rule for the multiplication of determinants

$$|\,g_{\mu\alpha}g^{\alpha\nu}\,| \;=\; |\,g_{\mu\alpha}\,| \;\times\; |\,g^{\alpha\nu}\,|.$$

On the other hand

$$|\,g_{\mu\alpha}g^{\alpha\nu}\,| \;=\; |\,\delta_{\mu}^{\nu}\,| \;=\; 1.$$

It therefore follows that

$$|\,g_{\mu\nu}\,| \;\times\; |\,g^{\mu\nu}\,| \;=\; 1 \quad . \qquad . \qquad . \quad (17)$$

The Volume Scalar.—We seek first the law of transformation of the determinant $g = |g_{\mu\nu}|$. In accordance with (11)

$$g' = \left| \frac{\partial x_\mu}{\partial x'_\sigma} \frac{\partial x}{\partial x'_\tau} g_{\mu\nu} \right|.$$

Hence, by a double application of the rule for the multiplication of determinants, it follows that

$$g' = \left| \frac{\partial x_\mu}{\partial x'_\sigma} \right| \cdot \left| \frac{\partial x_\nu}{\partial x'_\tau} \right| \cdot |g_{\mu\nu}| = \left| \frac{\partial x_\mu}{\partial x'_\sigma} \right|^2 g,$$

or

$$\sqrt{g'} = \left| \frac{\partial x_\mu}{\partial x'_\sigma} \right| \sqrt{g}.$$

On the other hand, the law of transformation of the element of volume

$$d\tau = \int dx_1 dx_2 dx_3 dx_4$$

is, in accordance with the theorem of Jacobi,

$$d\tau' = \left| \frac{\partial x'_\sigma}{\partial x_\mu} \right| d\tau.$$

By multiplication of the last two equations, we obtain

$$\sqrt{g'} d\tau' = \sqrt{g} d\tau \qquad . \qquad . \qquad . \quad (18).$$

Instead of $\sqrt{g}$, we introduce in what follows the quantity $\sqrt{-g}$, which is always real on account of the hyperbolic character of the space-time continuum. The invariant $\sqrt{-g}\,d\tau$ is equal to the magnitude of the four-dimensional element of volume in the "local" system of reference, as measured with rigid rods and clocks in the sense of the special theory of relativity.

Note on the Character of the Space-time Continuum.—Our assumption that the special theory of relativity can always be applied to an infinitely small region, implies that ds^2 can always be expressed in accordance with (1) by means of real quantities $dX_1 \ldots dX_4$. If we denote by $d\tau_0$ the "natural" element of volume dX_1, dX_2, dX_3, dX_4, then

$$d\tau_0 = \sqrt{-g}\,d\tau \qquad . \qquad . \qquad . \quad (18a)$$

If $\sqrt{-g}$ were to vanish at a point of the four-dimensional continuum, it would mean that at this point an infinitely small "natural" volume would correspond to a finite volume in the co-ordinates. Let us assume that this is never the case. Then g cannot change sign. We will assume that, in the sense of the special theory of relativity, g always has a finite negative value. This is a hypothesis as to the physical nature of the continuum under consideration, and at the same time a convention as to the choice of co-ordinates.

But if $-g$ is always finite and positive, it is natural to settle the choice of co-ordinates *a posteriori* in such a way that this quantity is always equal to unity. We shall see later that by such a restriction of the choice of co-ordinates it is possible to achieve an important simplification of the laws of nature.

In place of (18), we then have simply $d\tau' = d\tau$, from which, in view of Jacobi's theorem, it follows that

$$\left| \frac{\partial x'_\sigma}{\partial x_\mu} \right| = 1 \qquad . \qquad . \qquad . \qquad . \qquad (19)$$

Thus, with this choice of co-ordinates, only substitutions for which the determinant is unity are permissible.

But it would be erroneous to believe that this step indicates a partial abandonment of the general postulate of relativity. We do not ask " What are the laws of nature which are covariant in face of all substitutions for which the determinant is unity ? " but our question is " What are the generally covariant laws of nature ? " It is not until we have formulated these that we simplify their expression by a particular choice of the system of reference.

The Formation of New Tensors by Means of the Fundamental Tensor.—Inner, outer, and mixed multiplication of a tensor by the fundamental tensor give tensors of different character and rank. For example,

$$A^\mu = g^{u\sigma}A_\sigma,$$
$$A = g_{\mu\nu}A^{\mu\nu}.$$

The following forms may be specially noted :—

$$A^{\mu\nu} = g^{\mu\alpha}g^{\nu\beta}A_{\alpha\beta},$$
$$A_{\mu\nu} = g_{\mu\alpha}g_{\nu\beta}A^{\alpha\beta}$$

(the " complements " of covariant and contravariant tensors respectively), and

$$B_{\mu\nu} = g_{\mu\nu}g^{\alpha\beta}A_{\alpha\beta}.$$

We call $B_{\mu\nu}$ the reduced tensor associated with $A_{\mu\nu}$. Similarly,

$$B^{\mu\nu} = g^{\mu\nu}g_{\alpha\beta}A^{\alpha\beta}.$$

It may be noted that $g^{\mu\nu}$ is nothing more than the complement of $g_{\mu\nu}$, since

$$g^{\mu\alpha}g^{\nu\beta}g_{\alpha\beta} = g^{\mu\alpha}\delta^\nu_\alpha = g^{\mu\nu}.$$

§ 9. The Equation of the Geodetic Line.   The Motion of a Particle

As the linear element ds is defined independently of the system of co-ordinates, the line drawn between two points P and P' of the four-dimensional continuum in such a way that $\int ds$ is stationary—a geodetic line—has a meaning which also is independent of the choice of co-ordinates. Its equation is

$$\delta\int_P^{P'} ds = 0 \quad . \quad . \quad . \quad . \quad (20)$$

Carrying out the variation in the usual way, we obtain from this equation four differential equations which define the geodetic line ; this operation will be inserted here for the sake of completeness. Let λ be a function of the co-ordinates x_ν, and let this define a family of surfaces which intersect the required geodetic line as well as all the lines in immediate proximity to it which are drawn through the points P and P'. Any such line may then be supposed to be given by expressing its co-ordinates x_ν as functions of λ. Let the symbol δ indicate the transition from a point of the required geodetic to the point corresponding to the same λ on a neighbouring line. Then for (20) we may substitute

$$\left.\begin{array}{l}\int_{\lambda_1}^{\lambda_2}\delta w\,d\lambda = 0 \\[2mm] w^2 = g_{\mu\nu}\dfrac{dx_\mu}{d\lambda}\dfrac{dx_\nu}{d\lambda}\end{array}\right\} \quad . \quad . \quad . \quad (20a)$$

But since

$$\delta w = \frac{1}{w}\left\{\frac{1}{2}\frac{\partial g_{\mu\nu}}{\partial x_\sigma}\frac{dx_\mu}{d\lambda}\frac{dx_\nu}{d\lambda}\delta x_\sigma + g_{\mu\nu}\frac{dx_\mu}{d\lambda}\delta\left(\frac{dx_\nu}{d\lambda}\right)\right\},$$

and

$$\delta\left(\frac{dx_\nu}{d\lambda}\right) = \frac{d}{d\lambda}(\delta x_\nu),$$

we obtain from (20a), after a partial integration,

$$\int_{\lambda_1}^{\lambda_2}\kappa_\sigma\delta x_\sigma d\lambda = 0,$$

where

[14]
$$\kappa_\sigma = \frac{d}{d\lambda}\left\{\frac{g_{\mu\nu}}{w}\frac{dx_\mu}{d\lambda}\right\} - \frac{1}{2w}\frac{\partial g_{\mu\nu}}{\partial x_\sigma}\frac{dx_\mu}{d\lambda}\frac{dx_\nu}{d\lambda} \qquad . \qquad (20\text{b})$$

Since the values of δx_σ are arbitrary, it follows from this that

$$\kappa_\sigma = 0 \qquad . \qquad . \qquad . \qquad . \qquad (20\text{c})$$

are the equations of the geodetic line.

If ds does not vanish along the geodetic line we may choose the "length of the arc " s, measured along the geodetic line, for the parameter λ. Then $w = 1$, and in place of (20c) we obtain

[15]
$$g_{\mu\nu}\frac{d^2x_\mu}{ds^2} + \frac{\partial g_{\mu\nu}}{\partial x_\sigma}\frac{dx_\sigma}{ds}\frac{dx_\mu}{ds} - \frac{1}{2}\frac{\partial g_{\mu\nu}}{\partial x_\sigma}\frac{dx_\mu}{ds}\frac{dx_\nu}{ds} = 0$$

or, by a mere change of notation,

$$g_{a\sigma}\frac{d^2x_a}{ds^2} + [\mu\nu, \sigma]\frac{dx_\mu}{ds}\frac{dx_\nu}{ds} = 0 \qquad . \qquad . \qquad (20\text{d})$$

where, following Christoffel, we have written

$$[\mu\nu, \sigma] = \frac{1}{2}\left(\frac{\partial g_{\mu\sigma}}{\partial x_\nu} + \frac{\partial g_{\nu\sigma}}{\partial x_\mu} - \frac{\partial g_{\mu\nu}}{\partial x_\sigma}\right) \qquad . \qquad . \qquad (21)$$

Finally, if we multiply (20d) by $g^{\sigma\tau}$ (outer multiplication with respect to τ, inner with respect to σ), we obtain the equations of the geodetic line in the form

$$\frac{d^2x_\tau}{ds^2} + \{\mu\nu, \tau\}\frac{dx_\mu}{ds}\frac{dx_\nu}{ds} = 0 \qquad . \qquad . \qquad (22)$$

where, following Christoffel, we have set

$$\{\mu\nu, \tau\} = g^{\tau a}[\mu\nu, a] \qquad . \qquad . \qquad . \qquad (23)$$

§ 10. The Formation of Tensors by Differentiation

With the help of the equation of the geodetic line we can now easily deduce the laws by which new tensors can be formed from old by differentiation. By this means we are able for the first time to formulate generally covariant differential equations. We reach this goal by repeated application of the following simple law :—

If in our continuum a curve is given, the points of which are specified by the arcual distance s measured from a fixed point on the curve, and if, further, ϕ is an invariant function of space, then $d\phi/ds$ is also an invariant. The proof lies in this, that ds is an invariant as well as $d\phi$.

As

$$\frac{d\phi}{ds} = \frac{\partial\phi}{\partial x_\mu}\frac{dx_\mu}{ds}$$

therefore

$$\psi = \frac{\partial\phi}{dx_\mu}\frac{dx_\mu}{ds}$$

is also an invariant, and an invariant for all curves starting from a point of the continuum, that is, for any choice of the vector dx_μ. Hence it immediately follows that

$$A_\mu = \frac{\partial\phi}{\partial x_\mu} \qquad . \qquad . \qquad . \qquad . \quad (24)$$

is a covariant four-vector—the " gradient " of ϕ.

According to our rule, the differential quotient

$$\chi = \frac{d\psi}{ds}$$

taken on a curve, is similarly an invariant. Inserting the value of ψ, we obtain in the first place

$$\chi = \frac{\partial^2\phi}{\partial x_\mu \partial x_\nu}\frac{dx_\mu}{ds}\frac{dx_\nu}{ds} + \frac{\partial\phi}{\partial x_\mu}\frac{d^2x_\mu}{ds^2}.$$

The existence of a tensor cannot be deduced from this forthwith. But if we may take the curve along which we have differentiated to be a geodetic, we obtain on substitution for d^2x_ν/ds^2 from (22),

$$\chi = \left(\frac{\partial^2\phi}{\partial x_\mu \partial x_\nu} - \{\mu\nu, \tau\}\frac{\partial\phi}{\partial x_\tau}\right)\frac{dx_\mu}{ds}\frac{dx_\nu}{ds}.$$

Since we may interchange the order of the differentiations,

and since by (23) and (21) $\{\mu\nu, \tau\}$ is symmetrical in μ and ν, it follows that the expression in brackets is symmetrical in μ and ν. Since a geodetic line can be drawn in any direction from a point of the continuum, and therefore dx_μ/ds is a four-vector with the ratio of its components arbitrary, it follows from the results of § 7 that

$$A_{\mu\nu} = \frac{\partial^2\phi}{\partial x_\mu \partial x_\nu} - \{\mu\nu, \tau\}\frac{\partial\phi}{\partial x_\tau}. \quad . \quad . \quad (25)$$

is a covariant tensor of the second rank. We have therefore come to this result: from the covariant tensor of the first rank

$$A_\mu = \frac{\partial\phi}{\partial x_\mu}$$

we can, by differentiation, form a covariant tensor of the second rank

$$A_{\mu\nu} = \frac{\partial A_\mu}{\partial x_\nu} - \{\mu\nu, \tau\}A_\tau \quad . \quad . \quad (26)$$

We call the tensor $A_{\mu\nu}$ the " extension " (covariant derivative) of the tensor A_μ. In the first place we can readily show that the operation leads to a tensor, even if the vector A_μ cannot be represented as a gradient. To see this, we first observe that

$$\psi\frac{\partial\phi}{\partial x_\mu}$$

is a covariant vector, if ψ and ϕ are scalars. The sum of four such terms

$$S_\mu = \psi^{(1)}\frac{\phi\partial^{(1)}}{\partial x_\mu} + . + . + \psi^{(4)}\frac{\partial\phi^{(4)}}{\partial x_\mu},$$

is also a covariant vector, if $\psi^{(1)}$, $\phi^{(1)}$ . . . $\psi^{(4)}$, $\phi^{(4)}$ are scalars. But it is clear that any covariant vector can be represented in the form S_μ. For, if A_μ is a vector whose components are any given functions of the x_ν, we have only to put (in terms of the selected system of co-ordinates)

$$\begin{aligned}
\psi^{(1)} &= A_1, & \phi^{(1)} &= x_1, \\
\psi^{(2)} &= A_2, & \phi^{(2)} &= x_2, \\
\psi^{(3)} &= A_3, & \phi^{(3)} &= x_3, \\
\psi^{(4)} &= A_4, & \phi^{(4)} &= x_4,
\end{aligned}$$

in order to ensure that S_μ shall be equal to A_μ.

Therefore, in order to demonstrate that $A_{\mu\nu}$ is a tensor if *any* covariant vector is inserted on the right-hand side for A_μ, we only need show that this is so for the vector S_μ. But for this latter purpose it is sufficient, as a glance at the right-hand side of (26) teaches us, to furnish the proof for the case

$$A_\mu = \psi\,\frac{\partial\phi}{\partial x_\mu}.$$

Now the right-hand side of (25) multiplied by ψ,

$$\psi\,\frac{\partial^2\phi}{\partial x_\mu\partial x_\nu} - \{\mu\nu,\,\tau\}\psi\frac{\partial\phi}{\partial x_\tau}$$

is a tensor. Similarly

$$\frac{\partial\psi}{\partial x_\mu}\frac{\partial\phi}{\partial x_\nu}$$

being the outer product of two vectors, is a tensor. By addition, there follows the tensor character of

$$\frac{\partial}{\partial x_\nu}\Big(\psi\,\frac{\partial\phi}{\partial x_\mu}\Big) - \{\mu\nu,\,\tau\}\Big(\psi\,\frac{\partial\phi}{\partial x_\tau}\Big).$$

As a glance at (26) will show, this completes the demonstration for the vector

$$\psi\,\frac{\partial\phi}{\partial x_\mu}$$

and consequently, from what has already been proved, for any vector A_μ.

By means of the extension of the vector, we may easily define the "extension" of a covariant tensor of any rank. This operation is a generalization of the extension of a vector. We restrict ourselves to the case of a tensor of the second rank, since this suffices to give a clear idea of the law of formation.

As has already been observed, any covariant tensor of the second rank can be represented * as the sum of tensors of the

---

* By outer multiplication of the vector with arbitrary components A_{11}, A_{12}, A_{13}, A_{14} by the vector with components 1, 0, 0, 0, we produce a tensor with components

$$
\begin{array}{cccc}
A_{11} & A_{12} & A_{13} & A_{14} \\
0 & 0 & 0 & 0 \\
0 & 0 & 0 & 0 \\
0 & 0 & 0 & 0.
\end{array}
$$

By the addition of four tensors of this type, we obtain the tensor $A_{\mu\nu}$ with any assigned components.

type $A_\mu B_\nu$. It will therefore be sufficient to deduce the expression for the extension of a tensor of this special type. By (26) the expressions

$$\frac{\partial A_\mu}{\partial x_\sigma} - \{\sigma\mu, \tau\}A_\tau,$$

$$\frac{\partial B_\nu}{\partial x_\sigma} - \{\sigma\nu, \tau\}B_\tau,$$

are tensors. On outer multiplication of the first by B_ν, and of the second by A_μ, we obtain in each case a tensor of the third rank. By adding these, we have the tensor of the third rank

$$A_{\mu\nu\sigma} = \frac{\partial A_{\mu\nu}}{\partial x_\sigma} - \{\sigma\mu, \tau\}A_{\tau\nu} - \{\sigma\nu, \tau\}A_{\mu\tau} . \qquad . \quad (27)$$

where we have put $A_{\mu\nu} = A_\mu B_\nu$. As the right-hand side of (27) is linear and homogeneous in the $A_{\mu\nu}$ and their first derivatives, this law of formation leads to a tensor, not only in the case of a tensor of the type $A_\mu B_\nu$, but also in the case of a sum of such tensors, i.e. in the case of any covariant tensor of the second rank. We call $A_{\mu\nu\sigma}$ the extension of the tensor $A_{\mu\nu}$.

It is clear that (26) and (24) concern only special cases of extension (the extension of the tensors of rank one and zero respectively).

In general, all special laws of formation of tensors are included in (27) in combination with the multiplication of tensors.

§ 11. Some Cases of Special Importance

The Fundamental Tensor.—We will first prove some lemmas which will be useful hereafter. By the rule for the differentiation of determinants

$$dg = g^{\mu\nu}g\,dg_{\mu\nu} = - g_{\mu\nu}g\,dg^{\mu\nu} \qquad . \qquad . \quad (28)$$

The last member is obtained from the last but one, if we bear in mind that $g_{\mu\nu}g^{\mu'\nu} = \delta_\mu^{\mu'}$, so that $g_{\mu\nu}g^{\mu\nu} = 4$, and consequently

$$g_{\mu\nu}dg^{\mu\nu} + g^{\mu\nu}dg_{\mu\nu} = 0.$$

From (28), it follows that

$$\frac{1}{\sqrt{-g}}\frac{\partial\sqrt{-g}}{\partial x_\sigma} = \tfrac{1}{2}\frac{\partial \log(-g)}{\partial x_\sigma} = \tfrac{1}{2}g^{\mu\nu}\frac{\partial g_{\mu\nu}}{\partial x_\sigma} = \tfrac{1}{2}g_{\mu\nu}\frac{\partial g^{\mu\nu}}{\partial x_\sigma}. \quad (29)$$

Further, from $g_{\mu\sigma}g^{\nu\sigma} = \delta_\mu^\nu$, it follows on differentiation that

$$\left.\begin{aligned} g_{\mu\sigma}dg^{\nu\sigma} &= -g^{\nu\sigma}dg_{\mu\sigma}\\ g_{\mu\sigma}\frac{\partial g^{\nu\sigma}}{\partial x_\lambda} &= -g^{\nu\sigma}\frac{\partial g_{\mu\sigma}}{\partial x_\lambda} \end{aligned}\right\} \quad . \quad . \quad . \quad (30)$$

From these, by mixed multiplication by $g^{\sigma\tau}$ and $g_{\nu\lambda}$ respectively, and a change of notation for the indices, we have

$$\left.\begin{aligned} dg^{\mu\nu} &= -g^{\mu\alpha}g^{\nu\beta}\, dg_{\alpha\beta}\\ \frac{\partial g^{\mu\nu}}{\partial x_\sigma} &= -g^{\mu\alpha}g^{\nu\beta}\frac{\partial g_{\alpha\beta}}{\partial x_\sigma} \end{aligned}\right\} \quad . \quad . \quad . \quad (31)$$

and

$$\left.\begin{aligned} dg_{\mu\nu} &= -g_{\mu\alpha}g_{\nu\beta}\, dg^{\alpha\beta}\\ \frac{\partial g_{\mu\nu}}{\partial x_\sigma} &= -g_{\mu\alpha}g_{\nu\beta}\frac{\partial g^{\alpha\beta}}{\partial x_\sigma} \end{aligned}\right\} \quad . \quad . \quad . \quad (32)$$

The relation (31) admits of a transformation, of which we also have frequently to make use. From (21)

$$\frac{\partial g_{\alpha\beta}}{\partial x_\sigma} = [\alpha\sigma, \beta] + [\beta\sigma, \alpha] \quad . \quad . \quad . \quad (33)$$

Inserting this in the second formula of (31), we obtain, in view of (23)

$$\frac{\partial g^{\mu\nu}}{\partial x_\sigma} = -g^{\mu\tau}\{\tau\sigma, \nu\} - g^{\nu\tau}\{\tau\sigma, \mu\} \quad . \quad . \quad (34)$$ [16]

Substituting the right-hand side of (34) in (29), we have

$$\frac{1}{\sqrt{-g}}\frac{\partial\sqrt{-g}}{\partial x_\sigma} = \{\mu\sigma, \mu\} \quad . \quad . \quad (29\text{a})$$ [17]

The " Divergence " of a Contravariant Vector.—If we take the inner product of (26) by the contravariant fundamental tensor $g^{\mu\nu}$, the right-hand side, after a transformation of the first term, assumes the form

$$\frac{\partial}{\partial x_\nu}(g^{\mu\nu}A_\mu) - A_\mu\frac{\partial g^{\mu\nu}}{\partial x_\nu} - \tfrac{1}{2}g^{\tau\alpha}\Big(\frac{\partial g_{\mu\alpha}}{\partial x_\nu} + \frac{\partial g_{\nu\alpha}}{\partial x_\mu} - \frac{\partial g_{\mu\nu}}{\partial x_\alpha}\Big)g^{\mu\nu}A_\tau.$$ [18]

[19]

In accordance with (31) and (29), the last term of this expression may be written

[20]

$$\tfrac{1}{2}\frac{\partial g^{\tau\nu}}{\partial x_\nu}A_\tau + \tfrac{1}{2}\frac{\partial g^{\tau\mu}}{\partial x_\mu}A_\tau + \frac{1}{\sqrt{-g}}\frac{\partial\sqrt{-g}}{\partial x_a}g^{\mu\nu}A_\tau.$$

As the symbols of the indices of summation are immaterial, the first two terms of this expression cancel the second of the one above. If we then write $g^{\mu\nu}A_\mu = A^\nu$, so that A^ν like A_μ is an arbitrary vector, we finally obtain

$$\Phi = \frac{1}{\sqrt{-g}}\frac{\partial}{\partial x_\nu}(\sqrt{-g}A^\nu).\qquad .\qquad .\quad (35)$$

This scalar is the *divergence* of the contravariant vector A^ν.

The " Curl " of a Covariant Vector.—The second term in (26) is symmetrical in the indices μ and ν. Therefore $A_{\mu\nu} - A_{\nu\mu}$ is a particularly simply constructed antisymmetrical tensor. We obtain

$$B_{\mu\nu} = \frac{\partial A_\mu}{\partial x_\nu} - \frac{\partial A_\nu}{\partial x_\mu}\qquad .\qquad .\qquad .\quad (36)$$

Antisymmetrical Extension of a Six-vector.—Applying (27) to an antisymmetrical tensor of the second rank $A_{\mu\nu}$, forming in addition the two equations which arise through cyclic permutations of the indices, and adding these three equations, we obtain the tensor of the third rank

$$B_{\mu\nu\sigma} = A_{\mu\nu\sigma} + A_{\nu\sigma\mu} + A_{\sigma\mu\nu} = \frac{\partial A_{\mu\nu}}{\partial x_\sigma} + \frac{\partial A_{\nu\sigma}}{\partial x_\mu} + \frac{\partial A_{\sigma\mu}}{\partial x_\nu}\quad (37)$$

which it is easy to prove is antisymmetrical.

The Divergence of a Six-vector.—Taking the mixed product of (27) by $g^{\mu a}g^{\nu\beta}$, we also obtain a tensor. The first term on the right-hand side of (27) may be written in the form

$$\frac{\partial}{\partial x_\sigma}(g^{\mu a}g^{\nu\beta}A_{\mu\nu}) - g^{\mu a}\frac{\partial g^{\nu\beta}}{\partial x_\sigma}A_{\mu\nu} - g^{\nu\beta}\frac{\partial g^{\mu a}}{\partial x_\sigma}A_{\mu\nu}.$$

If we write $A_\sigma^{a\beta}$ for $g^{\mu a}g^{\nu\beta}A_{\mu\nu\sigma}$ and $A^{a\beta}$ for $g^{\mu a}g^{\nu\beta}A_{\mu\nu}$, and in the transformed first term replace

$$\frac{\partial g^{\nu\beta}}{\partial x_\sigma}\quad\text{and}\quad\frac{\partial g^{\mu a}}{\partial x_\sigma}$$

by their values as given by (34), there results from the right-
hand side of (27) an expression consisting of seven terms, of
which four cancel, and there remains

$$A_\sigma^{\alpha\beta} = \frac{\partial A^{\alpha\beta}}{\partial x_\sigma} + \{\sigma\gamma, \alpha\}A^{\gamma\beta} + \{\sigma\gamma, \beta\}A^{\alpha\gamma} . \qquad . \quad (38)$$

This is the expression for the extension of a contravariant
tensor of the second rank, and corresponding expressions for
the extension of contravariant tensors of higher and lower
rank may also be formed.

We note that in an analogous way we may also form the
extension of a mixed tensor :—

$$A_{\mu\sigma}^\alpha = \frac{\partial A_\mu^\alpha}{\partial x_\sigma} - \{\sigma\mu, \tau\}A_\tau^\alpha + \{\sigma\tau, \alpha\}A_\mu^\tau . \qquad . \quad (39)$$

On contracting (38) with respect to the indices β and σ
(inner multiplication by δ_β^σ), we obtain the vector

$$A^\alpha = \frac{\partial A^{\alpha\beta}}{\partial x_\beta} + \{\beta\gamma, \beta\}A^{\alpha\gamma} + \{\beta\gamma, \alpha\}A^{\gamma\beta}.$$

On account of the symmetry of $\{\beta\gamma, \alpha\}$ with respect to the indices β and γ, the third term on the right-hand side vanishes,
if $A^{\alpha\beta}$ is, as we will assume, an antisymmetrical tensor. The
second term allows itself to be transformed in accordance
with (29a). Thus we obtain

$$A^\alpha = \frac{1}{\sqrt{.-g}} \frac{\partial(\sqrt{-g}A^{\alpha\beta})}{\partial x_\beta} . \qquad . \quad . \quad (40)$$

This is the expression for the divergence of a contravariant
six-vector.

The Divergence of a Mixed Tensor of the Second Rank.—
Contracting (39) with respect to the indices α and σ, and
taking (29a) into consideration, we obtain

$$\sqrt{-g}A_\mu = \frac{\partial(\sqrt{-g}A_\mu^\sigma)}{\partial x_\sigma} - \{\sigma\mu, \tau\}\sqrt{-g}A_\tau^\sigma . \quad (41)$$

If we introduce the contravariant tensor $A^{\rho\sigma} = g^{\rho\tau}A_\tau^\sigma$ in the
last term, it assumes the form

$$- [\sigma\mu, \rho]\sqrt{-g}A^{\rho\sigma}.$$

If, further, the tensor $A^{\rho\sigma}$ is symmetrical, this reduces to

$$-\tfrac{1}{2}\sqrt{-g}\,\frac{\partial g_{\rho\sigma}}{\partial x_\mu}A^{\rho\sigma}.$$

Had we introduced, instead of $A^{\rho\sigma}$, the covariant tensor $A_{\rho\sigma} = g_{\rho\alpha}g_{\sigma\beta}A^{\alpha\beta}$, which is also symmetrical, the last term, by virtue of (31), would assume the form

$$\tfrac{1}{2}\sqrt{-g}\,\frac{\partial g^{\rho\sigma}}{\partial x_\mu}A_{\rho\sigma}.$$

In the case of symmetry in question, (41) may therefore be replaced by the two forms

$$\sqrt{-g}\,A_\mu = \frac{\partial(\sqrt{-g}\,A_\mu^\sigma)}{\partial x_\sigma} - \tfrac{1}{2}\frac{\partial g_{\rho\sigma}}{\partial x_\mu}\sqrt{-g}\,A^{\rho\sigma} \ . \quad (41a)$$

$$\sqrt{-g}\,A_\mu = \frac{\partial(\sqrt{-g}\,A_\mu^\sigma)}{\partial x_\sigma} + \tfrac{1}{2}\frac{\partial g^{\rho\sigma}}{\partial x_\mu}\sqrt{-g}\,A_{\rho\sigma} \ . \quad (41b)$$

which we have to employ later on.

§ 12. The Riemann-Christoffel Tensor

We now seek the tensor which can be obtained from the fundamental tensor *alone*, by differentiation. At first sight the solution seems obvious. We place the fundamental tensor of the $g_{\mu\nu}$ in (27) instead of any given tensor $A_{\mu\nu}$, and thus have a new tensor, namely, the extension of the fundamental tensor. But we easily convince ourselves that this extension vanishes identically. We reach our goal, however, in the following way. In (27) place

$$A_{\mu\nu} = \frac{\partial A_\mu}{\partial x_\nu} - \{\mu\nu,\ \rho\}A_\rho,$$

i.e. the extension of the four-vector A_μ. Then (with a somewhat different naming of the indices) we get the tensor of the third rank

$$A_{\mu\sigma\tau} = \frac{\partial^2 A_\mu}{\partial x_\sigma \partial x_\tau} - \{\mu\sigma,\ \rho\}\frac{\partial A_\rho}{\partial x_\tau} - \{\mu\tau,\ \rho\}\frac{\partial A_\rho}{\partial x_\sigma} - \{\sigma\tau,\ \rho\}\frac{\partial A_\mu}{\partial x_\rho}$$

$$+ \left[-\frac{\partial}{\partial x_\tau}\{\mu\sigma,\ \rho\} + \{\mu\tau,\ a\}\{a\sigma,\ \rho\} + \{\sigma\tau,\ a\}\{a\mu,\ \rho\} \right]A_\rho.$$

This expression suggests forming the tensor $A_{\mu\sigma\tau} - A_{\mu\tau\sigma}$. For, if we do so, the following terms of the expression for $A_{\mu\sigma\tau}$ cancel those of $A_{\mu\tau\sigma}$, the first, the fourth, and the member corresponding to the last term in square brackets; because all these are symmetrical in σ and τ. The same holds good for the sum of the second and third terms. Thus we obtain

$$A_{\mu\sigma\tau} - A_{\mu\tau\sigma} = B^{\rho}_{\mu\sigma\tau} A_{\rho} \qquad . \qquad . \qquad . \quad (42)$$

where

$$B^{\rho}_{\mu\sigma\tau} = -\frac{\partial}{\partial x_{\tau}}\{\mu\sigma, \rho\} + \frac{\partial}{\partial x_{\sigma}}\{\mu\tau, \rho\} - \{\mu\sigma, a\}\{a\tau, \rho\}$$

$$+ \{\mu\tau, a\}\{a\sigma, \rho\} \quad (43)$$

The essential feature of the result is that on the right side of (42) the A_{ρ} occur alone, without their derivatives. From the tensor character of $A_{\mu\sigma\tau} - A_{\mu\tau\sigma}$ in conjunction with the fact that A_{ρ} is an arbitrary vector, it follows, by reason of § 7, that $B^{\rho}_{\mu\sigma\tau}$ is a tensor (the Riemann-Christoffel tensor).

The mathematical importance of this tensor is as follows : If the continuum is of such a nature that there is a co-ordinate system with reference to which the $g_{\mu\nu}$ are constants, then all the $B^{\rho}_{\mu\sigma\tau}$ vanish. If we choose any new system of co-ordinates in place of the original ones, the $g_{\mu\nu}$ referred thereto will not be constants, but in consequence of its tensor nature, the transformed components of $B^{\rho}_{\mu\sigma\tau}$ will still vanish in the new system. Thus the vanishing of the Riemann tensor is a necessary condition that, by an appropriate choice of the system of reference, the $g_{\mu\nu}$ may be constants. In our problem this corresponds to the case in which,* with a suitable choice of the system of reference, the special theory of relativity holds good for a *finite* region of the continuum.

Contracting (43) with respect to the indices τ and ρ we obtain the covariant tensor of second rank

[21]

---

* The mathematicians have proved that this is also a *sufficient* condition.

where

$$G_{\mu\nu} = B^{\rho}_{\mu\nu\rho} = R_{\mu\nu} + S_{\mu\nu}$$

$$R_{\mu\nu} = -\frac{\partial}{\partial x_a}\{\mu\nu, a\} + \{\mu a, \beta\}\{\nu\beta, a\}$$

$$S_{\mu\nu} = \frac{\partial^2 \log\sqrt{-g}}{\partial x_\mu \partial x_\nu} - \{\mu\nu, a\}\frac{\partial \log\sqrt{-g}}{\partial x_a}$$

(44)

Note on the Choice of Co-ordinates.—It has already been observed in § 8, in connexion with equation (18a), that the choice of co-ordinates may with advantage be made so that $\sqrt{-g} = 1$. A glance at the equations obtained in the last two sections shows that by such a choice the laws of formation of tensors undergo an important simplification. This applies particularly to $G_{\mu\nu}$, the tensor just developed, which plays a fundamental part in the theory to be set forth. For this specialization of the choice of co-ordinates brings about the vanishing of $S_{\mu\nu}$, so that the tensor $G_{\mu\nu}$ reduces to $R_{\mu\nu}$.

On this account I shall hereafter give all relations in the simplified form which this specialization of the choice of co-ordinates brings with it. It will then be an easy matter to revert to the *generally* covariant equations, if this seems desirable in a special case.

C. Theory of the Gravitational Field

§ 13. Equations of Motion of a Material Point in the Gravitational Field. Expression for the Field-components of Gravitation

A freely movable body not subjected to external forces moves, according to the special theory of relativity, in a straight line and uniformly. This is also the case, according to the general theory of relativity, for a part of four-dimensional space in which the system of co-ordinates K_0, may be, and is, so chosen that they have the special constant values given in (4).

If we consider precisely this movement from any chosen system of co-ordinates K_1, the body, observed from K_1, moves, according to the considerations in § 2, in a gravitational field. The law of motion with respect to K_1 results without difficulty from the following consideration. With respect to K_0 the law of motion corresponds to a four-dimensional straight line, i.e. to a geodetic line. Now since the geodetic line is defined independently of the system of reference, its equations will also be the equation of motion of the material point with respect to K_1. If we set

$$\Gamma^{\tau}_{\mu\nu} = - \{\mu\nu, \tau\} \qquad . \qquad . \qquad . \quad (45)$$

the equation of the motion of the point with respect to K_1, becomes

$$\frac{d^2x_{\tau}}{ds^2} = \Gamma_{\mu\nu}{}^{\tau} \frac{dx_{\mu}}{ds} \frac{dx_{\nu}}{ds} \qquad . \qquad . \qquad . \quad (46)$$

We now make the assumption, which readily suggests itself, that this covariant system of equations also defines the motion of the point in the gravitational field in the case when there is no system of reference K_0, with respect to which the special theory of relativity holds good in a finite region. We have all the more justification for this assumption as (46) contains only *first* derivatives of the $g_{\mu\nu}$, between which even in the special case of the existence of K_0, no relations subsist.*

If the $\Gamma^{\tau}_{\mu\nu}$ vanish, then the point moves uniformly in a straight line. These quantities therefore condition the deviation of the motion from uniformity. They are the components of the gravitational field.

§ 14. The Field Equations of Gravitation in the Absence of Matter

We make a distinction hereafter between " gravitational field " and " matter " in this way, that we denote everything but the gravitational field as " matter." Our use of the word therefore includes not only matter in the ordinary sense, but the electromagnetic field as well.

Our next task is to find the field equations of gravitation in the absence of matter. Here we again apply the method

* It is only between the second (and first) derivatives that, by § 12, the relations $B^{\rho}_{\mu\sigma\tau} = 0$ subsist.

employed in the preceding paragraph in formulating the equations of motion of the material point. A special case in which the required equations must in any case be satisfied is that of the special theory of relativity, in which the $g_{\mu\nu}$ have certain constant values. Let this be the case in a certain finite space in relation to a definite system of co-ordinates K_0. Relatively to this system all the components of the Riemann tensor $B_{\mu\sigma\tau}^{\rho}$, defined in (43), vanish. For the space under consideration they then vanish, also in any other system of co-ordinates.

Thus the required equations of the matter-free gravitational field must in any case be satisfied if all $B_{\mu\sigma\tau}^{\rho}$ vanish. But this condition goes too far. For it is clear that, e.g., the gravitational field generated by a material point in its environment certainly cannot be "transformed away" by any choice of the system of co-ordinates, i.e. it cannot be transformed to the case of constant $g_{\mu\nu}$.

This prompts us to require for the matter-free gravitational field that the symmetrical tensor $G_{\mu\nu}$, derived from the tensor $B_{\mu\nu\tau}^{\rho}$, shall vanish. Thus we obtain ten equations for the ten quantities $g_{\mu\nu}$, which are satisfied in the special case of the vanishing of all $B_{\mu\nu\tau}^{\rho}$. With the choice which we have made of a system of co-ordinates, and taking (44) into consideration, the equations for the matter-free field are

$$\left. \begin{aligned} \frac{\partial \Gamma_{\mu\nu}^{\alpha}}{\partial x_{\alpha}} + \Gamma_{\mu\beta}^{\alpha} \Gamma_{\nu\alpha}^{\beta} &= 0 \\ \sqrt{-g} &= 1 \end{aligned} \right\} \qquad . \qquad . \qquad . \quad (47)$$

It must be pointed out that there is only a minimum of arbitrariness in the choice of these equations. For besides $G_{\mu\nu}$ there is no tensor of second rank which is formed from the $g_{\mu\nu}$ and its derivatives, contains no derivations higher than second, and is linear in these derivatives.[*]

These equations, which proceed, by the method of pure

---

[*] Properly speaking, this can be affirmed only of the tensor

$$G_{\mu\nu} + \lambda g_{\mu\nu} g^{\alpha\beta} G_{\alpha\beta},$$

where λ is a constant. If, however, we set this tensor $= 0$, we come back again to the equations $G_{\mu\nu} = 0$.

mathematics, from the requirement of the general theory of relativity, give us, in combination with the equations of motion (46), to a first approximation Newton's law of attraction, and to a second approximation the explanation of the motion of the perihelion of the planet Mercury discovered by Leverrier (as it remains after corrections for perturbation have been made). These facts must, in my opinion, be taken as a convincing proof of the correctness of the theory.

§ 15. The Hamiltonian Function for the Gravitational Field.   Laws of Momentum and Energy

To show that the field equations correspond to the laws of momentum and energy, it is most convenient to write them in the following Hamiltonian form :—

$$
\left.
\begin{aligned}
&\delta \int \mathrm{H} d\tau = 0 \\
&\mathrm{H} = g^{\mu\nu}\, \Gamma^{\alpha}_{\mu\beta}\, \Gamma^{\beta}_{\nu\alpha} \\
&\sqrt{-g} = 1
\end{aligned}
\right\}
\qquad . \qquad . \qquad . \quad (47\mathrm{a})
$$

where, on the boundary of the finite four-dimensional region of integration which we have in view, the variations vanish.

We first have to show that the form (47a) is equivalent to the equations (47). For this purpose we regard H as a function of the $g^{\mu\nu}$ and the $g^{\mu\nu}_{\sigma}\ (= \partial g^{\mu\nu}/\partial x_{\sigma})$.

Then in the first place

$$
\begin{aligned}
\delta\mathrm{H} &= \Gamma^{\alpha}_{\mu\beta}\Gamma^{\beta}_{\nu\alpha}\,\delta g^{\mu\nu} + 2g^{\mu\nu}\Gamma^{\alpha}_{\mu\beta}\,\delta\Gamma^{\beta}_{\nu\alpha} \\
&= -\,\Gamma^{\alpha}_{\mu\beta}\Gamma^{\beta}_{\nu\alpha}\,\delta g^{\mu\nu} + 2\Gamma^{\alpha}_{\mu\beta}\,\delta(g^{\mu\nu}\Gamma^{\beta}_{\nu\alpha}).
\end{aligned}
$$

But

$$
\delta\!\left(g^{\mu\nu}\Gamma^{\beta}_{\nu\alpha}\right) = -\tfrac{1}{2}\delta\!\left[\, g^{\mu\nu}g^{\beta\lambda}\!\left(\frac{\partial g_{\nu\lambda}}{\partial x_{\alpha}} + \frac{\partial g_{\alpha\lambda}}{\partial x_{\nu}} - \frac{\partial g_{\alpha\nu}}{\partial x_{\lambda}}\right)\right].
$$

[22]

The terms arising from the last two terms in round brackets are of different sign, and result from each other (since the denomination of the summation indices is immaterial) through interchange of the indices μ and β. They cancel each other in the expression for δH, because they are multiplied by the quantity $\Gamma^{\alpha}_{\mu\beta}$, which is symmetrical with respect to the indices μ and β. Thus there remains only the first term in round brackets to be considered, so that, taking (31) into account, we obtain

$$\delta H = - \Gamma^{\alpha}_{\mu\beta}\Gamma^{\beta}_{\nu\alpha}\delta g^{\mu\nu} + \Gamma^{\alpha}_{\mu\beta}\delta g^{\mu\beta}_{\alpha}.$$

Thus

[23]

$$\left.\begin{aligned} \frac{\partial H}{\partial g^{\mu\nu}} &= - \Gamma^{\alpha}_{\mu\beta}\Gamma^{\beta}_{\nu\alpha} \\ \frac{\partial H}{\partial g^{\mu\nu}_{\sigma}} &= \Gamma^{\sigma}_{\mu\nu} \end{aligned}\right\} \qquad . \quad . \quad . \quad (48)$$

Carrying out the variation in (47a), we get in the first place

$$\frac{\partial}{\partial x_{\alpha}}\left(\frac{\partial H}{\partial g^{\mu\nu}_{\alpha}}\right) - \frac{\partial H}{\partial g^{\mu\nu}} = 0, \quad . \quad . \quad . \quad (47b)$$

which, on account of (48), agrees with (47), as was to be proved.

If we multiply (47b) by $g^{\mu\nu}_{\sigma}$, then because

$$\frac{\partial g^{\mu\nu}_{\sigma}}{\partial x_{\alpha}} = \frac{\partial g^{\mu\nu}_{\alpha}}{\partial x_{\sigma}}$$

and, consequently,

$$g^{\mu\nu}_{\sigma}\frac{\partial}{\partial x_{\alpha}}\left(\frac{\partial H}{\partial g^{\mu\nu}_{\alpha}}\right) = \frac{\partial}{\partial x_{\alpha}}\left(g^{\mu\nu}_{\sigma}\frac{\partial H}{\partial g^{\mu\nu}_{\alpha}}\right) - \frac{\partial H}{\partial g^{\mu\nu}_{\alpha}}\frac{\partial g^{\mu\nu}_{\alpha}}{\partial x_{\sigma}},$$

we obtain the equation

$$\frac{\partial}{\partial x_{\alpha}}\left(g^{\mu\nu}_{\sigma}\frac{\partial H}{\partial g^{\mu\nu}_{\alpha}}\right) - \frac{\partial H}{\partial x_{\sigma}} = 0$$

or *

$$\left.\begin{aligned} \frac{\partial t^{\alpha}_{\sigma}}{\partial x_{\alpha}} &= 0 \\ - 2\kappa t^{\alpha}_{\sigma} &= g^{\mu\nu}_{\sigma}\frac{\partial H}{\partial g^{\mu\nu}_{\alpha}} - \delta^{\alpha}_{\sigma}H \end{aligned}\right\} \quad . \quad . \quad . \quad (49)$$

where, on account of (48), the second equation of (47), and (34)

[24]

$$\kappa t^{\alpha}_{\sigma} = \tfrac{1}{2}\delta^{\alpha}_{\sigma}g^{\mu\nu}\Gamma^{\lambda}_{\mu\beta}\Gamma^{\beta}_{\nu\lambda} - g^{\mu\nu}\Gamma^{\alpha}_{\mu\beta}\Gamma^{\beta}_{\nu\sigma} \quad . \quad . \quad (50)$$

* The reason for the introduction of the factor $- 2\kappa$ will be apparent later.

It is to be noticed that t_σ^a is not a tensor; on the other hand (49) applies to all systems of co-ordinates for which $\sqrt{-g} = 1$. This equation expresses the law of conservation of momentum and of energy for the gravitational field. Actually the integration of this equation over a three-dimensional volume V yields the four equations

$$\frac{d}{dx_4}\int t_\sigma^4 dV = \int (l t_\sigma^1 + m t_\sigma^2 + n t_\sigma^3) dS . \qquad . \quad (49a)$$

where l, m, n denote the direction-cosines of direction of the inward drawn normal at the element dS of the bounding surface (in the sense of Euclidean geometry). We recognize in this the expression of the laws of conservation in their usual form. The quantities t_σ^a we call the " energy components " of the gravitational field.

I will now give equations (47) in a third form, which is particularly useful for a vivid grasp of our subject. By multiplication of the field equations (47) by $g^{\nu\sigma}$ these are obtained in the " mixed " form. Note that

$$g^{\nu\sigma}\frac{\partial \Gamma_{\mu\nu}^a}{\partial x_a} = \frac{\partial}{\partial x_a}\left(g^{\nu\sigma}\Gamma_{\mu\nu}^a\right) - \frac{\partial g^{\nu\sigma}}{\partial x_a}\Gamma_{\mu\nu}^a,$$

which quantity, by reason of (34), is equal to

$$\frac{\partial}{\partial x_a}\left(g^{\nu\sigma}\Gamma_{\mu\nu}^a\right) - g^{\nu\beta}\Gamma_{\alpha\beta}^\sigma\Gamma_{\mu\nu}^a - g^{\sigma\beta}\Gamma_{\beta\alpha}^\nu\Gamma_{\mu\nu}^a,$$

or (with different symbols for the summation indices)

$$\frac{\partial}{\partial x_a}\left(g^{\sigma\beta}\Gamma_{\mu\beta}^a\right) - g^{\gamma\delta}\Gamma_{\gamma\beta}^\sigma\Gamma_{\delta\mu}^\beta - g^{\nu\sigma}\Gamma_{\mu\beta}^a\Gamma_{\nu\alpha}^\beta.$$

The third term of this expression cancels with the one arising from the second term of the field equations (47); using relation (50), the second term may be written

$$\kappa(t_\mu^\sigma - \tfrac{1}{2}\delta_\mu^\sigma t),$$

where $t = t_\alpha^a$. Thus instead of equations (47) we obtain

$$\left.\begin{array}{r}\dfrac{\partial}{\partial x_a}\left(g^{\sigma\beta}\Gamma_{\mu\beta}^a\right) = - \kappa(t_\mu^\sigma - \tfrac{1}{2}\delta_\mu^\sigma t) \\[2mm] \sqrt{-g} = 1 \end{array}\right\} \qquad . \quad . \quad (51)$$

§ 16. The General Form of the Field Equations of Gravitation

The field equations for matter-free space formulated in § 15 are to be compared with the field equation

$$\nabla^2 \phi = 0$$

of Newton's theory. We require the equation corresponding to Poisson's equation

$$\nabla^2 \phi = 4\pi\kappa\rho,$$

where ρ denotes the density of matter.

The special theory of relativity has led to the conclusion that inert mass is nothing more or less than energy, which finds its complete mathematical expression in a symmetrical tensor of second rank, the energy-tensor. Thus in the general theory of relativity we must introduce a corresponding energy-tensor of matter T_σ^α, which, like the energy-components t_σ [equations (49) and (50)] of the gravitational field, will have mixed character, but will pertain to a symmetrical covariant tensor.*

The system of equation (51) shows how this energy-tensor (corresponding to the density ρ in Poisson's equation) is to be introduced into the field equations of gravitation. For if we consider a complete system (e.g. the solar system), the total mass of the system, and therefore its total gravitating action as well, will depend on the total energy of the system, and therefore on the ponderable energy together with the gravitational energy. This will allow itself to be expressed by introducing into (51), in place of the energy-components of the gravitational field alone, the sums $t_\mu^\sigma + T_\mu^\sigma$ of the energy-components of matter and of gravitational field. Thus instead of (51) we obtain the tensor equation

$$\left.\begin{array}{l} \dfrac{\partial}{\partial x_\alpha}(g^{\sigma\beta}T_{\mu\beta}^\alpha) = -\kappa[(t_\mu^\sigma + T_\mu^\sigma) - \tfrac{1}{2}\delta_\mu^\sigma(t + T)], \\[1ex] \sqrt{-g} = 1 \end{array}\right\} \quad . \quad (52)$$

where we have set $T = T_\mu^\mu$ (Laue's scalar). These are the

---

* $g_{\alpha\tau}T_\sigma^\alpha = T_{\sigma\tau}$ and $g^{\sigma\beta}T_\sigma^\alpha = T^{\alpha\beta}$ are to be symmetrical tensors.

required general field equations of gravitation in mixed form. Working back from these, we have in place of (47)

$$\frac{\partial}{\partial x_a}\Gamma^a_{\mu\nu} + \Gamma^a_{\mu\beta}\Gamma^\beta_{\nu a} = -\kappa(T_{\mu\nu} - \tfrac{1}{2}g_{\mu\nu}T),$$
$$\sqrt{-g} = 1 \qquad \qquad \qquad \qquad (53)$$

It must be admitted that this introduction of the energy-tensor of matter is not justified by the relativity postulate alone. For this reason we have here deduced it from the requirement that the energy of the gravitational field shall act gravitatively in the same way as any other kind of energy. But the strongest reason for the choice of these equations lies in their consequence, that the equations of conservation of momentum and energy, corresponding exactly to equations (49) and (49a), hold good for the components of the total energy. This will be shown in § 17.

§ 17. The Laws of Conservation in the General Case

Equation (52) may readily be transformed so that the second term on the right-hand side vanishes. Contract (52) with respect to the indices μ and σ, and after multiplying the resulting equation by $\tfrac{1}{2}\delta^\sigma_\mu$, subtract it from equation (52). This gives

$$\frac{\partial}{\partial x_a}(g^{\sigma\beta}\Gamma^a_{\mu\beta} - \tfrac{1}{2}\delta^\sigma_\mu g^{\lambda\beta}\Gamma^a_{\lambda\beta}) = -\kappa(t^\sigma_\mu + T^\sigma_\mu). \qquad (52a)$$

On this equation we perform the operation $\partial/\partial x_\sigma$. We have

$$\frac{\partial^2}{\partial x_a \partial x_\sigma}\left(g^\sigma \Gamma^{\ a}_{\ \beta\mu}\right) = -\tfrac{1}{2}\frac{\partial^2}{\partial x_a \partial x_\sigma}\left[g^{\sigma\beta}g^{a\lambda}\left(\frac{\partial g_{\mu\lambda}}{\partial x_\beta} + \frac{\partial g_{\beta\lambda}}{\partial x_\mu} - \frac{\partial g_{\mu\beta}}{\partial x_\lambda}\right)\right].$$

The first and third terms of the round brackets yield contributions which cancel one another, as may be seen by interchanging, in the contribution of the third term, the summation indices a and σ on the one hand, and β and λ on the other. The second term may be re-modelled by (31), so that we have

$$\frac{\partial^2}{\partial x_a \partial x_\sigma}\left(g^{\sigma\beta}\Gamma^a_{\mu\beta}\right) = \tfrac{1}{2}\frac{\partial^3 g^{a\beta}}{\partial x_a \partial x_\beta \partial x_\mu} \qquad . \qquad . \quad (54)$$

The second term on the left-hand side of (52a) yields in the first place

$$- \tfrac{1}{2}\frac{\partial^2}{\partial x_a \partial x_\mu}\Big(g^{\lambda\beta}\Gamma^\alpha_{\lambda\beta}\Big)$$

or

$$\tfrac{1}{2}\frac{\partial^2}{\partial x_a \partial x_\mu}\Big[g^{\lambda\beta}g^{a\delta}\Big(\frac{\partial g_{\delta\lambda}}{\partial x_\beta} + \frac{\partial g_{\delta\beta}}{\partial x_\lambda} - \frac{\partial g_{\lambda\beta}}{\partial x_\delta}\Big)\Big].$$

With the choice of co-ordinates which we have made, the term deriving from the last term in round brackets disappears by reason of (29). The other two may be combined, and together, by (31), they give

$$- \tfrac{1}{2}\frac{\partial^3 g^{\alpha\beta}}{\partial x_a \partial x_\beta \partial x_\mu},$$

so that in consideration of (54), we have the identity

$$\frac{\partial^2}{\partial x_a \partial x_\sigma}\Big(g^{\rho\beta}\Gamma_{\mu\beta} - \tfrac{1}{2}\delta^\sigma_\mu g^{\lambda\beta}\Gamma^\alpha_{\lambda\beta}\Big) \equiv 0 \quad . \quad . \quad (55)$$

From (55) and (52a), it follows that

$$\frac{\partial(t^\sigma_\mu + T^\sigma_\mu)}{\partial x_\sigma} = 0. \quad . \quad . \quad . \quad (56)$$

Thus it results from our field equations of gravitation that the laws of conservation of momentum and energy are satisfied. This may be seen most easily from the consideration which leads to equation (49a); except that here, instead of the energy components t^σ of the gravitational field, we have to introduce the totality of the energy components of matter and gravitational field.

§ 18. The Laws of Momentum and Energy for Matter, as a Consequence of the Field Equations

Multiplying (53) by $\partial g^{\mu\nu}/\partial x_\sigma$, we obtain, by the method adopted in § 15, in view of the vanishing of

$$g_{\mu\nu}\frac{\partial g^{\mu\nu}}{\partial x_\sigma},$$

the equation

$$\frac{\partial t^\alpha_\sigma}{\partial x_a} + \tfrac{1}{2}\frac{\partial g^{\mu\nu}}{\partial x_\sigma}T_{\mu\nu} = 0,$$

[26]

or, in view of (56),

$$\frac{\partial T^\alpha_\sigma}{\partial x_\alpha} + \tfrac{1}{2}\frac{\partial g^{\mu\nu}}{\partial x_\sigma}T_{\mu\nu} = 0 \quad . \qquad . \qquad . \quad (57)$$

Comparison with (41b) shows that with the choice of system of co-ordinates which we have made, this equation predicates nothing more or less than the vanishing of divergence of the material energy-tensor. Physically, the occurrence of the second term on the left-hand side shows that laws of conservation of momentum and energy do not apply in the strict sense for matter alone, or else that they apply only when the $g^{\mu\nu}$ are constant, i.e. when the field intensities of gravitation vanish. This second term is an expression for momentum, and for energy, as transferred per unit of volume and time from the gravitational field to matter. This is brought out still more clearly by re-writing (57) in the sense of (41) as

$$\frac{\partial T^\alpha_\sigma}{\partial x_\alpha} = -\; \Gamma^\beta_{\alpha\sigma}T^\alpha_\beta \quad . \qquad . \qquad . \quad (57a)$$

The right side expresses the energetic effect of the gravitational field on matter.

Thus the field equations of gravitation contain four conditions which govern the course of material phenomena. They give the equations of material phenomena completely, if the latter is capable of being characterized by four differential equations independent of one another.*

D. Material Phenomena

The mathematical aids developed in part B enable us forthwith to generalize the physical laws of matter (hydrodynamics, Maxwell's electrodynamics), as they are formulated in the special theory of relativity, so that they will fit in with the general theory of relativity. When this is done, the general principle of relativity does not indeed afford us a further limitation of possibilities ; but it makes us acquainted with the influence of the gravitational field on all processes,

---

* On this question cf. H. Hilbert, Nachr. d. K. Gesellsch. d. Wiss. zu Göttingen, Math.-phys. Klasse, 1915, p. 3.

without our having to introduce any new hypothesis whatever.

Hence it comes about that it is not necessary to introduce definite assumptions as to the physical nature of matter (in the narrower sense). In particular it may remain an open question whether the theory of the electromagnetic field in conjunction with that of the gravitational field furnishes a sufficient basis for the theory of matter or not. The general postulate of relativity is unable on principle to tell us anything about this. It must remain to be seen, during the working out of the theory, whether electromagnetics and the doctrine of gravitation are able in collaboration to perform what the former by itself is unable to do.

§ 19. Euler's Equations for a Frictionless Adiabatic Fluid

Let p and ρ be two scalars, the former of which we call the " pressure," the latter the " density " of a fluid; and let an equation subsist between them. Let the contravariant symmetrical tensor

$$T^{\alpha\beta} = -g^{\alpha\beta}p + \rho\frac{dx_\alpha}{ds}\frac{dx_\beta}{ds} . \qquad . \qquad . \quad (58)$$

be the contravariant energy-tensor of the fluid. To it belongs the covariant tensor

$$T_{\mu\nu} = -g_{\mu\nu}p + g_{\mu\alpha}g_{\mu\beta}\frac{dx_\alpha}{ds}\frac{dx_\beta}{ds}\rho, \qquad . \qquad . \quad (58a)$$

as well as the mixed tensor *

$$T^\alpha_\sigma = -\delta^\alpha_\sigma p + g_{\sigma\beta}\frac{dx_\beta}{ds}\frac{dx_\alpha}{ds}\rho \qquad . \qquad . \quad (58b)$$

Inserting the right-hand side of (58b) in (57a), we obtain the Eulerian hydrodynamical equations of the general theory of relativity. They give, in theory, a complete solution of the problem of motion, since the four equations (57a), together

---

* For an observer using a system of reference in the sense of the special theory of relativity for an infinitely small region, and moving with it, the density of energy T^4_4 equals $\rho - p$. This gives the definition of ρ. Thus ρ is not constant for an incompressible fluid.

with the given equation between p and ρ, and the equation

$$g_{\alpha\beta}\frac{dx_\alpha}{ds}\frac{dx_\beta}{ds} = 1,$$

are sufficient, $g_{\alpha\beta}$ being given, to define the six unknowns

$$p,\ \rho,\ \frac{dx_1}{ds},\ \frac{dx_2}{ds},\ \frac{dx_3}{ds},\ \frac{dx_4}{ds}.$$

If the $g_{\mu\nu}$ are also unknown, the equations (53) are brought in. These are eleven equations for defining the ten functions $g_{\mu\nu}$, so that these functions appear over-defined. We must remember, however, that the equations (57a) are already contained in the equations (53), so that the latter represent only seven independent equations. There is good reason for this lack of definition, in that the wide freedom of the choice of co-ordinates causes the problem to remain mathematically undefined to such a degree that three of the functions of space may be chosen at will.*

§ 20. Maxwell's Electromagnetic Field Equations for Free Space

[28]

Let ϕ_ν be the components of a covariant vector—the electromagnetic potential vector. From them we form, in accordance with (36), the components $F_{\rho\sigma}$ of the covariant six-vector of the electromagnetic field, in accordance with the system of equations

$$F_{\rho\sigma} = \frac{\partial\phi_\rho}{\partial x_\sigma} - \frac{\partial\phi_\sigma}{\partial x_\rho} \qquad . \qquad . \qquad . \quad (59)$$

It follows from (59) that the system of equations

$$\frac{\partial F_{\rho\sigma}}{\partial x_\tau} + \frac{\partial F_{\sigma\tau}}{\partial x_\rho} + \frac{\partial F_{\tau\rho}}{\partial x_\sigma} = 0 \quad . \qquad . \qquad . \quad (60)$$

[29]

is satisfied, its left side being, by (37), an antisymmetrical tensor of the third rank. System (60) thus contains essentially four equations which are written out as follows:—

---

* On the abandonment of the choice of co-ordinates with $g = -1$, there remain *four* functions of space with liberty of choice, corresponding to the four arbitrary functions at our disposal in the choice of co-ordinates.

$$\left.\begin{array}{l} \dfrac{\partial F_{23}}{\partial x_4} + \dfrac{\partial F_{34}}{\partial x_2} + \dfrac{\partial F_{42}}{\partial x_3} = 0 \\[2mm] \dfrac{\partial F_{34}}{\partial x_1} + \dfrac{\partial F_{41}}{\partial x_3} + \dfrac{\partial F_{13}}{\partial x_4} = 0 \\[2mm] \dfrac{\partial F_{41}}{\partial x_2} + \dfrac{\partial F_{12}}{\partial x_4} + \dfrac{\partial F_{24}}{\partial x_1} = 0 \\[2mm] \dfrac{\partial F_{12}}{\partial x_3} + \dfrac{\partial F_{23}}{\partial x_1} + \dfrac{\partial F_{31}}{\partial x_2} = 0 \end{array}\right\} \qquad . \qquad . \quad (60a)$$

This system corresponds to the second of Maxwell's systems of equations. We recognize this at once by setting

$$\left.\begin{array}{ll} F_{23} = H_x, & F_{14} = E_x \\ F_{31} = H_y, & F_{24} = E_y \\ F_{12} = H_z, & F_{34} = E_z \end{array}\right\} \qquad . \qquad . \quad (61)$$

Then in place of (60a) we may set, in the usual notation of three-dimensional vector analysis,

$$\left.\begin{array}{l} -\dfrac{\partial H}{\partial t} = \operatorname{curl} E \\[2mm] \operatorname{div} H = 0 \end{array}\right\} \qquad . \qquad . \quad (60b)$$

We obtain Maxwell's first system by generalizing the form given by Minkowski. We introduce the contravariant six-vector associated with $F^{\alpha\beta}$

$$F^{\mu\nu} = g^{\mu\alpha} g^{\nu\beta} F_{\alpha\beta} \qquad . \qquad . \quad (62)$$

and also the contravariant vector J^μ of the density of the electric current. Then, taking (40) into consideration, the following equations will be invariant for any substitution whose invariant is unity (in agreement with the chosen co-ordinates) :—

$$\dfrac{\partial}{\partial x_\nu} F^{\mu\nu} = J^\mu \qquad . \qquad . \qquad . \quad (63)$$

Let

$$\left.\begin{array}{ll} F^{23} = H'_x, & F^{14} = -E'_x \\ F^{31} = H'_y, & F^{24} = -E'_y \\ F^{12} = H'_z, & F^{34} = -E'_z \end{array}\right\} \qquad . \qquad . \quad (64)$$

which quantities are equal to the quantities $H_x \dots E_z$ in the special case of the restricted theory of relativity ; and in addition

$$J^1 = j_x, \ J^2 = j_y, \ J^3 = j_z, \ J^4 = \rho,$$

we obtain in place of (63)

$$\left.\begin{array}{c} \dfrac{\partial E'}{\partial t} + j = \text{curl } H' \\[2mm] \text{div } E' = \rho \end{array}\right\} \qquad . \qquad . \qquad . \quad (63a)$$

The equations (60), (62), and (63) thus form the generalization of Maxwell's field equations for free space, with the convention which we have established with respect to the choice of co-ordinates.

The Energy-components of the Electromagnetic Field.— We form the inner product

$$\kappa_\sigma = F_{\sigma\mu}J^\mu \qquad . \qquad . \qquad . \qquad . \quad (65)$$

By (61) its components, written in the three-dimensional manner, are

$$\left.\begin{array}{c} \kappa_1 = \rho E_x + [j \, . \, H]^x \\[1mm] \cdot \qquad \cdot \qquad \cdot \qquad \cdot \\[1mm] \cdot \qquad \cdot \qquad \cdot \\[1mm] \kappa_4 = -\, (j E) \end{array}\right\} \qquad . \qquad . \qquad . \quad (65a)$$

κ_σ is a covariant vector the components of which are equal to the negative momentum, or, respectively, the energy, which is transferred from the electric masses to the electromagnetic field per unit of time and volume. If the electric masses are free, that is, under the sole influence of the electromagnetic field, the covariant vector κ_σ will vanish.

To obtain the energy-components T_σ^ν of the electromagnetic field, we need only give to equation $\kappa_\sigma = 0$ the form of equation (57). From (63) and (65) we have in the first place

$$\kappa_\sigma = F_{\sigma\mu}\frac{\partial F^{\mu\nu}}{\partial x_\nu} = \frac{\partial}{\partial x_\nu}(F_{\sigma\mu}F^{\mu\nu}) - F^{\mu\nu}\frac{\partial F_{\sigma\mu}}{\partial x_\nu}.$$

The second term of the right-hand side, by reason of (60), permits the transformation

$$F^{\mu\nu}\frac{\partial F_{\sigma\mu}}{\partial x_\nu} = -\tfrac{1}{2}F^{\mu\nu}\frac{\partial F_{\mu\nu}}{\partial x_\sigma} = -\tfrac{1}{2}g^{\mu\alpha}g^{\nu\beta}F_{\alpha\beta}\frac{\partial F_{\mu\nu}}{\partial x_\sigma},$$

which latter expression may, for reasons of symmetry, also be written in the form

$$- \tfrac{1}{4}\left[g^{\mu a}g^{\nu\beta}F_{a\beta}\frac{\partial F_{\mu\nu}}{\partial x_\sigma} + g^{\mu a}g^{\nu\beta}\frac{\partial F_{a\beta}}{\partial x_\sigma}F_{\mu\nu} \right].$$

But for this we may set

$$- \tfrac{1}{4}\frac{\partial}{\partial x_\sigma}(g^{\mu a}g^{\nu\beta}F_{a\beta}F_{\mu\nu}) + \tfrac{1}{4}F_{a\beta}F_{\mu\nu}\frac{\partial}{\partial x_\sigma}(g^{\mu a}g^{\nu\beta}).$$

The first of these terms is written more briefly

$$- \tfrac{1}{4}\frac{\partial}{\partial x_\sigma}(F^{\mu\nu}F_{\mu\nu}) ;$$

the second, after the differentiation is carried out, and after some reduction, results in

$$- \tfrac{1}{2}F^{\mu\tau}F_{\mu\nu}g^{\nu\rho}\frac{\partial g_{\sigma\tau}}{\partial x_\sigma}.$$

[31]

Taking all three terms together we obtain the relation

$$\kappa_\sigma = \frac{\partial T^\nu_\sigma}{\partial x_\nu} - \tfrac{1}{2}g^{\tau\mu}\frac{\partial g_{\mu\nu}}{\partial x_\sigma}T^\nu_\tau \ . \qquad . \qquad . \ (66)$$

where

$$T^\nu_\sigma = - F_{\sigma a}F^{\nu a} + \tfrac{1}{4}\delta^\nu_\sigma F_{a\beta}F^{a\beta}.$$

Equation (66), if κ_σ vanishes, is, on account of (30), equivalent to (57) or (57a) respectively. Therefore the T^ν_σ are the energy-components of the electromagnetic field. With the help of (61) and (64), it is easy to show that these energy-components of the electromagnetic field in the case of the special theory of relativity give the well-known Maxwell-Poynting expressions.

We have now deduced the general laws which are satisfied by the gravitational field and matter, by consistently using a system of co-ordinates for which $\sqrt{-g} = 1$. We have thereby achieved a considerable simplification of formulæ and calculations, without failing to comply with the requirement of general covariance; for we have drawn our equations from generally covariant equations by specializing the system of co-ordinates.

Still the question is not without a formal interest, whether with a correspondingly generalized definition of the energy-components of gravitational field and matter, even without specializing the system of co-ordinates, it is possible to formulate laws of conservation in the form of equation (56), and field equations of gravitation of the same nature as (52) or (52a), in such a manner that on the left we have a divergence (in the ordinary sense), and on the right the sum of the energy-components of matter and gravitation. I have found that in both cases this is actually so. But I do not think that the communication of my somewhat extensive reflexions on this subject would be worth while, because after all they do not give us anything that is materially new.

E

§ 21. Newton's Theory as a First Approximation

[32]

As has already been mentioned more than once, the special theory of relativity as a special case of the general theory is characterized by the $g_{\mu\nu}$ having the constant values (4). From what has already been said, this means complete neglect of the effects of gravitation. We arrive at a closer approximation to reality by considering the case where the $g_{\mu\nu}$ differ from the values of (4) by quantities which are small compared with 1, and neglecting small quantities of second and higher order. (First point of view of approximation.)

It is further to be assumed that in the space-time territory under consideration the $g_{\mu\nu}$ at spatial infinity, with a suitable choice of co-ordinates, tend toward the values (4) ; i.e. we are considering gravitational fields which may be regarded as generated exclusively by matter in the finite region.

It might be thought that these approximations must lead us to Newton's theory. But to that end we still need to approximate the fundamental equations from a second point of view. We give our attention to the motion of a material point in accordance with the equations (16). In the case of the special theory of relativity the components

$$\frac{dx_1}{ds}, \frac{dx_2}{ds}, \frac{dx_3}{ds}$$

may take on any values.   This signifies that any velocity

$$v = \sqrt{\left(\frac{dx_1}{dx_4}\right)^2 + \left(\frac{dx_2}{dx_4}\right)^2 + \left(\frac{dx_3}{dx_4}\right)^2}$$

may occur, which is less than the velocity of light *in vacuo*. If we restrict ourselves to the case which almost exclusively offers itself to our experience, of v being small as compared with the velocity of light, this denotes that the components

$$\frac{dx_1}{ds}, \frac{dx_2}{ds}, \frac{dx_3}{ds}$$

are to be treated as small quantities, while dx_4/ds, to the second order of small quantities, is equal to one. (Second point of view of approximation.)

Now we remark that from the first point of view of approximation the magnitudes $\Gamma^\tau_{\mu\nu}$ are all small magnitudes of at least the first order. A glance at (46) thus shows that in this equation, from the second point of view of approximation, we have to consider only terms for which $\mu = \nu = 4$. Restricting ourselves to terms of lowest order we first obtain in place of (46) the equations

$$\frac{d^2x_\tau}{dt^2} = \Gamma^\tau_{44}$$

where we have set $ds = dx_4 = dt$; or with restriction to terms which from the first point of view of approximation are of first order :—

$$\frac{d^2x_\tau}{dt^2} = [44, \tau] \quad (\tau = 1, 2, 3)$$

$$\frac{d^2x_4}{dt^2} = -[44, 4].$$

If in addition we suppose the gravitational field to be a quasi-static field, by confining ourselves to the case where the motion of the matter generating the gravitational field is but slow (in comparison with the velocity of the propagation of light), we may neglect on the right-hand side differentiations with respect to the time in comparison with those with respect to the space co-ordinates, so that we have

$$\frac{d^2 x_\tau}{dt^2} = -\tfrac{1}{2}\frac{\partial g_{44}}{\partial x_\tau} \quad (\tau = 1, 2, 3) \quad . \quad . \quad (67)$$

This is the equation of motion of the material point according to Newton's theory, in which $\tfrac{1}{2}g_{44}$ plays the part of the gravitational potential. What is remarkable in this result is that the component g_{44} of the fundamental tensor alone defines, to a first approximation, the motion of the material point.

We now turn to the field equations (53). Here we have to take into consideration that the energy-tensor of "matter" is almost exclusively defined by the density of matter in the narrower sense, i.e. by the second term of the right-hand side of (58) [or, respectively, (58a) or (58b)]. If we form the approximation in question, all the components vanish with the one exception of $T_{44} = \rho = T$. On the left-hand side of (53) the second term is a small quantity of second order; the first yields, to the approximation in question,

$$\frac{\partial}{\partial x_1}[\mu\nu, 1] + \frac{\partial}{\partial x_2}[\mu\nu, 2] + \frac{\partial}{\partial x_3}[\mu\nu, 3] - \frac{\partial}{\partial x_4}[\mu\nu, 4].$$

For $\mu = \nu = 4$, this gives, with the omission of terms differentiated with respect to time,

$$-\tfrac{1}{2}\left(\frac{\partial^2 g_{44}}{\partial x_1^2} + \frac{\partial^2 g_{44}}{\partial x_2^2} + \frac{\partial^2 g_{44}}{\partial x_3^2}\right) = -\tfrac{1}{2}\nabla^2 g_{44}.$$

The last of equations (53) thus yields

$$\nabla^2 g_{44} = \kappa\rho \quad . \quad . \quad . \quad . \quad (68)$$

The equations (67) and (68) together are equivalent to Newton's law of gravitation.

By (67) and (68) the expression for the gravitational potential becomes

$$-\frac{\kappa}{8\pi}\int\frac{\rho d\tau}{r} \quad . \quad . \quad . \quad . \quad (68a)$$

while Newton's theory, with the unit of time which we have chosen, gives

$$-\frac{K}{c^2}\int\frac{\rho d\tau}{r}$$

in which K denotes the constant $6·7 \times 10^{-8}$, usually called the constant of gravitation. By comparison we obtain

$$\kappa = \frac{8\pi K}{c^2} = 1·87 \times 10^{-27} \qquad . \qquad . \quad (69)$$

§ 22. Behaviour of Rods and Clocks in the Static Gravitational Field.   Bending of Light-rays.   Motion of the Perihelion of a Planetary Orbit

To arrive at Newton's theory as a first approximation we had to calculate only one component, g_{44}, of the ten $g_{\mu\nu}$ of the gravitational field, since this component alone enters into the first approximation, (67), of the equation for the motion of the material point in the gravitational field. From this, however, it is already apparent that other components of the $g_{\mu\nu}$ must differ from the values given in (4) by small quantities of the first order. This is required by the condition $g = -1$.

For a field-producing point mass at the origin of co-ordinates, we obtain, to the first approximation, the radially symmetrical solution

$$\left. \begin{aligned} g_{\rho\sigma} &= -\delta_{\rho\sigma} - a\frac{x_\rho x_\sigma}{r^3} \ (\rho,\ \sigma = 1, 2, 3) \\ g_{\rho 4} &= g_{4\rho} = 0 \qquad (\rho = 1, 2, 3) \\ g_{44} &= 1 - \frac{a}{r} \end{aligned} \right\} \qquad . \quad (70)$$

[33]

where $\delta_{\rho\sigma}$ is 1 or 0, respectively, accordingly as $\rho = \sigma$ or $\rho \neq \sigma$, and r is the quantity $+\sqrt{x_1^2 + x_2^2 + x_3^2}$. On account of (68a)

$$a = \frac{\kappa M}{4\pi}, \qquad . \qquad . \qquad . \quad (70a)$$

if M denotes the field-producing mass. It is easy to verify that the field equations (outside the mass) are satisfied to the first order of small quantities.

We now examine the influence exerted by the field of the mass M upon the metrical properties of space. The relation

$$ds^2 = g_{\mu\nu}dx_\mu dx_\nu.$$

always holds between the "locally" (§ 4) measured lengths and times ds on the one hand, and the differences of co-ordinates dx_ν on the other hand.

For a unit-measure of length laid " parallel " to the axis of x, for example, we should have to set $ds^2 = -1$; $dx_2 = dx_3 = dx_4 = 0$.   Therefore $-1 = g_{11}dx_1^2$. If, in addition, the unit-measure lies on the axis of x, the first of equations (70) gives

$$g_{11} = -\left(1 + \frac{a}{r}\right).$$

From these two relations it follows that, correct to a first order of small quantities,

$$dx = 1 - \frac{a}{2r} \qquad . \qquad . \qquad . \qquad (71)$$

The unit measuring-rod thus appears a little shortened in relation to the system of co-ordinates by the presence of the gravitational field, if the rod is laid along a radius.

In an analogous manner we obtain the length of co-ordinates in tangential direction if, for example, we set

$$ds^2 = -1; \quad dx_1 = dx_3 = dx_4 = 0; \quad x_1 = r, \ x_2 = x_3 = 0.$$

The result is

$$-1 = g_{22}dx_2^2 = -dx_2^2 \qquad . \qquad . \qquad (71a)$$

With the tangential position, therefore, the gravitational field of the point of mass has no influence on the length of a rod.

Thus Euclidean geometry does not hold even to a first approximation in the gravitational field, if we wish to take one and the same rod, independently of its place and orientation, as a realization of the same interval; although, to be sure, a glance at (70a) and (69) shows that the deviations to be expected are much too slight to be noticeable in measurements of the earth's surface.

Further, let us examine the rate of a unit clock, which is arranged to be at rest in a static gravitational field.   Here we have for a clock period $ds = 1$; $dx_1 = dx_2 = dx_3 = 0$ Therefore

$$1 = g_{44}dx_4^2;$$

$$dx_4 = \frac{1}{\sqrt{g_{44}}} = \frac{1}{\sqrt{(1 + (g_{44} - 1))}} = 1 - \tfrac{1}{2}(g_{44} - 1)$$

or

$$dx_4 = 1 + \frac{\kappa}{8\pi} \int \rho \frac{d\tau}{r} \qquad . \qquad . \qquad . \quad (72)$$

Thus the clock goes more slowly if set up in the neighbourhood of ponderable masses. From this it follows that the spectral lines of light reaching us from the surface of large stars must appear displaced towards the red end of the spectrum.*

We now examine the course of light-rays in the static gravitational field. By the special theory of relativity the velocity of light is given by the equation

$$- dx_1^2 - dx_2^2 - dx_3^2 + dx_4^2 = 0$$

and therefore by the general theory of relativity by the equation

$$ds^2 = g_{\mu\nu} dx_\mu dx_\nu = 0 \qquad . \qquad . \qquad . \quad (73)$$

If the direction, i.e. the ratio $dx_1 : dx_2 : dx_3$ is given, equation (73) gives the quantities

$$\frac{dx_1}{dx_4}, \frac{dx_2}{dx_4}, \frac{dx_3}{dx_4}$$

and accordingly the velocity

$$\sqrt{\left(\frac{dx_1}{dx_4}\right)^2 + \left(\frac{dx_2}{dx_4}\right)^2 + \left(\frac{dx_3}{dx_4}\right)^2} = \gamma$$

defined in the sense of Euclidean geometry. We easily recognize that the course of the light-rays must be bent with regard to the system of co-ordinates, if the $g_{\mu\nu}$ are not constant. If n is a direction perpendicular to the propagation of light, the Huyghens principle shows that the light-ray, envisaged in the plane (γ, n), has the curvature $- \partial\gamma/\partial n$.

We examine the curvature undergone by a ray of light passing by a mass M at the distance $\triangle$. If we choose the system of co-ordinates in agreement with the accompanying diagram, the total bending of the ray (calculated positively if

[35]

[34]

* According to E. Freundlich, spectroscopical observations on fixed stars of certain types indicate the existence of an effect of this kind, but a crucial test of this consequence has not yet been made.

concave towards the origin) is given in sufficient approximation by

$$B = \int_{-\infty}^{+\infty} \frac{\partial \gamma}{\partial x_1} dx_2,$$

while (73) and (70) give

$$\gamma = \sqrt{\left(-\frac{g_{44}}{g_{22}}\right)} = 1 - \frac{a}{2r}\left(1 + \frac{x_2^2}{r^2}\right).$$

Carrying out the calculation, this gives

$$B = \frac{2a}{\Delta} = \frac{\kappa M}{2\pi\Delta}. \qquad . \qquad . \qquad . \quad (74) \qquad [36]$$

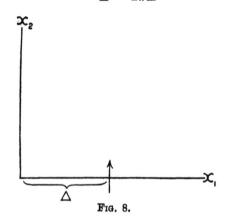

Fig. 8.

According to this, a ray of light going past the sun undergoes a deflexion of 1·7''; and a ray going past the planet Jupiter a deflexion of about ·02''.

If we calculate the gravitational field to a higher degree of approximation, and likewise with corresponding accuracy the orbital motion of a material point of relatively infinitely small mass, we find a deviation of the following kind from the Kepler-Newton laws of planetary motion. The orbital ellipse of a planet undergoes a slow rotation, in the direction of motion, of amount

$$\epsilon = 24\pi^3 \frac{a^2}{T^2 c^2 (1 - e^2)} \qquad . \qquad . \qquad . \quad (75)$$

per revolution. In this formula a denotes the major semi-axis, c the velocity of light in the usual measurement, e the eccentricity, T the time of revolution in seconds.*

Calculation gives for the planet Mercury a rotation of the orbit of 43″ per century, corresponding exactly to astronomical observation (Leverrier); for the astronomers have discovered in the motion of the perihelion of this planet, after allowing for disturbances by other planets, an inexplicable remainder of this magnitude.

[37]

*For the calculation I refer to the original papers: A. Einstein, Sitzungsber. d. Preuss. Akad. d. Wiss., 1915, p. 831; K. Schwarzschild, *ibid.*, 1916, p. 189.

Doc. 31

Appendix. Formulation of the Theory on the Basis of a Variational Principle

Not translated for this volume.

Doc. 32

Session of the physical-mathematical class on June 22, 1916   [p. 688]

Approximative Integration of the Field Equations of Gravitation
by A. Einstein

For the treatment of the special (not basic) problems in gravitational theory one can be satisfied with a first approximation of the $g_{\mu\nu}$. The same reasons as in the special theory of relativity make it advantageous to use the imaginary time variable $x_4 = it$. By "first approximation" we mean that the quantities $\gamma_{\mu\nu}$, defined by the equation

$$g_{\mu\nu} = -\delta_{\mu\nu} + \gamma_{\mu\nu}, \tag{1}$$

are small compared to 1, such that their squares and products are negligible compared with first powers; furthermore, they have a tensorial character under linear, orthogonal transformations. In addition, $\delta_{\mu\nu} = 1$ or $\delta_{\mu\nu} = 0$ resp. depending upon $\mu = \nu$ or $\mu \neq \nu$.

We shall show that these $\gamma_{\mu\nu}$ can be calculated in a manner analogous to that of retarded potentials in electrodynamics. From this follows next that gravitational fields propagate at the speed of light. Subsequent to this general solution we shall investigate gravitational waves and how they originate. It turned out that my suggested choice of a system of reference with the condition $g = |g_{\mu\nu}| = -1$ is not advantageous for the calculation of fields in first approximation. A note in a letter   [1] from the astronomer DE SITTER alerted me to his finding that a choice of reference system, different from the one I had previously given,[1] leads to a simpler expression of the gravitational field of a mass point at rest. I therefore take the generally invariant field equations as a basis in what follows.

§1. Integration of the Approximated Equations of the Gravitational Field   [p. 689]

The field equations in their covariant form are   [3]

---

[1]*Sitzungsber.* 47 (1915), p. 833.   [2]

$$
\left.
\begin{aligned}
R_{\mu\nu} + S_{\mu\nu} &= -\kappa\left(T_{\mu\nu} + \frac{1}{2} g_{\mu\nu} T\right) \\
R_{\mu\nu} &= -\sum_{\alpha} \frac{\partial}{\partial x_\alpha}\begin{Bmatrix} \mu\nu \\ \alpha \end{Bmatrix} + \sum_{\alpha\beta} \begin{Bmatrix} \mu\alpha \\ \beta \end{Bmatrix}\begin{Bmatrix} \nu\beta \\ \alpha \end{Bmatrix} \\
S_{\mu\nu} &= \frac{\partial^2 \log \sqrt{g}}{\partial x_\mu \partial x_\nu} - \sum_{\alpha} \begin{Bmatrix} \mu\nu \\ \alpha \end{Bmatrix} \frac{\partial \log \sqrt{g}}{\partial x_\alpha}
\end{aligned}
\right\}. \tag{1}
$$

{1}

The braces denote here the well-known CHRISTOFFEL symbols (of the second kind), $T_{\mu\nu}$ the covariant energy tensor of matter, T its associated scalar. The equations (1) produce, in the approximation that interests us, and after expansion, directly the equations

$$
\sum_{\alpha} \frac{\partial^2 \gamma_{\mu\alpha}}{\partial x_\nu \partial x_\alpha} + \sum_{\alpha} \frac{\partial^2 \gamma_{\nu\alpha}}{\partial x_\mu \partial x_\alpha} - \sum_{\alpha} \frac{\partial^2 \gamma_{\mu\nu}}{\partial x_\alpha^2} - \frac{\partial^2}{\partial x_\mu \partial x_\nu}\left(\sum_{\alpha} \gamma_{\alpha\alpha}\right) = -2\kappa\left(T_{\mu\nu} - \frac{1}{2}\delta_{\mu\nu}\sum T_{\alpha\alpha}\right). \tag{2}
$$

The last term on the left-hand side originates from the quantity $S_{\mu\nu}$, which can be solved by

$$
\gamma_{\mu\nu} = \gamma'_{\mu\nu} + \psi \delta_{\mu\nu}, \tag{3}
$$

when the $\gamma'_{\mu\nu}$ obey the additional condition

$$
\sum_{\nu} \frac{\partial \gamma'_{\mu\nu}}{\partial x_\nu} = 0. \tag{4}
$$

Substituting (3) into (2) one gets for the left-hand side

[4] {2}

$$
-\sum_{\alpha} \frac{\partial^2 \gamma'_{\mu\nu}}{\partial x_\alpha^2} - \frac{\partial^2}{\partial x_\mu \partial x_\nu}\left(\sum_{\alpha} \gamma'_{\alpha\alpha}\right) + 2\frac{\partial^2 \psi}{\partial x_\mu \partial x_\nu} - \delta_{\mu\nu}\sum_{\alpha} \frac{\partial^2 \psi}{\partial x_\alpha^2} - 4\frac{\partial^2 \psi}{\partial x_\mu \partial x_\nu}.
$$

The contributions from the second, third, and fifth terms vanish if ψ is chosen from the equation

$$
\sum_{\alpha} \gamma'_{\alpha\alpha} + 2\psi = 0, \tag{5}
$$

which we will do. Considering this, one gets in place of (2)

[p. 690]

$$
\sum_{\alpha} \frac{\partial^2}{\partial x_\alpha^2}\left(\gamma'_{\mu\nu} - \frac{1}{2}\delta_{\mu\nu}\sum_{\alpha} \gamma'_{\alpha\alpha}\right) = 2\kappa\left(T_{\mu\nu} - \frac{1}{2}\delta_{\mu\nu}\sum_{\alpha} T_{\alpha\alpha}\right)
$$

or

$$\sum_\alpha \frac{\partial^2}{\partial x_\alpha^2} \gamma'_{\mu\nu} = 2\kappa T_{\mu\nu}. \tag{6}$$

It is to be noted that equation (6) is an agreement with equation (4), because it is easily shown that under the accuracy we want to achieve the energy-momentum theorem for matter is expressed by the equation

$$\sum_\nu \frac{\partial T_{\mu\nu}}{\partial x_\nu} = 0. \tag{7}$$

If (6) is subjected to the operation $\sum_\nu \dfrac{\partial}{\partial x_\nu}$, the left-hand side vanishes not only due

to (4) but—as it must—also the right-hand side of (6) due to (7). We note that the equations

$$\gamma_{\mu\nu} = \gamma'_{\mu\nu} - \frac{1}{2}\delta_{\mu\nu}\sum_\alpha \gamma'_{\alpha\alpha} \tag{8}$$

$$\gamma'_{\mu\nu} = \gamma_{\mu\nu} - \frac{1}{2}\delta_{\mu\nu}\sum_\alpha \gamma_{\alpha\alpha} \tag{8a}$$

hold because of (3) and (5). Our problem is solved, since the $\gamma'_{\mu\nu}$ can be calculated in the manner of retarded potentials. It is

$$\gamma'_{\nu\mu} = -\frac{\kappa}{2\pi}\int \frac{T_{\mu\nu}(x_0,y_0,z_0,t-r)}{r}dV_0. \tag{9}$$

x, y, z, t denote here the real-valued coordinates x_1, x_2, x_3, $\dfrac{x_4}{i}$ and, specifically, they

denote without index the coordinates of the "field point," with index "0" those of the element of integration. dV_0 is the three-dimensional volume element of the space of

integration, r is the spatial distance $\sqrt{(x-x_0)^2 + (y-y_0)^2 + (z-z_0.)^2}$.   [5] {3}

For what follows, we also need the energy components of the gravitational field.

The simplest way is to get them directly from equations (6). Multiplying by $\dfrac{\partial \gamma'_{\mu\nu}}{\partial x_\sigma}$ [6] {4}

and summing over μ and ν, one obtains on the left-hand side after customary rearrangement

$$\sum_\alpha \frac{\partial}{\partial x_\alpha}\left[\sum_{\mu\nu}\frac{\partial \gamma'_{\mu\nu}}{\partial x_\sigma}\frac{\partial \gamma'_{\mu\nu}}{\partial x_\alpha} - \frac{1}{2}\delta_{\sigma\alpha}\sum_{\mu\nu\beta}\left(\frac{\partial \gamma'_{\mu\nu}}{\partial x_\beta}\right)^2\right]. \tag{[7] {5}}$$

The quantity in brackets obviously expresses the energy components $t_{\sigma\alpha}$ up to a   [p. 691] proportionality factor; the latter is found easily from calculating the right-hand side. The energy-momentum theorem of matter, without neglecting any terms, is

$$\sum_\sigma \frac{\partial \sqrt{-g}\, T_\mu^\sigma}{\partial x_\sigma} + \frac{1}{2}\sum_{\rho\sigma} \frac{\partial g^{\rho\sigma}}{\partial x_\mu} \sqrt{-g}\, T_{\rho\sigma} = 0.$$

Under the desired order of approximation this can be replaced by

$$\sum_\sigma \frac{\partial T_{\mu\sigma}}{\partial x_\sigma} + \frac{1}{2}\sum_{\rho\sigma} \frac{\partial g_{\rho\sigma}}{\partial x_\mu} T_{\rho\sigma} = 0. \tag{7a}$$

This formulation is by one order more precise than equation (7). It follows that the right-hand side of (6) yields under the here-considered modification,

[8]
$$- 4\kappa \sum_\nu \frac{\partial T_{\mu\nu}}{\partial x_\nu}.$$

The conservation theorem, therefore, appears as

$$\sum_\nu \frac{\partial (T_{\mu\nu} + t_{\mu\nu})}{\partial x_\nu} = 0, \tag{10}$$

where

$$t_{\mu\nu} = \frac{1}{4\kappa}\left[\sum_{\alpha\beta} \frac{\partial \gamma'_{\alpha\beta}}{\partial x_\mu} \frac{\partial \gamma'_{\alpha\beta}}{\partial x_\nu} - \frac{1}{2}\delta_{\mu\nu}\sum_{\alpha\beta\tau} \left(\frac{\partial \gamma'_{\alpha\beta}}{\partial x_\tau}\right)^2\right] \tag{11}$$

are the energy components of the gravitational field.

As the simplest example of application, we calculate the gravitational field of a mass point of mass M, resting at the point of origin of the coordinates. The energy tensor of matter, neglecting surface forces, is

$$T_{\mu\nu} = \rho \frac{dx_\mu}{ds} \frac{dx_\nu}{ds}, \tag{12}$$

considering hereby that in the first approximation the covariant energy tensor can be replaced by the contravariant one. The scalar ρ is the (naturally measured) mass density. It follows from (9) and (12) that all $\gamma'_{\mu\nu}$ except γ'_{44} vanish, and the latter component is

[9] {6}
$$\gamma'_{44} = \frac{\kappa}{2\pi} \frac{M}{r}. \tag{13}$$

[p. 692] With the help of (8) and (1) one obtains for the $g_{\mu\nu}$ the values

$$\left.\begin{array}{cccc} -1-\dfrac{\kappa}{4\pi}\dfrac{M}{r} & 0 & 0 & 0 \\[2em] 0 & -1-\dfrac{\kappa}{4\pi}\dfrac{M}{r} & 0 & 0 \\[2em] 0 & 0 & -1-\dfrac{\kappa}{4\pi}\dfrac{M}{r} & 0 \\[2em] 0 & 0 & 0 & -1+\dfrac{\kappa}{4\pi}\dfrac{M}{r} \end{array}\right\} \tag{14}$$

Herr DE SITTER sent me these values by letter; they differ from those which I previously gave only in the choice of the system of reference. They led me to the simple approximative solution given above. However, one has to keep in mind that the choice of coordinates which has been made here has no equivalent in the general case, as the $\gamma_{\mu\nu}$ and the $\gamma'_{\mu\nu}$ have tensorial character only with respect to linear, orthogonal substitutions, but not under general substitutions.

§2. Plane Gravitational Waves

It follows from equations (6) and (9) that gravitational fields always propagate with velocity 1, i.e., with the velocity of light. Plane gravitational waves, traveling along the positive X-axis, can therefore be found by setting

$$\gamma'_{\mu\nu} = \alpha_{\mu\nu}f(x_1 + ix_4) = \alpha_{\mu\nu}f(x - t). \tag{15}$$

The $\alpha_{\mu\nu}$ are constants here; f is a function of the argument $x - t$. If the space under consideration is free of matter, i.e., if the $T_{\mu\nu}$ vanish, equations (6) are satisfied by (15). Equations (4) yield the following relations between the $\alpha_{\mu\nu}$:

$$\left.\begin{array}{l} \alpha_{11} + i\alpha_{14} = 0 \\ \alpha_{12} + i\alpha_{24} = 0 \\ \alpha_{13} + i\alpha_{34} = 0 \\ \alpha_{14} + i\alpha_{44} = 0 \end{array}\right\} \tag{16}$$

Consequently, only 6 of the 10 constants $\alpha_{\mu\nu}$ can be chosen freely. We can superpose the most general type of wave, among those considered, from the following 6 types

$$\left.\begin{array}{lll} \text{a)}\begin{array}{l}\alpha_{11} + i\alpha_{14} = 0 \\ \alpha_{14} + i\alpha_{44} = 0\end{array} & \begin{array}{l}\text{b) } \alpha_{12} + i\alpha_{24} = 0 \\ \text{c) } \alpha_{13} + i\alpha_{34} = 0\end{array} & \begin{array}{l}\text{d) } \alpha_{22} \neq 0 \\ \text{e) } \alpha_{23} \neq 0 \\ \text{f) } \alpha_{33} \neq 0\end{array} \end{array}\right\} \tag{17}$$

[p. 693]   These statements can be understood such that for the conditions of each type the not explicitly named $\alpha_{\mu\nu}$ vanish; e.g., for type a) only α_{11}, α_{14}, and α_{44} differ from zero, etc. Type a) has the symmetry properties of a longitudinal wave; type b) and c) are transversal waves, while types d), e), and f) correspond to a new type of symmetry. Types b) and c) differ from one another only in the orientations of their y- and z-axes; and the same is true for types d), e), f), so that there are essentially only three wave types proper.

We are primarily interested in the energy transport of these waves, and this is measured by the energy current $\hat{s}_x = \frac{1}{i} t_{41}$. One gets from (11) for the individual types

$$\text{a) } \frac{1}{i} t_{41} = \frac{f'^2}{4x}(\alpha_{11}^2 + \alpha_{14}^2 + \alpha_{44}^2) = 0$$

$$\text{b) } \frac{1}{i} t_{41} = \frac{f'^2}{2x}(\alpha_{12}^2 + \alpha_{24}^2) = 0$$

$$\text{c) } \frac{1}{i} t_{41} = \frac{f'^2}{2x}(\alpha_{13}^2 + \alpha_{34}^2) = 0$$

[10] {7}

$$\text{d) } \frac{1}{i} t_{41} = \frac{f'^2}{4x}\alpha_{22}^2 = \frac{1}{4x}\left(\frac{\partial \gamma'_{22}}{\partial t}\right)^2$$

$$\text{e) } \frac{1}{i} t_{41} = \frac{f'^2}{4x}\alpha_{23}^2 = \frac{1}{4x}\left(\frac{\partial \gamma'_{23}}{\partial t}\right)^2$$

$$\text{f) } \frac{1}{i} t_{41} = \frac{f'^2}{4x}\alpha_{33}^2 = \frac{1}{4x}\left(\frac{\partial \gamma'_{33}}{\partial t}\right)^2.$$

It follows, therefore, that only waves of the last-named type do transport energy; and the energy transport of any plane wave is given by

$$I_z = \frac{1}{i} t_{41} = \frac{1}{4x}\left[\left(\frac{\partial \gamma'_{22}}{\partial t}\right)^2 + 2\left(\frac{\partial \gamma'_{23}}{\partial t}\right)^2 + \left(\frac{\partial \gamma'_{33}}{\partial t}\right)^2\right]. \tag{18}$$

§3. Energy Loss of Material Systems by Emissions of Gravitational Waves

Let the system whose radiation we want to investigate be permanently in the neighborhood of the origin of the coordinates. We consider the gravitational field, generated by the system, only in a field point whose distance R from the origin of the coordinates is large compared to the dimensions of the system. Let the field point be on the positive x-axis, i.e., let there be

$$x_1 = R, \quad x_2 = x_3 = 0.$$

Then the question arises if there is an energy-transporting wave radiation at the [p. 694]
field point, and whether it is directed along the positive x-axis. The considerations in
§2 show that such radiation at the field point can only be contributed by the
components g'_{22}, g'_{23}, g'_{33}. We need to calculate only these. From (9) we find

$$\gamma'_{22} = -\frac{\kappa}{2\pi} \int \frac{T_{22}(x_0, y_0, z_0, t - r)}{r} dV_0.$$

If the system is less spread out and its energy components are not changing too
fast, one can replace, without noticeable error, the argument $t - r$ by $t - R$, where
the latter is constant during integration. If, furthermore, $\dfrac{1}{r}$ is replaced by $\dfrac{1}{R}$, one

obtains an approximative equation that suffices in most cases, viz.,

$$\gamma'_{22} = -\frac{\kappa}{2\pi R} \int T_{22} dV_0, \tag{19}$$

where the integration is to be carried out in the conventional way, i.e., under constant
time argument. By means of (7), this expression can be replaced by one which is
more convenient for the computation with material systems. From

$$\frac{\partial T_{21}}{\partial x_1} + \frac{\partial T_{22}}{\partial x_2} + \frac{\partial T_{23}}{\partial x_3} + \frac{\partial T_{24}}{\partial x_4} = 0$$

it follows after multiplication by x_2, integration over the entire system, and partial
integration of the second term, that

$$-\int T_{22} dV + \frac{\partial}{\partial x_4}\left(\int T_{24} x_2 dV\right) = 0. \tag{20}$$

Furthermore, from

$$\frac{\partial T_{41}}{\partial x_1} + \frac{\partial T_{42}}{\partial x_2} + \frac{\partial T_{43}}{\partial x_3} + \frac{\partial T_{44}}{\partial x_4} = 0,$$

and after multiplication by $\dfrac{x_2^2}{2}$ one gets in an analogous manner

$$-\int T_{24} x_2 dV + \frac{\partial}{\partial x_4}\left(\int T_{44} \frac{x_2^2}{2} dV\right) = 0. \tag{21}$$

It follows from (20) and (21) that

$$\int T_{22} dV = \frac{\partial^2}{\partial x_4^2} \left(\int T_{44} \frac{x_2^2}{2} dV \right),$$

[p. 695] or, after introducing real-valued coordinates, and allowing oneself, as an approximation, to equate the energy density $(-T_{44})$ with the mass density ρ of arbitrarily
moving masses:

$$\int T_{22} dV = \frac{1}{2} \frac{\partial^2}{\partial t^2} \left(\int \rho y^2 dV \right). \tag{22}$$

One also has

$$\gamma'_{22} = - \frac{\kappa}{4\pi R} \frac{\partial^2}{\partial t^2} \left(\int \rho y^2 dV \right). \tag{23}$$

In an analogous manner one calculates

$$\gamma'_{33} = - \frac{\kappa}{4\pi R} \frac{\partial^2}{\partial t^2} \left(\int \rho z^2 dV \right) \tag{23a}$$

[11] {8}

$$\gamma'_{23} = - \frac{\kappa}{4\pi R} \frac{\partial^2}{\partial t^2} \left(\int \rho yz \, dV \right). \tag{23b}$$

The integrals in (23), (23a), and (23b) are nothing more than time-variable
momenta of inertia, and shall be denoted by the abbreviations J_{22}, J_{33}, J_{23}. The
intensity $\hat{\mathbf{s}}_x$ of the energy radiation from (18) then yields

{9}

$$\hat{\mathbf{s}}_x = \frac{\kappa}{64\pi^2 R^2} \left[\left(\frac{\partial^2 J_{22}}{\partial t^3} \right)^2 + 2 \left(\frac{\partial^3 J_{23}}{\partial t^3} \right)^2 + \left(\frac{\partial^3 J_{33}}{\partial t^3} \right)^2 \right]. \tag{20}$$

From this follows, furthermore, that the mean value of energy radiation in all
directions is given by

$$\frac{\kappa}{64\pi^2 R^3} \cdot \frac{2}{3} \sum_{\alpha\beta} \left(\frac{\partial^3 J_{\alpha\beta}}{\partial t^3} \right)^2,$$

where the summation of the indices 1 to 3 is to be extended over all 9 combinations,
the reason being that this expression is on the one hand invariant under spatial
rotations of the coordinate system (as is easily seen from the three-dimensional
tensorial character of $J_{\alpha\beta}$) and, on the other hand, it coincides with (20) in case of
radial symmetry $(J_{11} = J_{22} = J_{33}; J_{23} = J_{31} = J_{12} = 0)$. Consequently, one obtains the
radiation A of the system per unit time by multiplying with $4\pi R^2$:

$$A = \frac{\kappa}{24\pi} \sum_{\alpha\beta} \left(\frac{\partial^3 J_{\alpha\beta}}{\partial t^3} \right)^2 . \qquad (21) \qquad \{10\}$$

This expression would get an additional factor $\frac{1}{c^4}$ if we would measure time in seconds and energy in Erg. Considering furthermore that $\kappa = 1.87 \cdot 10^{-27}$, it is obvious that A has, in all imaginable cases, a practically vanishing value.

Nevertheless, due to the inneratomic movement of electrons, atoms would have [p. 696] to radiate not only electromagnetic but also gravitational energy, if only in tiny amounts. As this is hardly true in nature, it appears that quantum theory would have to modify not only MAXWELLIAN electrodynamics, but also the new theory of gravitation.

Supplement. There is a simple way to clarify the strange result that gravitational waves (types *a, b, c*), which transport no energy, could exist. The reason is that they [12] are not "real" waves but rather "apparent" waves, initiated by the use of a system of reference whose origin of coordinates is subject to wavelike jitters. This is easily seen in the following manner. If one selects from the beginning a coordinate system in the usual manner such that $\sqrt{-g} = 1$, one gets instead of (2) as field equations in the {11} absence of matter

$$\sum_\alpha \frac{\partial^2 \gamma_{\mu\alpha}}{\partial x_\nu \partial x_\alpha} + \frac{\partial^2 \gamma_{\nu\alpha}}{\partial x_\mu \partial x_\alpha} - \frac{\partial^2 \gamma_{\mu\nu}}{\partial x_\alpha^2} = 0.$$

Substituting into this equation

$$\gamma_{\mu\nu} = \alpha_{\mu\nu} f(x_1 + ix_4),$$

one obtains 10 equations between the constants $\alpha_{\mu\nu}$ from which it can be seen that *only* α_{22}, α_{33}, and α_{23} can differ from zero (where $\alpha_{22} + \alpha_{33} = 0$). With this choice of system of reference only the wave types (*d, e, f*) exist, which do transport energy. The other types of waves are eliminated by this choice of coordinates; in this sense they are not "real" waves.

Even though it turned out to be advantageous for this investigation not to restrain the choice of coordinate system for the calculation of a first approximation, our last result shows, nevertheless, that the choice of coordinates under the restriction $\sqrt{-g} = 1$ has a deep-seated physical justification. [13]

Additional notes by translator

{1}   The third equation has been corrected here to include a minus sign in front of the summation symbol; "∂" in the first term has been corrected to "∂^2."

{2}   "$\partial\gamma'_{\mu\nu}$" in the first term has been corrected to "$\partial^2\gamma'_{\mu\nu}$."

{3}   "$z - z$" has been corrected to "$z - z_0$."

{4}   "$g_{\mu\nu}$" has been corrected to "$g'_{\mu\nu}$."

{5}   An implied summation over α has been added; "$d_{\sigma\alpha}$" has been corrected to "$\delta_{\sigma\alpha}$."

{6} $-\dfrac{k}{2\pi}\dfrac{M}{r}$ has been corrected to $\dfrac{k}{2\pi}\dfrac{M}{r}$.

{7}   In expressions b) and c) a factor of 2 has been added on the right-hand side; d), e), and f) have been given the 41-component of t.

{8}   "y^2" has been corrected to "yz."

{9}   This is the second time in this paper that "(20)" has been assigned to an expression.

{10} This is the second time in this paper that "(21)" has been assigned to an expression.

{11} "$\sqrt{g}$" has been corrected to "$\sqrt{-g}$."

Doc. 33

Einstein's Memorial Lecture on Karl Schwarzschild

Not translated for this volume.

Doc. 34

[p. 318]
Emission and Absorption of Radiation in Quantum Theory

by A. Einstein

(Received on July 17, 1916)

[1] Sixteen years ago, when PLANCK created quantum theory by deriving his radiation formula, he took the following approach. He calculated the mean energy $\overline{E}$ of a resonator as a function of temperature according to his newly found quantum-theoretic basic principles, and determined from this the radiation density ρ as a function of frequency ν and temperature. He accomplished this by deriving—based upon electromagnetic considerations—a relation between radiation density and [2] resonator energy $\overline{E}$:

$$\overline{E} = \frac{c^3\rho}{8\pi\nu^2}.\tag{1}$$

His derivation was of unparalleled boldness, but found brilliant confirmation. Not only the radiation formula proper and the calculated value of the elementary quantum [3] in it was confirmed, but also the quantum-theoretically calculated value of $\overline{E}$ was [4] confirmed by later investigations on specific heat. In this manner, equation (1), originally found by electromagnetic reasoning, was also confirmed. However, it remained unsatisfactory that the electromagnetic-mechanical analysis, which led to [5] (1), is incompatible with quantum theory, and it is not surprising that PLANCK himself and all theoreticians who work on this topic incessantly tried to modify the theory [6] such as to base it on noncontradictory foundations.

[7] Since BOHR's theory of spectra has achieved its great successes, it seems no longer doubtful that the basic idea of quantum theory must be maintained. It so appears that the uniformity of the theory must be established such that the electromagneto-mechanical considerations, which led PLANCK to equation (1), are to be replaced by quantum-theoretical contemplations on the interaction between matter [p. 319] and radiation. In this endeavor I feel galvanized by the following consideration, [8] which is attractive both for its simplicity and generality.

§1. PLANCK's Resonator in a Field of Radiation

The behavior of a monochromatic resonator in a field of radiation, according to the classical theory, can be easily understood if one recalls the manner of treatment that [9] was first used in the theory of BROWNian movement. Let E be the energy of the resonator at a given moment in time; we ask for the energy after time τ has elapsed. Hereby, τ is assumed to be large compared to the period of oscillation of the resonator, but still so small that the percentage change of E during τ can be treated as infinitely small. Two kinds of change can be distinguished. First the change

$$\Delta_1 E = -AE\tau$$

effected by emission; and second, the change $\Delta_2 E$ caused by the work done by the electric field on the resonator. This second change increases with the radiation density and has a "chance"-dependent value and a "chance"-dependent sign. An electromagnetic, statistical consideration yields the mean-value relation

$$\overline{\Delta_2 E} = B\rho\tau.$$

The constants A and B can be calculated in known manner. We call $\Delta_1 E$ the   [10]
energy change due to emitted radiation, $\Delta_2 E$ the energy change due to incident radiation. Since the mean value of E, taken over many resonators, is supposed to be independent of time, there has to be

$$\overline{E + \Delta_1 E + \Delta_2 E} = \overline{E}$$

or

$$\overline{E} = \frac{B}{A}\rho.$$

One obtains relation (1) if one calculates B and A for the monochromatic resonator in the known way with the help of electromagnetism and mechanics.

We now want to undertake corresponding considerations, but on a quantum-theoretical basis and without specialized suppositions about the interaction between   [p. 320]
radiation and those structures which we want to call "molecules."

§2. Quantum Theory and Radiation

We consider a gas of identical molecules that are in static equilibrium with thermal radiation. Let each molecule be able to assume only a discrete sequence Z_1, Z_2, etc., of states with energy values ε_1, ε_2, respectively. Then it follows in known manner and in analogy to statistical mechanics, or directly from BOLTZMANN's principle, or finally from thermodynamic considerations, that the probability W_n of state Z_n (or the relative number of molecules which were in state Z_n) is given by

$$W_n = p_n e^{\frac{-\varepsilon_n}{kT}} \tag{2}$$

where K is the well-known BOLTZMANN constant. p_n is the statistical "weight" of state Z_n, i.e., a constant that is characteristic of the quantum state of the molecule but independent of the gas temperature T.

We shall now assume that a molecule can go from state Z_n to state Z_m by absorbing radiation of the distinct frequency $\nu = \nu_{nm}$; and likewise from state Z_m to state Z_n by emitting such radiation. The radiation energy involved is $\varepsilon_m - \varepsilon_n$. In general, this is possible for any combination of two indices m and n. With respect to any of these elementary processes there must be a statistical equilibrium in thermal equilibrium. Therefore, we can confine ourselves to a single elementary process belonging to a distinct pair of indices (n,m).

At the thermal equilibrium, as many molecules per time unit will change from state Z_n to state Z_m under absorption of radiation, as molecules will go from state Z_m to state Z_n with emission of radiation. We shall state simple hypotheses about these transitions, where our guiding principle is the limiting case of classical theory, as it has been briefly outlined above.

We shall distinguish here also two types of transitions:

[p. 321]   a) *Emission of Radiation.* This will be a transition from state Z_m to state Z_n with emission of the radiation energy $\varepsilon_m - \varepsilon_n$. This transition will take place without external influence. One can hardly imagine it to be other than similar to radioactive reactions. The number of transitions per time unit will have to be put at

$$A_m^n N_m,$$

where A_m^n is a constant that is characteristic of the combination of the states Z_m and   [11]
Z_n, and N_m is the number of molecules in state Z_m.

b) *Incidence of Radiation.* Incidence is determined by the radiation within which the molecule resides; let it be proportional to the radiation density ρ of the effective frequency. In case of the resonator it may cause a loss in energy as well as an increase in energy; that is, in our case, it may cause a transition $Z_n \to Z_m$ as well as a transition $Z_m \to Z_n$. The number of transitions $Z_n \to Z_m$ per unit time is then

$$B_n^m N_n \rho,$$

and the number of transitions $Z_m \to Z_n$ is to be expressed as

$$B_m^n N_m \rho,$$

where B_n^m, B_m^n are constants related to the combination of states Z_n, Z_m.

As a condition for the statistical equilibrium between the reactions $Z_n \to Z_m$ and $Z_m \to Z_n$ one finds, therefore, the equation

$$A_m^n N_m + B_m^n N_m \rho = B_n^m N_n \rho. \tag{3}$$

Equation (2), on the other hand, yields

$$\frac{N_n}{N_m} = \frac{p_n}{p_m} e^{\frac{\varepsilon_m - \varepsilon_n}{kT}}. \tag{4}$$

From (3) and (4) follows

$$A_m^{\ n} p_m = \rho \left(B_n^{\ m} p_n e^{\frac{\varepsilon_m - \varepsilon_n}{kT}} - B_m^{\ n} p_m \right). \tag{5}$$

ρ is the radiation density of that frequency which is emitted with the transition $Z_n \rightarrow Z_m$ and is absorbed with $Z_m \rightarrow Z_n$. Our equation shows the relation between T and ρ at this frequency. If we postulate that ρ must approach infinity with ever [p. 322] increasing T, then we necessarily have

$$B_n^{\ m} p_n = B_m^{\ n} p_m. \tag{6}$$

Introducing the abbreviation

$$\frac{A_m^{\ n}}{B_m^{\ n}} = \alpha_{mn}, \tag{7}$$

one finds

$$\rho = \frac{\alpha_{mn}}{e^{\frac{\varepsilon_m - \varepsilon_n}{kT}} - 1}. \tag{5a}$$

This is PLANCK's relation between ρ and T with the constants left indeterminate. The constants $A_m^{\ n}$ and $B_m^{\ n}$ could be calculated directly if we possessed a modified version of electrodynamics and mechanics that is in compliance with the quantum hypothesis.   [12]

The fact that ρ must be a universal function of T and ν implies that α_{mn} and $\varepsilon_m - \varepsilon_n$ cannot depend upon the specific constitution of the molecule, but only upon the effective frequency ν. From WIEN's law follows furthermore that α_{mn} must be proportional to the third power, and $\varepsilon_m - \varepsilon_n$ to the first power of ν. Consequently, one has   [13]

$$\varepsilon_m - \varepsilon_n = h\nu, \tag{8}$$

where h is a constant.

While the three hypotheses concerning emission and incidence of radiation lead to PLANCK's radiation formula, I am of course very willing to admit that this does not elevate them to confirmed results. But the simplicity of the hypotheses, the generality with which the analysis can be carried out so effortlessly, and the natural connection to PLANCK's linear oscillator (as a limiting case of classical electrodynamics and mechanics) seem to make it highly probable that these are basic traits of a future theoretical representation. The postulated statistical law of emission is nothing but RUTHERFORD's law of radioactive decay, and the law expressed by (8), in conjunction with (5a), is identical with the second basic hypothesis in BOHR's theory of spectra—this too speaks in favor of the theory presented here.

[p. 323]

§3. Remark on the Photochemical Law of Equivalence

[14]

The photochemical law of equivalence falls in line with our train of thoughts in the following manner. Let there be a gas of such low temperature that the thermal radiation of frequency ν, which leads from state Z_m to state Z_n, does not practically occur.

According to (2) and (5a), the state Z_m will be quite rare compared to state Z_n, and we shall assume that almost all gas molecules are in state Z_n. Aside from the previously considered process $Z_m \to Z_n$, let the molecule in state Z_m also have the capability of another elementary "chemical" process, e.g., monomolecular dissociation. Let us furthermore assume that the reaction rate of this dissociation is large compared to the rate of occurrence of the reaction $Z_m \to Z_n$.

What will happen now if we irradiate the gas with the effective frequency? Under absorption of the radiation energy $\varepsilon_m - \varepsilon_n = h\nu$, molecules will continually go from state Z_n to state Z_m. Only a very small fraction of these molecules will return to state Z_n by emission or absorption. Most, by far, will suffer chemical dissociation, corresponding to the postulated higher reaction rate of this process. This means that per dissociating molecule, we will practically find that the radiation energy $h\nu$ has been absorbed, just as the law of equivalence demands.

The essence of this interpretation is that molecular dissociation is achieved by the absorption of light via the quantum state Z_m, but not directly without this intermediate state. In consequence, one need not distinguish between a chemically effective and a chemically ineffective absorption of radiation. The absorption of light and the chemical process appear as independent processes.

Doc. 35

"Preface" to Erwin Freundlich, *The Foundations of Einstein's Theory of Gravitation*

Not translated for this volume.

Doc. 36

[p. 375]
[1]

Review of H. A. Lorentz, *Statistical Theories in Thermodynamics: Five Lectures . . .*

Whoever has studied mathematical theories has had the following, embarrassing experience: he verifies every step in a deduction with diligence and eagerness, and at the end of his efforts he understands *nothing*. He did not get the guiding idea of the whole concept because the author himself suppressed it, either from an incapacity to phrase it concisely, or worse, from an almost comical coquettishness—as the insightful would say—which was especially popular in the past. This evil can only be overcome by unrestrained openness of the author, who should not shy away from familiarizing the reader even with his incomplete guiding ideas if they have furthered his own work. There is hardly a field in theoretical physics where this commandment is more difficult to fulfill than in statistical mechanics. Every knowledgeable reader will agree with me that Gibbs, in his pioneering book about this very topic, has sinned quite a lot against this commandment. Many have read it, many have verified it—and did *not* understand it. Lorentz tackled this evil in his first three lectures by displaying the foundations of the theory in a stunningly simple mathematical form, such that the guiding ideas are sharply focused upon.

In doing so, he puts Boltzmann's principle at the forefront, and thoroughly discusses the question of how the probability W in Boltzmann's equation $S = \kappa \lg W$ is to be defined. Thereby he uses the definition "W = phase integral" and demonstrates that the definition which has been suggested by this referee, "W = frequency of occurrence as a function of time," is essentially the same. The author explains on this occasion what made him refrain from the second, more descriptive definition; and I specifically want to direct the reader's attention to this point.

The last two lectures deal mainly with the theory of Brownian movement and with fluctuations. The last lecture is a masterly application of this latter theory to Planck's radiation formula. As is well known, there emerge statistical properties of radiation that cannot be represented by the undulation theory. The fact that these relations found H. A. Lorentz's interest is a special pleasure to this reviewer. Every physicist can learn from this illuminating booklet.

A. Einstein, Berlin-Charlottenburg

[p. 376]
[2]
[3]

Doc. 37
Author's Summary of *The Foundation*
of the General Theory of Relativity

Not translated for this volume.

[p. 47]

On the Quantum Theory of Radiation

[1]

by A. Einstein

The formal similarity between the curve of the chromatic distribution of thermal radiation and the Maxwellian distribution law of velocities is so striking that it could not have been hidden for long. As a matter of fact, W. Wien was already led by this

[2]   similarity to a farther-reaching determination of his radiation formula in his theoretically important paper, where he derives his displacement law

$$\rho = \nu^3 f\left(\frac{\nu}{T}\right). \tag{1}$$

As is well known, he found in this pursuit the formula

$$\rho = \alpha \nu^3 e^{\frac{-h\nu}{kT}}, \tag{2}$$

[3]   which is still recognized today as a limiting law for large values of $\frac{\nu}{T}$ (Wien's

[4]   radiation formula). We know today that an analysis that is based upon classical

[5]   mechanics and electrodynamics cannot provide a usable radiation formula; the classical theory leads instead, by necessity, to Rayleigh's formula

$$\rho = \frac{k\alpha}{h} \nu^2 T. \tag{3}$$

When Planck, in his fundamental investigation, based his radiation formula

$$\rho = \alpha \nu^3 \frac{1}{e^{\frac{h\nu}{kT}} - 1} \tag{4}$$

upon the postulate of discrete elements of energy (upon which quantum theory

[6]   developed in quick succession), Wien's investigation—which had led to equation (2)—was, naturally, quickly forgotten.

Recently I was able to find a derivation of Planck's radiation formula[1] which I based upon the fundamental postulate of quantum theory, and which is also related

[p. 48]   to the original considerations of Wien such that the relation between Maxwell's curve and the chromatic distribution curve comes to the fore. This derivation deserves attention not only because of its simplicity, but especially because it seems to clarify

---

[7]   [1]*Verh. d. deutschen physikal. Gesellschaft* 13/14 (1916), p. 318. The considerations in the quoted paper are repeated in the present investigation.

somewhat the still unclear processes of emission and absorption of radiation by matter. I made a few hypotheses about the emission and absorption of radiation by molecules, which suggested themselves from a quantum-theoretic point of view, and thus was able to show that molecules under quantum theoretically distributed states at temperature equilibrium are in dynamical equilibrium with Planck's radiation. By this procedure, Planck's formula (4) followed in an amazingly simple and general manner. It resulted from the condition that the distribution of molecules over their states of the inner energy, which quantum theory demands, must be the sole result of absorption and emission of radiation.

If the hypotheses which I introduced about the interaction between radiation and matter are correct, they must provide more than merely the correct statistical distribution of the *inner* energy of the molecules. Because, during absorption and emission of radiation there occurs also a transfer of *momentum* upon the molecules. This transfer effects a certain distribution of velocities of the molecules, by way of the mere interaction between radiation and the molecules. This distribution must be identical to the one which results from the mutual collision of the molecules, i.e., it must be identical with the Maxwell distribution. One has to demand that the mean kinetic energy of a molecule (per degree of freedom) in a Planck field of radiation of temperature T is equal to $\frac{kT}{2}$; this must hold independently of the absorbed or emitted frequencies and the nature of the molecules considered. We want to show in the present paper that this far-reaching demand is indeed satisfied in a quite general manner. In this way our simple hypotheses about the elementary processes of emission and absorption gain new support.

In order to achieve the result mentioned above, we have to supplement to some [p. 49] extent the hypotheses upon which our earlier work was based because they relate only to the exchange of *energy*. There arises the question: Does the molecule receive an impulse when it absorbs or emits the energy ε? For example, let us look at emission from the point of view of classical electrodynamics. When a body emits the radiation ε it suffers a recoil (momentum) $\frac{\varepsilon}{c}$ if the entire amount of radiation energy ε is emitted in the same direction. If, however, the emission is a spatially symmetric process, e.g., a spherical wave, no recoil at all occurs. This alternative also plays a role in the quantum theory of radiation. When a molecule absorbs or emits the energy ε in the form of radiation during the transition between quantum theoretically possible states, then this elementary process can be viewed either as a completely or partially directed one in space, or also as a symmetrical (nondirected) one. *It turns out that we arrive at a theory that is free of contradictions, only if we interpret those elementary processes as completely directed processes.* Herein lies the main result of the [8] following considerations.

§1. The Fundamental Hypothesis of Quantum Theory.
Canonical Distribution of States

According to quantum theory, a molecule of a distinct kind can—aside from orientation and translatory movement—take only a discrete sequence of states $Z_1, Z_2, ..., Z_n, ...$ with (inner) energies $\varepsilon_1, \varepsilon_2, ..., \varepsilon_n, ...$ If these molecules are part of a gas of temperature T, then the relative occurrence W_n of these states Z_n is given by the canonical partition of states of statistical mechanics according to the formula,

$$W_n = p_n e^{\frac{-\varepsilon_n}{kT}}. \tag{5}$$

In this formula $k = \dfrac{R}{N}$ is the well-known Boltzmann constant; p_n can be called the

[p. 50] statistical "weight" of the state and is a number characteristic of the molecule, namely its n-th quantum state, and it is independent of T. Formula (5) can be derived from Boltzmann's principle or in a purely thermodynamic manner. Equation (5) expresses the farthest-reaching generalization of Maxwell's law of velocity distribution.

The most recent principal advances of quantum theory deal with the theoretical determination of quantum theoretically possible states Z_n and their weights p_n. For our present principal investigation a closer determination of these quantum states is not required.

§2. Hypotheses Concerning the Energy Exchange by Radiation

Let Z_n and Z_m be two quantum theoretically possible states of a gas molecule, and their respective energies ε_n and ε_m shall satisfy the inequality

$$\varepsilon_m > \varepsilon_n.$$

The molecule shall be able to change from state Z_n into state Z_m under absorption of the radiation energy $\varepsilon_m - \varepsilon_n$; likewise, a change from state Z_m into state Z_n shall be possible under emission of radiation energy. The radiation absorbed or emitted by the molecule shall be of frequency ν, which is characteristic of the index combination (m,n).

For the laws that determine this transition we introduce several hypotheses that are obtained from the known circumstances of the classical theory of Planck's resonator, and transferred to the still unknown ones of quantum theory.

a) *"Ausstrahlung" (Spontaneous Emission)*. An oscillating Planck resonator   [9]
radiates energy in a known manner according to Hertz, independent of whether it is
excited by an external field of not. Correspondingly, a molecule shall be able to
change from state Z_m into state Z_n under emission of the energy $\varepsilon_m - \varepsilon_n$ with
frequency ν without external causes. The probability dW that this actually occurs
during the time element dt shall be

$$dW = A_m^{\,n} dt, \qquad (A)$$

where $A_m^{\,n}$ is a constant characteristic of the index combination under consideration. [p. 51]

The assumed statistical law corresponds to a radioactive reaction, and the
assumed elementary process to a reaction where only γ-radiation is emitted. One does
not need to assume that this process requires no time; the time need only be
negligible compared to the times during which the molecule is in states Z_1, etc.

b) *"Einstrahlung" (Induced Radiation)*. In a field of radiation, a Planck resonator
changes its energy because the electromagnetic field of the radiation transfers work
upon the resonator. Depending upon the phases of the resonator and the oscillating
field, this amount of work can be positive or negative. In taking account of this, we
introduce the following quantum-theoretic hypotheses. Under the action of the   [10]
radiation density ρ of frequency ν, the molecule can go from state Z_n to Z_m by
absorbing the radiation energy $\varepsilon_m - \varepsilon_n$ according to the probability law

$$dW = B_n^{\,m} \rho \, dt. \qquad (B)$$

In a similar manner the transition $Z_m \to Z_n$ is possible under the action of
radiation, whereby the radiation energy $\varepsilon_m - \varepsilon_n$ is set free according to the
probability law

$$dW = B_m^{\,n} \rho \, dt. \qquad (B')$$

$B_n^{\,m}$ and $B_m^{\,n}$ are constants. Both processes are called "changes of state by
'Einstrahlung' (induced radiation)."

The question is now what momentum is transferred upon the molecule during
these changes of state. We begin with the induced processes. If a beam of radiation
with a certain direction does work upon a Planck resonator, the beam loses the
corresponding energy. According to the momentum theorem, this transfer of energy
corresponds to a transfer of momentum from the beam to the resonator. The latter
suffers the action of a force in the direction of the ray of the beam. If the transfer of
energy is negative, the action of the force upon the resonator is also in the opposite
direction. In case of the quantum hypothesis it obviously means the following. If the [p. 52]
process $Z_n \to Z_m$ takes place due to induced radiation, then the molecule receives

a momentum $\dfrac{(\varepsilon_m - \varepsilon_n)}{c}$ in the direction in which the beam propagates. The transferred momentum has the same magnitude but opposite direction if the process of induced radiation effects a transition $Z_m \rightarrow Z_n$. In case the resonator is simultaneously exposed to several beams, we assume that the entire energy $\varepsilon_m - \varepsilon_n$ of an elementary process is taken from (or given to) only *one* beam of rays; thus, the transfer of the momentum $\dfrac{(\varepsilon_m - \varepsilon_n)}{c}$ upon the molecule occurs also in this case.

If the loss of energy occurs through spontaneous emission, the Planck resonator does, as a whole, not receive a transfer of momentum because under the classical theory the emission of radiation occurs in the form of a spherical wave. But we noted already that we can only obtain a quantum theory that is free of contradictions if we assume that the process of emission of radiation is also a directed one. Then every process of emission of radiation ($Z_m \rightarrow Z_n$) transfers a momentum of magnitude $\dfrac{(\varepsilon_m - \varepsilon_n)}{c}$ upon the molecule. For an isotropic molecule we have to assume equal probability for all directions of emission. We arrive at the same statement for nonisotropic molecules if the orientation changes over time according to the laws of chance. By the way, such assumption also has to be made for the statistical laws (B) and (B′) of induced radiation; otherwise the constants B_n^m and B_m^n could become dependent upon direction, which can be avoided by assuming isotropy or pseudo-isotropy of the molecule (by forming mean values over time).

§3. Derivation of Planck's Law of Radiation

We now ask which effective radiation density ρ must prevail in order to ensure that the energy exchange between radiation and molecules, according to the statistical laws (A), (B), and (B′), does not disturb the distribution of molecules given by equation (5). For this, it is necessary and sufficient that, per time unit, on average as many elementary processes of type (B) occur as of types (A) and (B′) together. This condition provides—due to (5), (A), (B), and (B′) for the elementary processes that correspond to the index combination (m,n)—the following equation:

[p. 53]

$$p_n e^{-\frac{\varepsilon_n}{kT}} B_n^m \rho = p_m e^{-\frac{\varepsilon_n}{kT}} (B_m^n \rho + A_m^n).$$

If furthermore ρ shall go to infinity with increasing T—and we shall assume this to be the case—then the constants B_n^m and B_m^n must satisfy the relation

$$p_n B_n^{\,m} = p_m B_m^{\,n}.$$ (6)

We then get, as a condition of dynamic equilibrium,

$$\rho = \cfrac{\cfrac{A_m^{\,n}}{B_m^{\,n}}}{e^{\frac{\varepsilon_m - \varepsilon_n}{kT}} - 1}.$$ (7)

This is how the radiation density depends upon temperature according to Planck's law. It now follows immediately from Wien's displacement law that necessarily

$$\frac{A_m^{\,n}}{B_m^{\,n}} = \alpha \nu^3$$ (8)

and $\varepsilon_m - \varepsilon_n = h\nu$, (9)

where α and h are universal constants. In order to obtain the numerical value of constant α one would have to have an exact theory of electrodynamic and mechanical processes. For the time being we must use Rayleigh's limiting case of high temperatures, for which the classical theory applies in the limit.

Equation (9) constitutes, as is well known, the second major rule in Bohr's theory of spectra, and after its completion by Sommerfeld and Epstein, one can claim it as [11] a secured part of our science. Implicitly, it also contains the photochemical law of [12] equivalence, as I have shown.

§4. Methods of Calculating the Motion of Molecules in a Field of Radiation

We now turn to the investigation of the motion which our molecules execute under the influence of radiation. In doing this we use a method which is well known from [p. 54] the theory of Brownian movement, and which I have used repeatedly for calculations of movements in a domain of radiation. In order to simplify the calculation, we carry [13] it out only for the case where movement occurs just in one direction, i.e., in the X-direction of the coordinate system. We also confine ourselves to the calculation of the mean value of the kinetic energy of the translational motion; i.e., we skip the proof that the velocities v are distributed according to Maxwell's law. Let the mass M of the molecule be sufficiently large that higher powers of $\frac{v}{c}$ can be neglected against the lower ones. The molecule can then be treated with ordinary mechanics. Furthermore, we can calculate without any real loss of generality, as if the states m

and n are the only ones the molecule can assume.

The momentum Mv of a molecule undergoes two changes in the short time span τ. Even though the radiation is of the same nature in all directions, the molecule will—due to its own movement—suffer a force from the radiation that counteracts its movement. Let this force equal Rv, where R denotes a constant that is to be calculated later. This force would bring the molecule to rest if the irregularities of the radiative actions would not force the molecule to receive a momentum Δ of changing sign and magnitude during the time τ. This nonsystematic influence will maintain a certain movement of the molecule, counter to the one previously mentioned. At the end of the short time τ considered, the momentum of the molecule will have the value

$$Mv - Rv\tau + \Delta.$$

Since the distribution of the velocities should remain constant over time, the mean value of the quantity above must equal the mean, absolute value of Mv. The mean values of the squares of both quantities must be equal to each other when extended over long times or over a large number of molecules:

$$\overline{(Mv - Rv\tau + \Delta)^2} = \overline{(Mv)^2}.$$

[p. 55]
[14] {1} Since the systematic influence of v upon [the molecule has been especially taken into account] we can neglect the mean value $\overline{v\Delta}$. Development of the left-hand side of the equation yields

$$\overline{\Delta^2} = 2RM\overline{v^2}\tau. \tag{10}$$

[15] The mean value v^2 which the radiation of temperature T generates among the molecules by interacting with them must be as large as the mean value $\overline{v^2}$ that the gas molecules attain—according to the gas laws—at the kinetic gas temperature T, because otherwise the presence of the molecules would disturb the thermal equilibrium between the thermal radiation and any gas of the same temperature. Consequently there must be

$$\frac{\overline{Mv^2}}{2} = \frac{kT}{2}. \tag{11}$$

Equation (10), therefore, becomes

$$\frac{\overline{\Delta^2}}{\tau} = 2RkT. \tag{12}$$

The investigation now has to be carried out as follows. With a given radiation $(\rho(v))$ the values of $\overline{\Delta^2}$ and R can be calculated by means of our hypotheses about the interaction between the radiation and the molecules. Substituting the results into

(12) must satisfy the equation identically if ρ is expressed as a function of ν and T [16]
from Planck's equation (4).

§5. Calculation of R

Let a molecule of the kind under consideration be moving along the X-axis of the
coordinate system K with the uniform velocity v. We ask for the mean value of
momentum per time unit, which is transferred from the radiation to the molecule. In
order to be able to calculate this, we have to judge the radiation from a coordinate
system K' that is at rest relative to the molecule. For we formulated our hypotheses
about emission and absorption only for molecules at rest. The transformation to the
system K' has been done several times in the literature, with particular accuracy in
MOSENGEIL's Berlin dissertation. However, for the sake of completeness I will repeat   [17]
here these simple considerations.

The radiation is isotropical relative to K, i.e., the radiation of frequency range $d\nu$ [p. 56]
per unit volume attributed to an infinitesimal solid angle $d\kappa$ about the direction of the
beam is

$$\rho d\nu \frac{d\kappa}{4\pi}, \tag{13}$$

where ρ depends only upon ν but not upon the direction. To this particular radiation
corresponds, relative to the coordinate system K', another particular radiation that is
also characterized by a frequency range $d\nu'$ and a solid angle $d\kappa'$. The volume
density of this particular radiation is

$$\rho'(\nu',\phi')d\nu' \frac{d\kappa'}{4\pi}. \tag{13'}$$

This defines ρ'. It is direction-dependent and commonly defined by an angle ϕ'
with the X'-axis and an angle ψ' between the Y'-Z'-projection and the Y'-axis. These
angles correspond to the angles ϕ and ψ which fix, in an analogous manner, the
direction of $d\kappa$ relative to K.

Next, it is clear that the same law of transformation must obtain between (13)
and (13') as it does for the squares of the amplitudes A^2 and A'^2 of a plane wave
of corresponding direction. In the approximation we want, we therefore have

$$\frac{\rho'(\nu',\phi')d\nu'd\kappa'}{\rho(\nu)d\nu d\kappa} = 1 - 2\frac{v}{c}\cos\phi \tag{14}$$

or

{2}
$$\rho'(\nu',\phi') = \rho(\nu)\frac{d\nu}{d\nu'}\,\frac{d\kappa}{d\kappa'}\,(1 - 2\frac{\nu}{c}\cos\phi).\tag{14'}$$

Furthermore, the theory of relativity provides, with the desired approximation, the formulas

$$\nu = \nu'(1 - \frac{v}{c}\cos\phi)\tag{15}$$

$$\cos\phi' = \cos\phi - \frac{v}{c} + \frac{v}{c}\cos^2\phi\tag{16}$$

$$\psi' = \psi.\tag{17}$$

From (15) follows, with corresponding approximation,

$$\nu = \nu'(1 + \frac{v}{c}\cos\phi').$$

[p. 57]   Therefore, and likewise in the desired approximation,

$$\rho(\nu) = \rho(\nu' + \frac{v}{c}\nu'\cos\phi')$$

or

$$\rho(\nu) = \rho(\nu') + \frac{\partial\rho}{\partial\nu}(\nu')\cdot\frac{v}{c}\nu'\cos\phi'.\tag{18}$$

According to (15), (16), and (17) there is in addition

$$\frac{d\nu}{d\nu'} = 1 + \frac{v}{c}\cos\phi'$$

$$\frac{d\kappa}{d\kappa'} = \frac{\sin\phi\,d\phi\,d\psi}{\sin\phi'\,d\phi'\,d\psi'} = \frac{d(\cos\phi)}{d(\cos\phi')} = 1 - 2\frac{v}{c}\cos\phi'.$$

Due to these two relations and (18), (14') becomes

$$\rho'(\nu',\phi') = \left[(\rho)_{\nu'} + \frac{v}{c}\nu'\cos\phi'\left(\frac{\partial\rho}{\partial\nu}\right)_{\nu'}\right](1 - 3\frac{v}{c}\cos\phi').\tag{19}$$

With the help of (19) and our hypotheses on the spontaneous emission and the induced processes of the molecule, we can easily calculate the mean value of momentum per time unit which is transferred to the molecule. However, before we do this we have to say something to justify the method used. One could object that equations (14), (15), (16) are based upon Maxwell's theory of the electromagnetic field, a theory that is incompatible with quantum theory. But this objection touches the form more than the essence of the matter. Because, in whichever way the theory

of electromagnetic processes may develop, it will certainly retain Doppler's principle
and the law of aberration, and with it also equations (15) and (16). Furthermore, the
validity of the energy relation (14) extends beyond that of the theory of undulation;
according to the theory of relativity, the transformation law applies, for example, also
to the energy density of a mass that moves with the (quasi) speed of light and is of
infinitesimally small density when seen at rest. Equation (19) can therefore claim
validity for any theory of radiation.—   [18]

The radiation in the solid angle $d\kappa'$ would, according to (B), give rise to

$$B_n^m \rho(\nu',\phi')\frac{d\kappa'}{4\pi}$$

induced elementary processes per second of type $Z_n \rightarrow Z_m$ if the molecule would
immediately be brought back into state Z_n after each such elementary process. In
reality, however, it remains in state Z_n according to (5)—for a time per second of   [p. 58]

$$\frac{1}{S}p_n e^{-\frac{\varepsilon_n}{kT}},$$

where we used the abbreviation

$$S = p_n e^{-\frac{\varepsilon_n}{kT}} + p_m e^{-\frac{\varepsilon_m}{kT}}. \qquad (20)$$

The number of processes is therefore in actual fact

$$\frac{1}{S}p_n e^{-\frac{\varepsilon_n}{kT}} B_n^m \rho(\nu',\phi')\frac{d\kappa'}{4\pi}.$$

During each one of these elementary processes the momentum $\dfrac{\varepsilon_m - \varepsilon_n}{c}\cos\phi'$

is transferred to the atom in the direction of the positive X'-axis. In an analogous
manner we found—based upon (B′)—that the corresponding number of induced
elementary processes of type $Z_m \rightarrow Z_n$ per second is

$$\frac{1}{S}p_m e^{-\frac{\varepsilon_m}{kT}} B_m^n \rho(\nu',\phi')\frac{d\kappa'}{4\pi}$$

and in this case the momentum $- \dfrac{\varepsilon_m - \varepsilon_n}{c}\cos\phi'$ is transferred to the molecule with

each such elementary process. Considering (6) and (9), the total momentum per time
unit that is transferred to the molecule and caused through induced processes per time
unit is

[19]
$$\frac{hv'}{cS}\,p_n B_n^{\,m}(e^{-\frac{\varepsilon_n}{kT}} - e^{-\frac{\varepsilon_m}{kT}})\int \rho(v',\phi')\cos\phi'\,\frac{d\kappa'}{4\pi},$$

where the integration is to be extended over all solid elementary angles. Executing this integration and using (19), one finds

$$-\frac{hv}{c^2 S}\left(\rho - \frac{1}{3}v\frac{\partial\rho}{\partial v}\right)p_n B_n^{\,m}(e^{-\frac{\varepsilon_n}{kT}} - e^{-\frac{\varepsilon_m}{kT}})\cdot v.$$

The effective frequency is here again denoted by v (instead of v').

This expression represents, however, the entire momentum which in the mean is transferred per time unit to a molecule moving at velocity v. It is clear that the elementary processes of spontaneous emission which occur without the interaction of [p. 59] radiation cannot have a preferential direction when viewed from the system K', and therefore in the mean are not able to transfer any momentum to the molecule. The end result of our deliberation is therefore

$$R = \frac{hv}{c^2 S}\left(\rho - \frac{1}{3}v\frac{\partial\rho}{\partial v}\right)p_n B_n^{\,m}e^{-\frac{\varepsilon_n}{kT}}(1 - e^{-\frac{hv}{kT}}). \tag{21}$$

§6. Calculation of $\overline{\Delta^2}$

It is much simpler to calculate the effect the irregularities of the elementary process have upon the mechanical behavior of the molecule, because, with the degree of approximation we considered sufficient from the beginning, one can base the calculations upon a molecule at rest.

Let some event cause the transfer of a momentum λ to a molecule in the X-direction. In the various cases these momenta shall be of diverse signs and diverse magnitudes. Let λ be subject to a law of statistics such that the mean value of λ vanishes. Now let $\lambda_1, \lambda_2, \ldots$ be the values of the momenta of several mutually independent causes, all transferred upon the molecule and all acting in the X-direction such that the total of the momenta Δ which are transferred is given by

$$\Delta = \sum \lambda_v.$$

[20]   If the mean value $\overline{\lambda_v}$ for the individual λ_v vanishes, one has

$$\overline{\Delta^2} = \Sigma\overline{\lambda_v^2}. \tag{22}$$

If the mean values $\overline{\lambda_v^2}$ of the individual momenta are equal among each other $(=\overline{\lambda^2})$ and l is the total number of momenta-producing causes, one gets the relation

$$\overline{\Delta^2} = l\overline{\lambda^2}. \tag{22a}$$

According to our hypotheses, each elementary process, induced or spontaneous, transfers to the molecule the momentum

$$\lambda = \frac{h\nu}{c}\,\phi,$$

where ϕ is the angle between the X-axis and a direction subject to the laws of chance. One therefore gets

$$\overline{\lambda^2} = \frac{1}{3}\left(\frac{h\nu}{c}\right)^2. \tag{23}$$

We can use (22a) because we assume all elementary processes that occur to be [p. 60] mutually independent events. l is then the total number of elementary events occurring during time τ. This is twice as large as the number of induced processes $Z_n \to Z_m$ during τ. Consequently,

$$l = \frac{2}{S}\,p_n B_n^m\, e^{-\frac{\varepsilon_n}{kT}}\, \rho\tau. \tag{24}$$

From (23), (24), and (22) follows

$$\frac{\overline{\Delta^2}}{\tau} = \frac{2}{3S}\left(\frac{h\nu}{c}\right)^2 p_n B_n^m\, e^{-\frac{\varepsilon_n}{kT}}\, \rho. \tag{25}$$

§7. Result

In order to show that, according to our basic hypotheses, the momenta exerted by the radiation upon the molecules do not disturb the thermodynamic equilibrium, we only have to insert the values of $\dfrac{\overline{\Delta^2}}{\tau}$ and R, which we calculated in (25) and (21), after we replace the value

$$\left(\rho - \frac{1}{3}\nu\frac{\partial\rho}{\partial\nu}\right)\left(1 - e^{-\frac{h\nu}{kT}}\right)$$

in (21) by $\dfrac{\rho}{3}\dfrac{h\nu}{kT}$ according to (4). It shows immediately that our fundamental equation (12) is satisfied.—

With this, our deliberations come to a close. They provide strong support for the hypotheses in §2 about the interaction between matter and radiation by means of processes of absorption and emission or, respectively, spontaneous and induced radiation processes. I was led to these hypotheses from a desire to postulate in the simplest manner the quantum-theoretical behavior of molecules in a manner analogous to the classical theory of Planck's resonator. From the general assumption of quanta for matter, Bohr's second rule (equation 9) and Planck's radiation formula followed effortlessly.

The result, referring to the momenta transferred to molecules by spontaneous and induced radiation processes, appears to me to be the most important one. If we were to modify one of our postulates about momenta, a violation of equation (12) would

[p. 61] be the consequence. To agree with this equation—which is demanded by the theory of heat—in a way other than by our assumptions, seems hardly possible. Therefore, we can consider the following as being rather certainly proven.

If a beam of radiation effects the targeted molecule to either accept or reject the quantity of energy $h\nu$ in the form of radiation by an elementary process (induced radiation process), then there is always a transfer of momentum $\dfrac{h\nu}{c}$ to the molecule, specifically in the direction of propagation of the beam when energy is absorbed by the molecule, in the opposite direction if the molecule releases the energy. If the molecule is exposed to the action of several directed beams of radiation, then always only one of them takes part in an induced elementary process; only this beam alone determines the direction of the momentum that is transferred to this molecule.

If the molecule suffers a loss of energy in the amount of $h\nu$ without external stimulation, i.e., by emitting the energy in the form of radiation (spontaneous emission), then this process too is a *directional* one. There is no emission of radiation in the form of spherical waves. The molecule suffers a recoil in the amount of $\dfrac{h\nu}{c}$ during this elementary process of emission of radiation; the direction of the recoil is, at the present state of theory, determined by "chance."

The properties of the elementary processes that are demanded by equation (12) let the establishment of a quantumlike theory of radiation appear as almost unavoidable. The weakness of the theory is, on the one hand, that it does not bring us closer to a link-up with the undulation theory; on the other hand, it also leaves

[21] time of occurrence and direction of the elementary processes a matter of "chance." Nevertheless, I fully trust in the reliability of the road taken.

Still one general remark should find its place here. Nearly all theories of thermal radiation are based upon a consideration of the interaction between radiation and molecules. But in general one is satisfied with a consideration of the *energy* exchange without consideration of the exchange of *momentum*. One feels easily justified to do so, because the smallness of the momenta transferred by radiation almost always pushes them into the background of reality when compared to other [p. 62] motion-generating causes. But in *theoretical* investigations these small effects are definitely as important as the more prominently appearing *energy* transfers by radiation, because energy and momenta are always intimately linked together. Therefore, a theory can only be viewed as justified if it shows that the momenta that are transferred from radiation onto matter lead to motions as they are demanded by the theory of heat.

Additional notes by translator

{1}   The English translation of the missing German text, which is quoted in note [14], has been included here in brackets.

{2}   " $\dfrac{dk}{dk}$ " has been corrected to " $\dfrac{d\kappa}{d\kappa}$."

Doc. 39

[1]

Elementary Theory of Water Waves and of Flight

by A. Einstein, Berlin-Wilmersdorf

What accounts for the carrying capacity of the wings[*] of our flying machines and of the birds soaring through the air in their flight? There is a widespread lack of clarity on this question. I must confess that I could not find anywhere in the specialized literature even the simplest answer. I hope, therefore, to give some readers pleasure when I try to remedy this deficiency with the following short consideration on the theory of the motion of liquids.

Let an incompressible liquid with negligible inner friction stream in the direction of the arrows through a pipe that is tapering off to the right (Fig. 1). We ask for the distribution of the pressure in the pipe. Since the same quantity of fluid per second

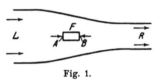

Fig. 1.

must flow through every cross section, the velocity of flow q must be the lowest at the largest cross sections and the fastest at the smallest cross sections. The velocities of the particles of the liquid will therefore be smallest at L in Fig. 1 and will continuously increase toward R. This acceleration of the particles of the liquid can only be effected by the pressure forces that act upon them. In order for the cylindrical liquid particle F to execute an accelerated movement toward the right, its rear-end surface A must be under a higher pressure than its front-end surface B. The pressure at A exceeds the pressure at B. Repeating this conclusion, it follows that the pressure within the pipe decreases steadily from L to R. The same distribution of pressure (decrease of pressure from L to R) is found in an analogous consideration if the direction of flow of the liquid is reversed.

In generalizing, we can state the following long-known theorem of the hydrodynamics of fluids without friction. When we follow the path of a liquid particle in a stationary flow, we find the pressure p always larger where the velocity q is lower and vice versa. This theorem is quantitatively expressed for noncompressible liquids by the well-known equation

---

[*]*Translator's note.* Instead of the basic term *Auftrieb* = wing lift, Einstein uses the extremely complex concept of *Tragfähigkeit der Flügel* = carrying capacity of wings, which linguistically and falsely suggests in German a rather elementary quantity—at least to the layman. The R. v. Mises lectures in *Fluglehre*, which in part date back to 1913, represent the state of the art at the time of Einstein's article; they were later incorporated into the *Theory of Flight* (Dover reprint, 1945 & 1959).

$$p = \text{const} - \frac{1}{2}\rho q^2,$$

where ρ is the density of the liquid.

Next, we consider several generally known examples of this theorem. The outflow of a liquid that is contained under pressure in a vessel (*Toricelli*). The pressure at J (Fig. 2) is higher and the velocity is lower than at A, such that

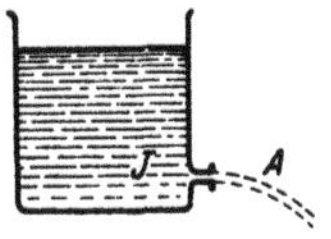

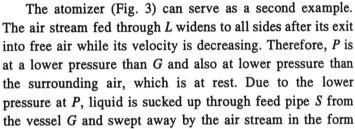

Fig. 2.

$$p + \frac{1}{2}\rho q^2$$

is constant during the outflow process.

The atomizer (Fig. 3) can serve as a second example. The air stream fed through L widens to all sides after its exit into free air while its velocity is decreasing. Therefore, P is at a lower pressure than G and also at lower pressure than the surrounding air, which is at rest. Due to the lower pressure at P, liquid is sucked up through feed pipe S from the vessel G and swept away by the air stream in the form of little droplets. (The fact that it is an air stream and not a stream of an incompressible liquid does not essentially change our considerations.)

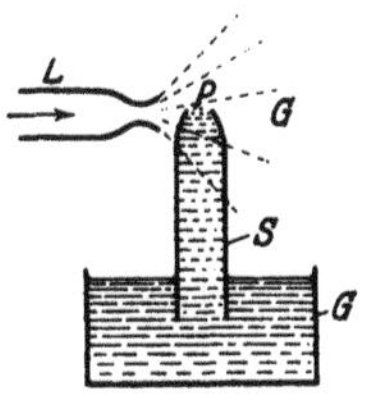

Fig. 3.

After these preparations, we turn to a consideration of water waves. Let W (Fig. 4) be a cylindrical and wave-shaped rigid wall, perpendicular to the plane of the paper; on one side it forms the boundary of a fluid that flows from left to right. We ask for the pressure forces exerted by the fluid toward the wall. Obviously, the cross section offered to the streaming fluid is larger at the B positions than at the positions T. The liquid at B will therefore flow slower, and at T faster, than at any locations deeper inside the fluid, i.e., farther removed from the wall W. The streaming fluid will therefore generate a higher pressure at B and a lower one at T. Consequently, the fluid will press against the wall such as to enlarge the already existing outward-pointing bulges of the wall. Therefore, the flow could not maintain itself under a free surface of the fluid, that is, if the wall would be infinitely moldable and stretchable.[1]

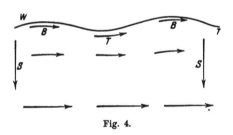

Fig. 4.

However, all these considerations are, like our previous ones, based on the assumption that there are no other causes to generate pressure in the fluid except the

---

[1]It is well known that fluttering flags can be understood based on these considerations.

flow itself. But if gravity acts in the direction of the arrows *S*, it will generate additional pressure forces in the liquid and they will increase toward the bottom. If gravity alone would act, there would be lesser pressure at points *B* than at points *T*.

Consequently, flow and gravity generate pressure differences opposite in sign between *B* and *T*, and it is obvious that the velocity of the fluid flow can be chosen such that the pressure differences between *B* and *T*, due to the two causes, vanish. In that case we can remove the wall *W* without disturbing the motion of the liquid. We then have a fluid flow with wave-like curved surface such as we often observe behind an obstacle in a stream. We notice this when we stand at a bridge-pier, looking downstream and watching the water behind the pier.

Imagine that the entire process is described by an observer who moves with the internal velocity of the liquid stream to the right; then we have the ordinary case of water waves before us. For the observer deep down, the liquid is at rest and the mountains *B* and valleys *T* propagate with constant velocity toward the left.

The possibility of the wave-process is therefore based upon the fact that both the statistically and dynamically produced pressure differences at various heights just balance one another.

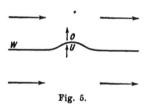

Fig. 5.

The carrying capacity of a wing has a very similar explanation. Let an air stream or fluid flow tangential to a solid, cylindrical wall *W* (Fig. 5), the latter perpendicular to the plane of the paper, and with an upward-directed bulge. If this bulge were missing there would be no force on the surface apart from the action of the unavoidable friction. The bulge, however, will affect the fluid flow above and below the wall, whereby pressures are generated.

For the lower flow the bulge represents a local increase in cross section, which means a slowdown in the flow and, thus, higher pressure at *U*. On top, however, the bulge contracts the cross section, increases the flow velocity locally, and causes diminished pressure at *O*. The dynamic pressure forces caused by the flow generate an upward-directed force upon the wall. In order to generate this force, one obviously needs to realize only a large enough portion of the wall as is necessary to cause an effective bulge in the fluid flow. We then have the supporting wing of a flying machine or of a bird soaring without moving its wings.

[2]

One sees from this simple analysis that flying requires the exertion of work only insofar as unavoidable obstacles of friction must be overcome. Without friction, a bird could fly any horizontal paths without doing any work.

Doc. 40

5. On Friedlich Kottler's Paper:

[p. 639]

"On Einstein's Equivalence Hypothesis and Gravitation"[1]

by A. Einstein

Especially noteworthy among the papers that take a critical look at the general theory of relativity are those by KOTTLER because this colleague has truly penetrated deeply into the spirit of the theory. I want to discuss here the most recent of his works.

[2]

Kottler claims I had abandoned in my later papers the "principle of equivalence" which I did introduce in order to unify the concepts of "inertial mass" and "gravitational mass." This opinion must be based upon the fact that we both do not denote the same thing as "the principle of equivalence"; because in my opinion my theory rests exclusively upon this principle. Therefore I repeat the following:

[3]

1. The Limiting Case of the Special Theory of Relativity. Let a finite space-time-like domain be without a gravitational field; i.e., let it be possible to introduce a system of reference K ("Galilean system") relative to which the following is true for the domain. As is usually presupposed in the special theory of relativity, let the coordinates be directly measurable in known manner by means of a unit measuring stick, and the times by a unit clock. Relative to this system an isolated material point shall move uniformly in a straight line, just as it was postulated by Galileo.

2. The Principle of Equivalence. Starting from the limiting case of the special theory of relativity, one may ask if in the domain under consideration an observer, who is uniformly accelerated relative to K, must necessarily judge his state as accelerated, or whether he has an option left—according to the (approximately) known laws of nature—to interpret his state as "at rest." Or, to phrase it more precisely: Do the laws of nature, known to us in some approximation, allow us to consider a reference system K' as being at rest if it is in uniform acceleration with respect to K? Or, somewhat more generally: Can the principle of relativity be extended such as to encompass reference systems that are in (uniform) accelerated motion relative to one another? The answer is: insofar as we really know the laws of nature, nothing prevents us from considering a system K' as at rest, provided we assume a gravitational field (homogeneous in first approximation) relative to K'. Because in a homogeneous gravitational field, as with regard to our system K', all bodies fall with the same acceleration independent of their physical nature. I call "principle of equivalence" the assumption that K' can be treated with all rigor as

[p. 640]

---

[1]*Ann. d. Phys.* 50 (1916), p. 955.

[1]

being at rest, such that no law of nature fails to be satisfied relative to K'.

3. The Gravitational Field Not Only Kinematically Caused. The previous consideration can also be inverted. Let the system K' with the gravitational field, which we considered above, be the original system. One can then introduce a new system K that is accelerated with respect to K' such that (isolated) masses move uniformly in straight lines (within the domain of consideration). But one must *not* go beyond this and say: If K' is a reference system with an *arbitrary* gravitational field, then one can always find a system K relative to which isolated masses move uniformly in straight lines, i.e., relative to which no gravitational field exists. The absurdity of such a hypothesis is plainly obvious. If the gravitational field relative to K' is, for example, that of a mass point at rest, then not even the most refined trick of transformation can transform the field away in the entire neighborhood of the mass point. Therefore, one may never maintain that a gravitational field could be explained, so to speak, by pure kinematics; a "kinematic, nondynamic interpretation of gravitation" is not possible. By mere transformation from a Galilean system into another one by means of an acceleration transformation, we do not learn about *arbitrary* gravitational fields but only about some of a very special kind; but these too must—of course—obey the same laws as all other fields of gravitation. This is again just another formulation of the principle of equivalence (specialized in its application to gravitation).

[4]

[p. 641]

A theory of gravitation violates the principle of equivalence—in the sense as I understand it—only if the equations of gravitation are not satisfied in *any* system K' of reference which moves nonuniformly relative to a Galilean system. It is evident that this accusation cannot be raised against my theory of *generally* covariant equations, because the equations are here satisfied in every system of reference. *The postulate for general covariance of the equations embraces the principle of equivalence as a special case.*

4. Are the Forces of a Field of Gravitation "Real" Forces? KOTTLER censures that I interpret the second term in the equations of motion

$$\frac{d^2 x_v}{ds^2} + \sum_{\alpha\beta} \begin{Bmatrix} \alpha\beta \\ v \end{Bmatrix} \frac{dx_\alpha}{ds} \frac{dx_\beta}{ds} = 0$$

as an expression of the influence of the gravitational field upon the mass point, and the first term more or less as the expression of the Galilean inertia. Allegedly this would introduce "real forces of the gravitational field" and this would not comply with the spirit of the principle of equivalence. My answer to this is that this equation as a whole is generally covariant, and therefore is quite in compliance with the hypothesis of equivalence. The naming of the parts, which I have introduced, is in principle meaningless and only meant to appeal to our physical habit of thinking.

This is especially true of the concepts

$$\Gamma^{\nu}_{\alpha\beta} = - \begin{Bmatrix} \alpha\beta \\ \nu \end{Bmatrix}$$

(components of the gravitational field) and t^{ν}_{σ} (energy components of the gravitational field). The introduction of these names is in principle unnecessary, but at least for the time being and in order to maintain the continuity of ideas, I do not think they [p. 642] are worthless—and that is why I introduced these quantities even though they do not have tensorial character. The principle of equivalence, however, is always satisfied when equations are covariant.

5. It is true that the price I had to pay for the general covariance of the equations was the abandonment of ordinary time measurement and Euclidean measurement of space. KOTTLER believes he can do without this sacrifice. But already in the case of the system K', considered by him, which is accelerated relative to a Galilean system in the sense of BORN, one has to abandon the ordinary measurement of time. The [5] point of view of the theory of relativity suggests then, obviously, also the need to abandon the ordinary measurement of space. Herr KOTTLER will certainly convince himself of this necessity if he tries to execute in a general manner his intended theoretical plans. [6]

October 1916

[p. 1111]

[1]

Doc. 41

Hamilton's Principle and the General Theory of Relativity

by A. Einstein

H. A. LORENTZ and D. HILBERT have recently succeeded[1] in presenting the theory of general relativity in a particularly comprehensive form by deriving its equations from a single variational principle. The same shall be done in this paper. My aim here is to present the fundamental connections as transparently and comprehensively [3] as the principle of general relativity allows. In contrast to HILBERT's presentation, I shall make as few assumptions about the constitution of matter as possible. On the other hand, and in contrast to my own very recent treatment of the subject matter, the [4] choice of a system of coordinates shall remain completely free.

§1. The Variational Principle and the Field Equations of Gravitation and Matter

The gravitational field shall be described as usual by the tensor[2] of the $g_{\mu\nu}$ (or $g^{\mu\nu}$ resp.); matter (inclusive of the electromagnetic field) by an arbitrary number of space-time functions $q_{(\rho)}$ whose invariance-theoretical character we ignore. Furthermore, let $\mathfrak{H}$ be a function of the

{1}
$$g^{\mu\nu},\ g_\sigma^{\mu\nu}\left(=\frac{\partial g^{\mu\nu}}{\partial x_\sigma}\right) \text{ and } g_{\sigma\tau}^{\mu\nu}\left(=\frac{\partial^2 g^{\mu\nu}}{\partial x_\sigma \partial x_\tau}\right),\ \text{the } q_{(\rho)} \text{ and } q_{(\rho)\alpha}\left(=\frac{\partial q_{(\rho)}}{\partial x_\alpha}\right).$$

The variational principle

$$\delta\left\{\int \mathfrak{H}\, d\tau\right\} = 0 \tag{1}$$

[p. 1112] then provides as many differential equations as there are functions $g_{\mu\nu}$ and $q_{(\rho)}$, which are to be determined, provided we agree that the $g^{\mu\nu}$ and $q_{(\rho)}$ are to be varied [5] independently of each other such that at the boundaries of integration the $\delta q_{(\rho)}$, $\delta g^{\mu\nu}$,

and $\dfrac{\partial \delta g_{\mu\nu}}{\partial x_\sigma}$ all vanish.

[6] {2}   We shall now assume $\mathfrak{H}$ to be linear in the $g_{\sigma\tau}^{\mu\nu}$ such that the coefficients of the $g_{\sigma\tau}^{\mu\nu}$

---

[1]Four papers by H. A. LORENTZ in volumes 1915 and 1916 of the *Publikationer d.*

[2]   *Koninkl. Akad. van Wetensch. te Amsterdam*; D. HILBERT, *Gött. Nachr.* 1915, Heft. 3.

[2]For the time being, the tensorial character of the $g_{\mu\nu}$ is not used.

depend only upon the $g^{\mu\nu}$. The variational principle (1) can then be replaced by one more convenient for us. With suitable partial integration one gets

$$\int \mathfrak{H} d\tau = \int \overset{*}{\mathfrak{H}} d\tau + F, \tag{2}$$

where F is an integral extended over the boundaries of the domain under consideration, while the quantity $\overset{*}{\mathfrak{H}}$ depends only upon the $g^{\mu\nu}$, $g_\sigma^{\mu\nu}$, $q_{(\rho)}$, $q_{(\rho)\alpha}$ but no longer upon $g_{\sigma\tau}^{\mu\nu}$. For the variation of interest to us one gets from (2)

$$\delta\left\{\int \mathfrak{H} d\tau\right\} = \delta\left\{\int \overset{*}{\mathfrak{H}} d\tau\right\}, \tag{3}$$

whereupon we can replace the variational principle (1) with the more convenient one

$$\delta\left\{\int \overset{*}{\mathfrak{H}} d\tau\right\} = 0. \tag{1a}$$

By executing the variation after the $g^{\mu\nu}$ and the $q_{(\rho)}$ one obtains for the field equations of gravitation and matter the equations[3]

$$\frac{\partial}{\partial x_\alpha}\left(\frac{\partial \overset{*}{\mathfrak{H}}}{\partial g_\alpha^{\mu\nu}}\right) - \frac{\partial \overset{*}{\mathfrak{H}}}{\partial g^{\mu\nu}} = 0 \tag{4}$$

$$\frac{\partial}{\partial x_\alpha}\left(\frac{\partial \overset{*}{\mathfrak{H}}}{\partial q_{(\rho)\alpha}}\right) - \frac{\partial \overset{*}{\mathfrak{H}}}{\partial q_{(\rho)}} = 0. \tag{5}$$

§2. Separate Existence of the Gravitational Field

The energy components cannot be split into two separate parts such that one belongs to the gravitational field and the other to matter, unless one makes special assumptions in which manner $\mathfrak{H}$ should depend upon the $g^{\mu\nu}$, $g_\sigma^{\mu\nu}$, $q_{(\rho)}$, $q_{(\rho)\alpha}$. In order to bring about this property of the theory we assume

$$\mathfrak{H} = \mathfrak{G} + \mathfrak{M}, \tag{6}$$

where $\mathfrak{G}$ depends only upon $g^{\mu\nu}$, $g_\sigma^{\mu\nu}$, $g_{\sigma\tau}^{\mu\nu}$ and $\mathfrak{M}$ only upon $g^{\mu\nu}$, $q_{(\rho)}$, $q_{(\rho)\alpha}$. [p. 1113]
Equations (4), (5) then take the form {3}

---

[3]As an abbreviation, the summation signs are omitted in the formulas. A summation has to be carried out over the indices that occur twice in a term. For example, in

(4) $\dfrac{\partial}{\partial x_\alpha}\left(\dfrac{\partial \overset{*}{\mathfrak{H}}}{\partial g_\alpha^{\mu\nu}}\right)$ denotes the term $\displaystyle\sum_\alpha \dfrac{\partial}{\partial x_\alpha}\left(\dfrac{\partial \overset{*}{\mathfrak{H}}}{\partial g_\alpha^{\mu\nu}}\right).$

$$\frac{\partial}{\partial x_\alpha}\left(\frac{\partial \mathfrak{G}^*}{\partial g_\alpha^{\mu\nu}}\right) - \frac{\partial \mathfrak{G}^*}{\partial g^{\mu\nu}} = \frac{\partial \mathfrak{M}}{\partial g^{\mu\nu}} \tag{7}$$

$$\frac{\partial}{\partial x_\alpha}\left(\frac{\partial \mathfrak{M}}{\partial q_{(\rho)\alpha}}\right) - \frac{\partial \mathfrak{M}}{\partial q_{(\rho)}} = 0. \tag{8}$$

$\mathfrak{G}^*$ is here in the same relation to $\mathfrak{G}$ as $\mathfrak{H}^*$ is to $\mathfrak{H}$.

It must be noted that equations (8) or (5) respectively would have to be replaced by others if we would assume that $\mathfrak{M}$ or $\mathfrak{H}$ resp. would depend upon higher than the first derivatives of $q_{(\rho)}$. Similarly, one could imagine the $q_{(\rho)}$ not as mutually independent but rather as connected to each other by further, conditional equations. All this is irrelevant for the following development, since it is solely based upon equation (7), which is obtained by varying our integral after the $g^{\mu\nu}$.

[7] {4}

§3. Properties of the Field Equations of Gravitation Based on the Theory of Invariants

We now introduce the assumption that

$$ds^2 = g_{\mu\nu}\,dx_\mu dx_\nu \tag{9}$$

is an invariant. This fixes the transformational character of the $g_{\mu\nu}$. We make no presuppositions about the transformational character of the $q_{(\rho)}$ which describe matter. However, the functions $H = \dfrac{\mathfrak{H}}{\sqrt{-g}}$ and $G = \dfrac{\mathfrak{G}}{\sqrt{-g}}$ and $M = \dfrac{\mathfrak{M}}{\sqrt{-g}}$ shall be invariants under arbitrary substitutions of the space-time coordinates. From these suppositions follows the general covariance of equations (7) and (8), which have been derived from (1). It follows furthermore that (up to a constant factor) G is equal to the scalar of the RIEMANN tensor of curvature, because there is no other invariant with the properties demanded for G.[4] With this, $\mathfrak{G}^*$, and hence the left-hand side of the field equation (7) is completely determined.[5]

[p. 1114]   The postulate of general relativity entails certain properties of the function $\mathfrak{G}^*$

---

[4]This is the reason why the requirements of general relativity led to a quite distinct theory of gravitation.

[5]Execution of the partial integration yields

$$\mathfrak{G}^* = \sqrt{-g}\,g^{\mu\nu}\left[\begin{Bmatrix}\mu\alpha\\\beta\end{Bmatrix}\begin{Bmatrix}\mu\beta\\\alpha\end{Bmatrix} - \begin{Bmatrix}\mu\nu\\\alpha\end{Bmatrix}\begin{Bmatrix}\alpha\beta\\\beta\end{Bmatrix}\right].$$

which we shall now derive. For this purpose we carry out an infinitesimal transformation of the coordinates by setting

$$x'_\nu = x_\nu + \Delta x_\nu,$$ (10)

where the Δx_ν are arbitrarily eligible, infinitesimally small functions of the coordinates. x'_ν are the coordinates of the world point in the new system, the same point whose coordinates were x_ν in the original system. Just as for the coordinates, there is a transformation law for any other quantity ψ, of the type

$$\psi' = \psi + \Delta\psi,$$

where $\Delta\psi$ must always be expressible in terms of the Δx_ν. From the covariant properties of the $g^{\mu\nu}$ one derives easily for the $g^{\mu\nu}$ and the $g_\sigma^{\mu\nu}$ the transformation laws:

$$\Delta g^{\mu\nu} = g^{\mu\alpha}\frac{\partial \Delta x_\nu}{\partial x_\alpha} + g^{\nu\alpha}\frac{\partial \Delta x_\mu}{\partial x_\alpha}$$ (11)

$$\Delta g_\sigma^{\mu\nu} = \frac{\partial(\Delta g^{\mu\nu})}{\partial x_\sigma} - g_\alpha^{\mu\nu}\frac{\partial \Delta x_\alpha}{\partial x_\sigma}.$$ (12)

$\Delta\mathfrak{G}^*$ can be calculated with the help of (11) and (12), since $\mathfrak{G}^*$ depends only upon the $g^{\mu\nu}$ and the $g_\sigma^{\mu\nu}$. Thus, one gets the equation [8] {5}

$$\sqrt{-g}\,\Delta\left(\frac{\mathfrak{G}^*}{\sqrt{-g}}\right) = S_\sigma^\nu\frac{\partial \Delta x_\sigma}{\partial x_\nu} + 2\frac{\partial \mathfrak{G}^*}{\partial g_\alpha^{\mu\nu}}g^{\mu\nu}\frac{\partial^2 \Delta x_\sigma}{\partial x_\nu \partial x_\alpha},$$ (13)

where we used the abbreviation

$$S_\sigma^\nu = 2\frac{\partial \mathfrak{G}^*}{\partial g^{\mu\sigma}}g^{\mu\nu} + 2\frac{\partial \mathfrak{G}^*}{\partial g_\alpha^{\mu\sigma}}g_\alpha^{\mu\nu} + \mathfrak{G}^*\delta_\sigma^\nu - \frac{\partial \mathfrak{G}^*}{\partial g_\nu^{\mu\alpha}}g_\sigma^{\mu\alpha}.$$ (14) [9]

From these two equations we draw two conclusions that are important in the following. We know $\dfrac{\mathfrak{G}}{\sqrt{-g}}$ to be an invariant under arbitrary substitutions but not $\dfrac{\mathfrak{G}^*}{\sqrt{-g}}$. It is, however, easy to show that the latter quantity is invariant under *linear* substitutions of the coordinates. Consequently, the right-hand side of (13) must always vanish when all $\dfrac{\partial^2 \Delta x_\sigma}{\partial x_\nu \partial x_\alpha}$ do. Then it follows that $\mathfrak{G}^*$ must satisfy the identity

$$S_\sigma^\nu \equiv 0.$$ (15)

If we furthermore choose the Δx_ν such that they differ from zero only inside the domain considered, but vanish in an infinitesimal neighborhood of the boundary, then [p. 1115] the value of the integral in equation (2) extended over the boundary does not change during the transformation. We therefore have

$$\Delta(F) = 0$$

and thus[6]

$$\Delta\left\{\int \mathfrak{G}d\tau\right\} = \Delta\left\{\int \overset{*}{\mathfrak{G}}d\tau\right\}.$$

But the left-hand side of the equation must vanish since both $\dfrac{\mathfrak{G}}{\sqrt{-g}}$ and $\sqrt{-g}\,d\tau$ are

[10] {6} invariants. Consequently, the right-hand side vanishes also. Due to (13), (14), and (15) we next get the equation

$$\int \frac{\partial \overset{*}{\mathfrak{G}}}{\partial g_\alpha^{\mu\sigma}} g^{\mu\nu} \frac{\partial^2 \Delta x_\sigma}{\partial x_\nu \partial x_\alpha} d\tau = 0. \tag{16}$$

Rearranging after twofold partial integration, and considering the free choice of the Δx_σ, one has the identity

$$\frac{\partial^2}{\partial x_\nu \partial x_\alpha}\left(\frac{\partial \overset{*}{\mathfrak{G}}}{\partial g_\alpha^{\mu\sigma}} g^{\mu\nu}\right) \equiv 0. \tag{17}$$

[11] {7} We now have to draw conclusions from the two identities (15) and (17), which follow from the invariance of $\dfrac{\mathfrak{G}}{\sqrt{-g}}$, i.e., from the postulate of general relativity.

The field equations (7) of gravitation can be transformed first by mixed multiplication with $g^{\mu\nu}$. One obtains then (also exchanging the indices σ and ν) as an equivalent of the field equations (7) the equations

$$\frac{\partial}{\partial x_\alpha}\left(\frac{\partial \overset{*}{\mathfrak{G}}}{\partial g_\alpha^{\mu\sigma}} g^{\mu\nu}\right) = -(\mathfrak{T}_\sigma^\nu + \mathfrak{t}_\sigma^\nu), \tag{18}$$

where we put

$$\mathfrak{T}_\sigma^\nu = -\frac{\partial \mathfrak{M}}{\partial g^{\mu\sigma}} g^{\mu\nu} \tag{19}$$

$$\mathfrak{t}_\sigma^\nu = -\left(\frac{\partial \overset{*}{\mathfrak{G}}}{\partial g_\alpha^{\mu\sigma}} g_\alpha^{\mu\nu} + \frac{\partial \overset{*}{\mathfrak{G}}}{\partial g^{\mu\sigma}} g^{\mu\nu}\right) = \frac{1}{2}\left(\overset{*}{\mathfrak{G}} \delta_\sigma^\nu - \frac{\partial \overset{*}{\mathfrak{G}}}{\partial g_\nu^{\mu\alpha}} g_\sigma^{\mu\alpha}\right). \tag{20}$$

The latter expression for $\mathfrak{t}_\sigma^\nu$ is justified by (14) and (15). After differentiation of (18) with respect to x_n and summation over ν, with consideration of (17), follows

---

[6]Introducing $\mathfrak{G}$ and $\mathfrak{G}^*$ instead of $\mathfrak{H}$ and $\mathfrak{H}^*$.

$$\frac{\partial}{\partial x_\nu} \, (\mathfrak{T}_\sigma^\nu + \mathfrak{t}_\sigma^\nu) = 0. \tag{21}$$

Equation (21) expresses the conservation of the momentum and the energy. We [p. 1116] call $\mathfrak{T}_\sigma^\nu$ the components of the energy of matter, $\mathfrak{t}_\sigma^\nu$ the components of the energy of the gravitational field.

From the field equations (7) of gravitation follows (after multiplication by $g_\sigma^{\mu\nu}$, summation over μ and ν, and on account of (20))

$$\frac{\partial \mathfrak{t}_\sigma^\nu}{\partial x_\nu} + \frac{1}{2} g_\sigma^{\mu\nu} \frac{\partial \mathfrak{M}}{\partial g^{\mu\nu}} = 0,$$

or, taking (19) and (21) into account,

$$\frac{\partial \mathfrak{T}_\sigma^\nu}{\partial x_\nu} + \frac{1}{2} g_\sigma^{\mu\nu} \mathfrak{T}_{\mu\nu} = 0, \tag{22} \quad [12] \ \{8\}$$

where $\mathfrak{T}_{\mu\nu}$ denotes the quantities $g_{\nu\sigma} \mathfrak{T}_\mu^\sigma$. These are four equations that the energy components of matter have to satisfy.

It is to be emphasized that the (generally covariant) conservation theorems (21) and (22) have been derived—using also the postulate of general covariance (relativity)—from the field equations (7) of gravitation *alone*, without use of the field equations (8) for material processes.

[13]

Additional notes by translator

In his footnote 1), just prior to equations (4) and (5), Einstein introduces into tensor calculus, in a formal manner, the rule of abbreviated writing of summations, which is now generally known as the Einstein summation convention. It was introduced in Doc. 30, p. 296.

{1}   The "q" with 2 subscripts and 2 superscripts has been corrected here to "g"; the 2nd derivative of "q" inside the following parenthesis has also been corrected to "g," both with indices as indicated. Editorial notes [6] and [7] relate to similar typesetting errors.

{2}   "$q_{\sigma\tau}^{\mu\nu}$" has been corrected here to "$g_{\sigma\tau}^{\mu\nu}$."

{3}   "(4a)" has been corrected to "(5)."

{4}   "$q^{\mu\nu}$" has been corrected here to "$g^{\mu\nu}$".

{5}   "(13)" and "(14)" have been corrected here to "(11)" and "(12)."

{6}   "(14), (15), (16)" have been corrected here to "(13), (14), (15)."

{7}   "(16)" has been corrected to "(15)."

{8}   "—" preceding the factor $\frac{1}{2}$ has been corrected here to "+".

Doc. 42

On the Special and General Theory of Relativity
(A Popular Account)

Relativity

The Special
and the General Theory

by Albert Einstein

AUTHORIZED TRANSLATION BY
ROBERT W. LAWSON

Preface

The present book is intended, as far as possible, to give an exact insight into the theory of Relativity to those readers who, from a general scientific and philosophical point of view, are interested in the theory, but who are not conversant with the mathematical apparatus of theoretical physics. The work presumes a standard of education corresponding to that of a university matriculation examination, and, despite the shortness of the book, a fair amount of patience and force of will on the part of the reader. The author has spared himself no pains in his endeavour to present the main ideas in the simplest and most intelligible form, and on the whole, in the sequence and connection in which they actually originated. In the interest of clearness, it appeared to me inevitable that I should repeat myself frequently, without paying the slightest attention to the elegance of the presentation. I adhered scrupulously to the precept of that brilliant theoretical physicist L. Boltzmann, according to whom matters of elegance ought to be left to the tailor and to the cobbler. I make no pretence of having withheld from the reader diffculties which are inherent to the subject. On the other hand, I have purposely treated the empirical physical foundations of the theory in a "step-motherly" fashion, so that readers unfamiliar with physics may not feel like the wanderer who was unable to see the forest for trees. May the book bring some one a few happy hours of suggestive thought!

[4]   *December 1916*   A. EINSTEIN

Part I

The Special Theory of Relativity

ONE

Physical Meaning of Geometrical Propositions

In your schooldays most of you who read this book made acquaintance with the noble building of Euclid's geometry, and you remember—perhaps with more respect than love—the magnificent structure, on the lofty staircase of which you were chased about for uncounted hours by conscientious teachers. By reason of your past experience, you would certainly regard everyone with disdain who should pronounce even the most out-of-the-way proposition of this science to be untrue. But perhaps this feeling of proud certainty would leave you immediately if some one were to ask you: "What, then, do you mean by the assertion that these propositions are true?" Let us proceed to give this question a little consideration.

Geometry sets out from certain conceptions such as "plane," "point," and "straight line," with which we are able to associate more or less definite ideas, and from certain simple propositions (axioms) which, in virtue of these ideas, we are inclined to accept as "true." Then, on the basis of a logical process, the justification of which we feel ourselves compelled to admit, all remaining propositions are shown to follow from

4   *Relativity*

those axioms, *i.e.* they are proven. A proposition is then correct
("true") when it has been derived in the recognised manner
from the axioms. The question of the "truth" of the individual
geometrical propositions is thus reduced to one of the "truth"
of the axioms. Now it has long been known that the last question is not only unanswerable by the methods of geometry, but
that it is in itself entirely without meaning. We cannot ask
whether it is true that only one straight line goes through two
points. We can only say that Euclidean geometry deals with
things called "straight lines," to each of which is ascribed the
property of being uniquely determined by two points situated
on it. The concept "true" does not tally with the assertions of
pure geometry, because by the word "true" we are eventually
in the habit of designating always the correspondence with a
"real" object; geometry, however, is not concerned with the
relation of the ideas involved in it to objects of experience, but
only with the logical connection of these ideas among themselves.

It is not difficult to understand why, in spite of this, we feel
constrained to call the propositions of geometry "true." Geometrical ideas correspond to more or less exact objects in
nature, and these last are undoubtedly the exclusive cause of
the genesis of those ideas. Geometry ought to refrain from
such a course, in order to give to its structure the largest
possible logical unity. The practice, for example, of seeing in
a "distance" two marked positions on a practically rigid body
is something which is lodged deeply in our habit of thought.
We are accustomed further to regard three points as being
situated on a straight line, if their apparent positions can be

Physical Meaning of Geometrical Propositions   5

made to coincide for observation with one eye, under suitable choice of our place of observation.

If, in pursuance of our habit of thought, we now supplement the propositions of Euclidean geometry by the single proposition that two points on a practically rigid body always correspond to the same distance (line-interval), independently of any changes in position to which we may subject the body, the propositions of Euclidean geometry then resolve themselves into propositions on the possible relative position of practically rigid bodies.[1] Geometry which has been supplemented in this way is then to be treated as a branch of physics. We can now legitimately ask as to the "truth" of geometrical propositions interpreted in this way, since we are justified in asking whether these propositions are satisfied for those real things we have associated with the geometrical ideas. In less exact terms we can express this by saying that by the "truth" of a geometrical proposition in this sense we understand its validity for a construction with ruler and compasses.

Of course the conviction of the "truth" of geometrical propositions in this sense is founded exclusively on rather incomplete experience. For the present we shall assume the "truth" of the geometrical propositions, then at a later stage (in the general theory of relativity) we shall see that this "truth" is limited, and we shall consider the extent of its limitation.

---

[1] It follows that a natural object is associated also with a straight line. Three points A, B and C on a rigid body thus lie in a straight line when, the points A and C being given, B is chosen such that the sum of the distances A B and B C is as short as possible. This incomplete suggestion will suffice for our present purpose.

TWO

The System of Co-ordinates

On the basis of the physical interpretation of distance which has been indicated, we are also in a position to establish the distance between two points on a rigid body by means of measurements. For this purpose we require a "distance" (rod S) which is to be used once and for all, and which we employ as a standard measure. If, now, A and B are two points on a rigid body, we can construct the line joining them according to the rules of geometry; then, starting from A, we can mark off the distance S time after time until we reach B. The number of these operations required is the numerical measure of the distance $A B$. This is the basis of all measurement of length.[1]

Every description of the scene of an event or of the position of an object in space is based on the specification of the point on a rigid body (body of reference) with which that event or object coincides. This applies not only to scientific descrip-

---

[1] Here we have assumed that there is nothing left over, *i.e.* that the measurement gives a whole number. This difficulty is got over by the use of divided measuring-rods, the introduction of which does not demand any fundamentally new method.

6

The System of Co-ordinates   7

tion, but also to everyday life. If I analyse the place specification "Trafalgar Square, London,"[1] I arrive at the following result. The earth is the rigid body to which the specification of place refers; "Trafalgar Square, London," is a well-defined point, to which a name has been assigned, and with which the event coincides in space.[2]

This primitive method of place specification deals only with places on the surface of rigid bodies, and is dependent on the existence of points on this surface which are distinguishable from each other. But we can free ourselves from both of these limitations without altering the nature of our specification of position. If, for instance, a cloud is hovering over Trafalgar Square, then we can determine its position relative to the surface of the earth by erecting a pole perpendicularly on the Square, so that it reaches the cloud. The length of the pole measured with the standard measuring-rod, combined with the specification of the position of the foot of the pole, supplies us with a complete place specification. On the basis of this illustration, we are able to see the manner in which a refinement of the conception of position has been developed.

(a) We imagine the rigid body, to which the place specification is referred, supplemented in such a manner that the object whose position we require is reached by the completed rigid body.

---

[1] I have chosen this as being more familiar to the English reader than the "Potsdamer Platz, Berlin," which is referred to in the original. (R. W. L.)

[2] It is not necessary here to investigate further the significance of the expression "coincidence in space." This conception is sufficiently obvious to ensure that differences of opinion are scarcely likely to arise as to its applicability in practice.

(b) In locating the position of the object, we make use of a number (here the length of the pole measured with the measuring-rod) instead of designated points of reference.

(c) We speak of the height of the cloud even when the pole which reaches the cloud has not been erected. By means of optical observations of the cloud from different positions on the ground, and taking into account the properties of the propagation of light, we determine the length of the pole we should have required in order to reach the cloud.

From this consideration we see that it will be advantageous if, in the description of position, it should be possible by means of numerical measures to make ourselves independent of the existence of marked positions (possessing names) on the rigid body of reference. In the physics of measurement this is attained by the application of the Cartesian system of co-ordinates.

This consists of three plane surfaces perpendicular to each other and rigidly attached to a rigid body. Referred to a system of co-ordinates, the scene of any event will be determined (for the main part) by the specification of the lengths of the three perpendiculars or co-ordinates (x, y, z) which can be dropped from the scene of the event to those three plane surfaces. The lengths of these three perpendiculars can be determined by a series of manipulations with rigid measuring-rods performed according to the rules and methods laid down by Euclidean geometry.

In practice, the rigid surfaces which constitute the system of co-ordinates are generally not available; furthermore, the magnitudes of the co-ordinates are not actually determined by

[5]

The System of Co-ordinates   9

constructions with rigid rods, but by indirect means. If the results of physics and astronomy are to maintain their clearness, the physical meaning of specifications of position must always be sought in accordance with the above considerations.[1]

We thus obtain the following result: Every description of events in space involves the use of a rigid body to which such events have to be referred. The resulting relationship takes for granted that the laws of Euclidean geometry hold for "distances," the "distance" being represented physically by means of the convention of two marks on a rigid body.

---

[1] A refinement and modification of these views does not become necessary until we come to deal with the general theory of relativity, treated in the second part of this book.

THREE

Space and Time in Classical Mechanics

The purpose of mechanics is to describe how bodies change their position in space with "time." I should load my conscience with grave sins against the sacred spirit of lucidity were I to formulate the aims of mechanics in this way, without serious reflection and detailed explanations. Let us proceed to disclose these sins.

It is not clear what is to be understood here by "position" and "space." I stand at the window of a railway carriage which is travelling uniformly, and drop a stone on the embankment, without throwing it. Then, disregarding the influence of the air resistance, I see the stone descend in a straight line. A pedestrian who observes the misdeed from the footpath notices that the stone falls to earth in a parabolic curve. I now ask: Do the "positions" traversed by the stone lie "in reality" on a straight line or on a parabola? Moreover, what is meant here by motion "in space"? From the considerations of the previous section the answer is self-evident. In the first place we entirely shun the vague word "space," of which, we must honestly acknowledge,

10

Space and Time in Classical Mechanics   *11*

we cannot form the slightest conception, and we replace it by
"motion relative to a practically rigid body of reference." The
positions relative to the body of reference (railway carriage or
embankment) have already been defined in detail in the preceding section. If instead of "body of reference" we insert
"system of co-ordinates," which is a useful idea for mathematical description, we are in a position to say: The stone traverses
a straight line relative to a system of co-ordinates rigidly attached to the carriage, but relative to a system of co-ordinates
rigidly attached to the ground (embankment) it describes a parabola. With the aid of this example it is clearly seen that there
is no such thing as an independently existing trajectory (lit.
"path-curve"[1]), but only a trajectory relative to a particular
body of reference.

[6]

In order to have a *complete* description of the motion, we
must specify how the body alters its position *with time*; *i.e.* for
every point on the trajectory it must be stated at what time the
body is situated there. These data must be supplemented by
such a definition of time that, in virtue of this definition, these
time-values can be regarded essentially as magnitudes (results
of measurements) capable of observation. If we take our stand
on the ground of classical mechanics, we can satisfy this requirement for our illustration in the following manner. We
imagine two clocks of identical construction; the man at the
railway-carriage window is holding one of them, and the man
on the footpath the other. Each of the observers determines

---

[1] That is, a curve along which the body moves.

12   *Relativity*

the position on his own reference-body occupied by the stone
at each tick of the clock he is holding in his hand. In this
connection we have not taken account of the inaccuracy involved by the finiteness of the velocity of propagation of light.
With this and with a second difficulty prevailing here we shall
have to deal in detail later.

FOUR

The Galileian System of Co-ordinates

As is well known, the fundamental law of the mechanics of Galilei-Newton, which is known as the *law of inertia*, can be stated thus: A body removed sufficiently far from other bodies continues in a state of rest or of uniform motion in a straight line. This law not only says something about the motion of the bodies, but it also indicates the reference-bodies or systems of co-ordinates, permissible in mechanics, which can be used in mechanical description. The visible fixed stars are bodies for which the law of inertia certainly holds to a high degree of approximation. Now if we use a system of co-ordinates which is rigidly attached to the earth, then, relative to this system, every fixed star describes a circle of immense radius in the course of an astronomical day, a result which is opposed to the statement of the law of inertia. So that if we adhere to this law we must refer these motions only to systems of co-ordinates relative to which the fixed stars do not move in a

13

circle. A system of co-ordinates of which the state of motion is such that the law of inertia holds relative to it is called a "Galileian system of co-ordinates." The laws of the mechanics of Galilei-Newton can be regarded as valid only for a Galileian system of co-ordinates.

FIVE

The Principle of Relativity
(in the Restricted Sense)

In order to attain the greatest possible clearness, let us return to our example of the railway carriage supposed to be travelling uniformly. We call its motion a uniform translation ("uniform" because it is of constant velocity and direction, "translation" because although the carriage changes its position relative to the embankment yet it does not rotate in so doing). Let us imagine a raven flying through the air in such a manner that its motion, as observed from the embankment, is uniform and in a straight line. If we were to observe the flying raven from the moving railway carriage, we should find that the motion of the raven would be one of different velocity and direction, but that it would still be uniform and in a straight line. Expressed in an abstract manner we may say: If a mass m is moving uniformly in a straight line with respect to a co-ordinate system K, then it will also be moving uniformly and in a straight line relative to a second co-ordinate system K', provided that the latter is executing a uniform translatory motion with respect to K. In accordance with the discussion contained in the preceding section, it follows that:

15

16   *Relativity*

[7]   If K is a Galileian co-ordinate system, then every other
co-ordinate system K' is a Galileian one, when, in relation to
K, it is in a condition of uniform motion of translation. Relative to K' the mechanical laws of Galilei-Newton hold good
exactly as they do with respect to K.

We advance a step farther in our generalisation when we
express the tenet thus: If, relative to K, K' is a uniformly
moving co-ordinate system devoid of rotation, then natural
phenomena run their course with respect to K' according to
exactly the same general laws as with respect to K. This statement is called the *principle of relativity* (in the restricted sense).

As long as one was convinced that all natural phenomena
were capable of representation with the help of classical mechanics, there was no need to doubt the validity of this principle of relativity. But in view of the more recent development
of electrodynamics and optics it became more and more evident that classical mechanics affords an insufficient foundation for the physical description of all natural phenomena. At
this juncture the question of the validity of the principle of
relativity became ripe for discussion, and it did not appear
impossible that the answer to this question might be in the
negative.

Nevertheless, there are two general facts which at the outset speak very much in favour of the validity of the principle
of relativity. Even though classical mechanics does not supply
us with a sufficiently broad basis for the theoretical presentation of all physical phenomena, still we must grant it a considerable measure of "truth," since it supplies us with the
actual motions of the heavenly bodies with a delicacy of detail

little short of wonderful. The principle of relativity must therefore apply with great accuracy in the domain of *mechanics*. But that a principle of such broad generality should hold with such exactness in one domain of phenomena, and yet should be invalid for another, is *a priori* not very probable.

We now proceed to the second argument, to which, moreover, we shall return later. If the principle of relativity (in the restricted sense) does not hold, then the Galileian co-ordinate systems K, K', K'', etc., which are moving uniformly relative to each other, will not be *equivalent* for the description of natural phenomena. In this case we should be constrained to believe that natural laws are capable of being formulated in a particularly simple manner, and of course only on condition that, from amongst all possible Galileian co-ordinate systems, we should have chosen *one* (K_0) of a particular state of motion as our body of reference. We should then be justified (because of its merits for the description of natural phenomena) in calling this system "absolutely at rest," and all other Galileian systems K "in motion." If, for instance, our embankment were the system K_0, then our railway carriage would be a system K, relative to which less simple laws would hold than with respect to K_0. This diminished simplicity would be due to the fact that the carriage K would be in motion (*i.e.* "really") with respect to K_0. In the general laws of nature which have been formulated with reference to K, the magnitude and direction of the velocity of the carriage would necessarily play a part. We should expect, for instance, that the note emitted by an organ-pipe placed with its axis parallel to the direction of travel would be different from that emitted if the axis of the pipe were placed perpendicular

18   *Relativity*

[8] to this direction. Now in virtue of its motion in an orbit round the sun, our earth is comparable with a railway carriage travelling with a velocity of about 30 kilometres per second. If the principle of relativity were not valid we should therefore expect that the direction of motion of the earth at any moment would enter into the laws of nature, and also that physical systems in their behaviour would be dependent on the orientation in space with respect to the earth. For owing to the alteration in direction of the velocity of revolution of the earth in the course of a year, the earth cannot be at rest relative to the hypothetical system K_0 throughout the whole year. However, the most careful observations have never revealed such anisotropic properties in terrestrial physical space, *i.e.* a physical non-equivalence of different directions. This is very powerful argument in favour of the principle of relativity.

SIX

The Theorem of the Addition of Velocities Employed in Classical Mechanics

Let us suppose our old friend the railway carriage to be travelling along the rails with a constant velocity v, and that a man traverses the length of the carriage in the direction of travel with a velocity w. How quickly or, in other words, with what velocity W does the man advance relative to the embankment during the process? The only possible answer seems to result from the following consideration: If the man were to stand still for a second, he would advance relative to the embankment through a distance v equal numerically to the velocity of the carriage. As a consequence of his walking, however, he traverses an additional distance w relative to the carriage, and hence also relative to the embankment, in this second, the distance w being numerically equal to the velocity with which he is walking. Thus in total he covers the distance $W = v + w$ relative to the embankment in the second considered. We shall see later that this result, which expresses the theorem of the addition of velocities employed in classical mechanics, cannot be maintained; in other words, the law that we have just written down does not hold in reality. For the time being, however, we shall assume its correctness.

SEVEN

The Apparent Incompatibility of the
Law of Propagation of Light
with the Principle of Relativity

There is hardly a simpler law in physics than that according to which light is propagated in empty space. Every child at school knows, or believes he knows, that this propagation takes place in straight lines with a velocity c = 300,000 km./sec. At all events we know with great exactness that this velocity is the same for all colours, because if this were not the case, the minimum of emission would not be observed simultaneously for different colours during the eclipse of a fixed star by its dark neighbour. By means of similar considerations based on observations of double stars, the Dutch astronomer De Sitter was also able to show that the velocity of propagation of light cannot depend on the velocity [9] of motion of the body emitting the light. The assumption that this velocity of propagation is dependent on the direction "in space" is in itself improbable.

In short, let us assume that the simple law of the constancy of the velocity of light c (in vacuum) is justifiably believed by the child at school. Who would imagine that this simple law has plunged the conscientiously thoughtful physicist into the

21

greatest intellectual difficulties? Let us consider how these difficulties arise.

Of course we must refer the process of the propagation of light (and indeed every other process) to a rigid reference-body (co-ordinate system). As such a system let us again choose our embankment. We shall imagine the air above it to have been removed. If a ray of light be sent along the embankment, we see from the above that the tip of the ray will be transmitted with the velocity c relative to the embankment. Now let us suppose that our railway carriage is again travelling along the railway lines with the velocity v, and that its direction is the same as that of the ray of light, but its velocity of course much less. Let us inquire about the velocity of propagation of the ray of light relative to the carriage. It is obvious that we can here apply the consideration of the previous section, since the ray of light plays the part of the man walking along relatively to the carriage. The velocity W of the man relative to the embankment is here replaced by the velocity of light relative to the embankment. w is the required velocity of light with respect to the carriage, and we have

$$w = c - v.$$

The velocity of propagation of a ray of light relative to the carriage thus comes out smaller than c.

But this result comes into conflict with the principle of relativity set forth in Section 5. For, like every other general law of nature, the law of the transmission of light *in vacuo* must, according to the principle of relativity, be the same for the railway carriage as reference-body as when the rails are the body of reference. But, from our above consideration, this would appear to be impossible. If every ray of light is propagated relative to the embankment with the velocity c, then for this reason it would appear that another law of propagation of light must necessarily hold with respect to the carriage—a result contradictory to the principle of relativity.

In view of this dilemma there appears to be nothing else for it than to abandon either the principle of relativity or the simple law of the propagation of light *in vacuo*. Those of you who have carefully followed the preceding discussion are almost sure to expect that we should retain the principle of relativity, which appeals so convincingly to the intellect because it is so natural and simple. The law of the propagation of light *in vacuo* would then have to be replaced by a more complicated law comformable to the principle of relativity. The development of theoretical physics shows, however, that we cannot pursue this course. The epoch-making theoretical investigations of H. A. Lorentz on the electrodynamical and optical phenomena connected with moving bodies show that experience in this domain leads conclusively to a theory of electromagnetic phenomena, of which the law of the constancy of the velocity of light *in vacuo* is a necessary conse-

[10]   quence. Prominent theoretical physicists were therefore more inclined to reject the principle of relativity, in spite of the fact that no empirical data had been found which were contradictory to this principle.

At this juncture the theory of relativity entered the arena. As a result of an analysis of the physical conceptions of time

24   *Relativity*

and space, it became evident that *in reality there is not the least incompatibility between the principle of relativity and the law of propagation of light*, and that by systematically holding fast to both these laws a logically rigid theory could be arrived at. This theory has been called the *special theory of relativity* to distinguish it from the extended theory, with which we shall deal later. In the following pages we shall present the fundamental ideas of the special theory of relativity.   [11]

EIGHT

On the Idea of
Time in Physics

Lightning has struck the rails on our railway embankment at two places *A* and *B* far distant from each other. I make the additional assertion that these two lightning flashes occurred simultaneously. If I ask you whether there is sense in this statement, you will answer my question with a decided "Yes." But if I now approach you with the request to explain to me the sense of the statement more precisely, you find after some consideration that the answer to this question is not so easy as it appears at first sight.

After some time perhaps the following answer would occur to you: "The significance of the statement is clear in itself and needs no further explanation; of course it would require some consideration if I were to be commissioned to determine by observations whether in the actual case the two events took place simultaneously or not." I cannot be satisfied with this answer for the following reason. Supposing that as a result of ingenious consideration an able meteorologist were to discover that the lightning must always strike the places *A* and *B* simultaneously, then we should be faced with the task of

testing whether or not this theoretical result is in accordance
with the reality. We encounter the same difficulty with all
physical statements in which the conception "simultaneous"
plays a part. The concept does not exist for the physicist until
he has the possibility of discovering whether or not it is fulfilled in an actual case. We thus require a definition of simultaneity such that this definition supplies us with the method
by means of which, in the present case, he can decide by
experiment whether or not both the lightning strokes occurred
simultaneously. As long as this requirement is not satisfied, I
allow myself to be deceived as a physicist (and of course the
same applies if I am not a physicist), when I imagine that I am
able to attach a meaning to the statement of simultaneity. (I
would ask the reader not to proceed farther until he is fully
convinced on this point.)

After thinking the matter over for some time you then offer
the following suggestion with which to test simultaneity. By
measuring along the rails, the connecting line *AB* should be
measured up and an observer placed at the mid-point *M* of the
distance *AB*. This observer should be supplied with an arrangement (*e.g.* two mirrors inclined at 90°) which allows him
visually to observe both places *A* and *B* at the same time. If
the observer perceives the two flashes of lightning at the same
time, then they are simultaneous.

I am very pleased with this suggestion, but for all that I
cannot regard the matter as quite settled, because I feel constrained to raise the following objection: "Your definition
would certainly be right, if only I knew that the light by means
of which the observer at *M* perceives the lightning flashes

travels along the length $A \longrightarrow M$ with the same velocity as along the length $B \longrightarrow M$. But an examination of this supposition would only be possible if we already had at our disposal the means of measuring time. It would thus appear as though we were moving here in a logical circle."

After further consideration you cast a somewhat disdainful glance at me—and rightly so—and you declare: "I maintain my previous definition nevertheless, because in reality it assumes absolutely nothing about light. There is only *one* demand to be made of the definition of simultaneity, namely, that in every real case it must supply us with an empirical decision as to whether or not the conception that has to be defined is fulfilled. That my definition satisfies this demand is indisputable. That light requires the same time to traverse the path $A \longrightarrow M$ as for the path $B \longrightarrow M$ is in reality neither a *supposition nor a hypothesis* about the physical nature of light, but a *stipulation* which I can make of my own freewill in order to arrive at a definition of simultaneity."

It is clear that this definition can be used to give an exact meaning not only to *two* events, but to as many events as we care to choose, and independently of the positions of the scenes of the events with respect to the body of reference[1] (here the railway embankment). We are thus led also to a definition of "time" in physics. For this purpose we suppose

---

[1] We suppose further, that, when three events A, B and C occur in different places in such a manner that A is simultaneous with B, and B is simultaneous with C (simultaneous in the sense of the above definition), then the criterion for the simultaneity of the pair of events A, C is also satisfied. This assumption is a physical hypothesis about the law of propagation of light; it must certainly be fulfilled if we are to maintain the law of the constancy of the velocity of light *in vacuo*.

that clocks of identical construction are placed at the points *A*, *B* and *C* of the railway line (co-ordinate system), and that they are set in such a manner that the positions of their pointers are simultaneously (in the above sense) the same. Under these conditions we understand by the "time" of an event the reading (position of the hands) of that one of these clocks which is in the immediate vicinity (in space) of the event. In this manner a time-value is associated with every event which is essentially capable of observation.

This stipulation contains a further physical hypothesis, the validity of which will hardly be doubted without empirical evidence to the contrary. It has been assumed that all these clocks go *at the same rate* if they are of identical construction. Stated more exactly: When two clocks arranged at rest in different places of a reference-body are set in such a manner that a *particular* position of the pointers of the one clock is *simultaneous* (in the above sense) with the *same* position of the pointers of the other clock, then identical "settings" are always simultaneous (in the sense of the above definition).

[12]

NINE

The Relativity of
Simultaneity

Up to now our considerations have been referred to a particular body of reference, which we have styled a "railway embankment." We suppose a very long train travelling along the rails with the constant velocity v and in the direction indicated in Fig. 1. People travelling in this train

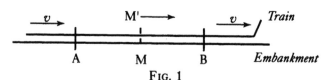

FIG. 1

will with advantage use the train as a rigid reference-body (co-ordinate system); they regard all events in reference to the train. Then every event which takes place along the line also takes place at a particular point of the train. Also the definition of simultaneity can be given relative to the train in exactly the same way as with respect to the embankment. As a natural consequence, however, the following question arises:

Are two events (*e.g.* the two strokes of lightning A and B) which are simultaneous *with reference to the railway embankment*

29

also simultaneous *relatively to the train?* We shall show directly that the answer must be in the negative.

When we say that the lightning strokes A and B are simultaneous with respect to the embankment, we mean: the rays of light emitted at the places A and B, where the lightning occurs, meet each other at the mid-point M of the length $A \longrightarrow B$ of the embankment. But the events A and B also correspond to positions A and B on the train. Let M' be the mid-point of the distance $A \longrightarrow B$ on the travelling train. Just when the flashes[1] of lightning occur, this point M' naturally coincides with the point M, but it moves towards the right in the diagram with the velocity v of the train. If an observer sitting in the position M' in the train did not possess this velocity, then he would remain permanently at M, and the light rays emitted by the flashes of lightning A and B would reach him simultaneously, *i.e.* they would meet just where he is situated. Now in reality (considered with reference to the railway embankment) he is hastening towards the beam of light coming from B, whilst he is riding on ahead of the beam of light coming from A. Hence the observer will see the beam of light emitted from B earlier than he will see that emitted from A. Observers who take the railway train as their reference-body must therefore come to the conclusion that the lightning flash B took place earlier than the lightning flash A. We thus arrive at the important result:

Events which are simultaneous with reference to the embankment are not simultaneous with respect to the train, and

---

[1] As judged from the embankment.

vice versa (relativity of simultaneity). Every reference-body (co-ordinate system) has its own particular time; unless we are told the reference-body to which the statement of time refers, there is no meaning in a statement of the time of an event.

Now before the advent of the theory of relativity it had always tacitly been assumed in physics that the statement of time had an absolute significance, *i.e.* that it is independent of the state of motion of the body of reference. But we have just seen that this assumption is incompatible with the most natural definition of simultaneity; if we discard this assumption, then the conflict between the law of the propagation of light *in vacuo* and the principle of relativity (developed in Section 7) disappears.

We were led to that conflict by the considerations of Section 6, which are now no longer tenable. In that section we concluded that the man in the carriage, who traverses the distance *w per second* relative to the carriage, traverses the same distance also with respect to the embankment *in each second* of time. But, according to the foregoing considerations, the time required by a particular occurrence with respect to the carriage must not be considered equal to the duration of the same occurrence as judged from the embankment (as reference-body). Hence it cannot be contended that the man in walking travels the distance *w* relative to the railway line in a time which is equal to one second as judged from the embankment.

Moreover, the considerations of Section 6 are based on yet a second assumption, which, in the light of a strict consideration, appears to be arbitrary, although it was always tacitly made even before the introduction of the theory of relativity.

TEN

On the Relativity
of the Conception
of Distance

Let us consider two particular points on the train[1] travelling along the embankment with the velocity v, and inquire as to their distance apart. We already know that it is necessary to have a body of reference for the measurement of a distance, with respect to which body the distance can be measured up. It is the simplest plan to use the train itself as reference-body (co-ordinate system). An observer in the train measures the interval by marking off his measuring-rod in a straight line (*e.g.* along the floor of the carriage) as many times as is necessary to take him from the one marked point to the other. Then the number which tells us how often the rod has to be laid down is the required distance.

It is a different matter when the distance has to be judged from the railway line. Here the following method suggests itself. If we call A' and B' the two points on the train whose distance apart is required, then both of these points are mov-

---

[1] *E.g.* the middle of the first and of the twentieth carriage.

32

ing with the velocity v along the embankment. In the first place we require to determine the points A and B of the embankment which are just being passed by the two points A' and B' at a particular time t—judged from the embankment. These points A and B of the embankment can be determined by applying the definition of time given in Section 8. The distance between these points A and B is then measured by repeated application of the measuring-rod along the embankment.

A priori it is by no means certain that this last measurement will supply us with the same result as the first. Thus the length of the train as measured from the embankment may be different from that obtained by measuring in the train itself. This circumstance leads us to a second objection which must be raised against the apparently obvious consideration of Section 6. Namely, if the man in the carriage covers the distance w in a unit of time—*measured from the train*—then this distance—*as measured from the embankment*—is not necessarily also equal to w.

ELEVEN

The Lorentz Transformation

The results of the last three sections show that the apparent incompatibility of the law of propagation of light with the principle of relativity (Section 7) has been derived by means of a consideration which borrowed two unjustifiable hypotheses from classical mechanics; these are as follows:

(1) The time-interval (time) between two events is independent of the condition of motion of the body of reference.

(2) The space-interval (distance) between two points of a rigid body is independent of the condition of motion of the body of reference.

If we drop these hypotheses, then the dilemma of Section 7 disappears, because the theorem of the addition of velocities derived in Section 6 becomes invalid. The possibility presents itself that the law of the propagation of light *in vacuo* may be compatible with the principle of relativity, and the question arises: How have we to modify the considerations of Section 6 in order to remove the apparent disagreement between these

34

two fundamental results of experience? This question leads to a general one. In the discussion of Section 6 we have to do with places and times relative both to the train and to the embankment. How are we to find the place and time of an event in relation to the train, when we know the place and time of the event with respect to the railway embankment? Is there a thinkable answer to this question of such a nature that the law of transmission of light *in vacuo* does not contradict the principle of relativity? In other words: Can we conceive of a relation between place and time of the individual events relative to both reference-bodies, such that every ray of light possesses the velocity of transmission c relative to the embankment and relative to the train? This question leads to a quite definite positive answer, and to a perfectly definite transformation law for the space-time magnitudes of an event when changing over from one body of reference to another.

Before we deal with this, we shall introduce the following incidental consideration. Up to the present we have only considered events taking place along the embankment, which had mathematically to assume the function of a straight line. In the manner indicated in Section 2 we can imagine this reference-body supplemented laterally and in a vertical direction by means of a framework of rods, so that an event which takes place anywhere can be localised with reference to this framework. Similarly, we can imagine the train travelling with the velocity v to be continued across the whole of space, so that every event, no matter how far off it may be, could also be localised with respect to the second framework. Without committing any fundamental error, we can disregard the fact that in reality these frameworks would continually interfere with each other, owing to the impenetrability of solid bodies. In every such framework we imagine three surfaces perpendicular to each other marked out, and designated as "co-ordinate planes" ("co-ordinate system"). A co-ordinate system K then corresponds to the embankment, and a co-ordinate system K' to the train. An event, wherever it may have taken place, would be fixed in space with respect to K by the three perpendiculars x, y, z on the co-ordinate planes, and with regard to time by a time-value t. Relative to K', *the same event* would be fixed in respect of space and time by corresponding values x', y', z', t', which of course are not identical with x, y, z, t. It has already been set forth in detail how these magnitudes are to be regarded as results of physical measurements.

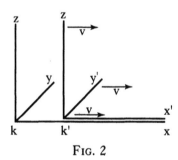

FIG. 2

Obviously our problem can be exactly formulated in the following manner. What are the values x', y', z', t', of an event with respect to K', when the magnitudes x, y, z, t, of the same event with respect to K are given? The relations must be so chosen that the law of the transmission of light *in vacuo* is satisfied for one and the same ray of light (and of course for

every ray) with respect to K and K'. For the relative orientation in space of the co-ordinate systems indicated in the diagram (Fig. 2), this problem is solved by means of the
equations:

$$x' = \frac{x - vt}{\sqrt{1 - \frac{v^2}{c^2}}}$$

$$y' = y$$

$$z' = z$$

$$t' = \frac{t - \frac{v}{c^2} \cdot x}{\sqrt{1 - \frac{v^2}{c^2}}}$$

This system of equations is known as the "Lorentz transfor-

[15]   mation."[1]

If in place of the law of transmission of light we had taken
as our basis the tacit assumptions of the older mechanics as to
the absolute character of times and lengths, then instead of
the above we should have obtained the following equations:

$$x' = x - vt$$

$$y' = y$$

$$z' = z$$

$$t' = t.$$

[1] A simple derivation of the Lorentz transformation is given in Appendix 1.

This system of equations is often termed the "Galilei transformation." The Galilei transformation can be obtained from
the Lorentz transformation by substituting an infinitely large
value for the velocity of light c in the latter transformation.

Aided by the following illustration, we can readily see that,
in accordance with the Lorentz transformation, the law of the
transmission of light *in vacuo* is satisfied both for the reference-
body K and for the reference-body K'. A light-signal is sent   [16]
along the positive x-axis, and this light-stimulus advances in
accordance with the equation.

$$x = ct,$$

i.e. with the velocity c. According to the equations of the
Lorentz transformation, this simple relation between x and t
involves a relation between x' and t'. In point of fact, if we
substitute for x the value ct in the first and fourth equations of
the Lorentz transformation, we obtain:

$$x' = \frac{(c - v)t}{\sqrt{1 - \dfrac{v^2}{c^2}}}$$

$$t' = \frac{\left(1 - \dfrac{v}{c}\right)t}{\sqrt{1 - \dfrac{v^2}{c^2}}},$$

from which, by division, the expression

$$x' = ct'$$

immediately follows. If referred to the system K', the propagation of light takes place according to this equation. We thus see that the velocity of transmission relative to the reference-body K' is also equal to c. The same result is obtained for rays of light advancing in any other direction whatsoever. Of course this is not surprising, since the equations of the Lorentz transformation were derived conformably to this point of view.

TWELVE

The Behaviour of
Measuring-Rods and
Clocks in Motion

I place a metre-rod in the x'-axis of K' in such a manner that one end (the beginning) coincides with the point $x' = 0$, whilst the other end (the end of the rod) coincides with the point $x' = 1$. What is the length of the metre-rod relatively to the system K? In order to learn this, we need only ask where the beginning of the rod and the end of the rod lie with respect to K at a particular time t of the system K. By means of the first equation of the Lorentz transformation the values of these two points at the time $t = 0$ can be shown to be   [17]

$$x_{\text{(beginning of rod)}} = 0 \sqrt{1 - \frac{v^2}{c^2}}$$

$$x_{\text{(end of rod)}} = 1.\sqrt{1 - \frac{v^2}{c^2}},$$

the distance between the points being $\sqrt{1 - \frac{v^2}{c^2}}$. But the metre-

40

rod is moving with the velocity v relative to K. It therefore follows that the length of a rigid metre-rod moving in the direction of its length with a velocity v is $\sqrt{1 - v^2/c^2}$ of a metre. The rigid rod is thus shorter when in motion than when at rest, and the more quickly it is moving, the shorter is the rod. For the velocity $v = c$ we should have $\sqrt{1 - v^2/c^2} = 0$, and for still greater velocities the square-root becomes imaginary. From this we conclude that in the theory of relativity the velocity c plays the part of a limiting velocity, which can neither be reached nor exceeded by any real body.

Of course this feature of the velocity c as a limiting velocity also clearly follows from the equations of the Lorentz transformation, for these become meaningless if we choose values of v greater than c.

If, on the contrary, we had considered a metre-rod at rest in the x-axis with respect to K, then we should have found that the length of the rod as judged from K' would have been $\sqrt{1 - v^2/c^2}$; this is quite in accordance with the principle of relativity which forms the basis of our considerations.

A priori it is quite clear that we must be able to learn something about the physical behaviour of measuring-rods and clocks from the equations of transformation, for the magnitudes x, y, z, t, are nothing more nor less than the results of measurements obtainable by means of measuring-rods and clocks. If we had based our considerations on the Galileian transformation we should not have obtained a contraction of the rod as a consequence of its motion.

Let us now consider a seconds-clock which is permanently

situated at the origin ($x' = 0$) of K'. $t' = 0$ and $t' = 1$ are two
successive ticks of this clock. The first and fourth equations of
the Lorentz transformation give for these two ticks:

$$t = 0$$

and

$$t = \frac{1}{\sqrt{1 - \dfrac{v^2}{c^2}}}$$

As judged from K, the clock is moving with the velocity v;
as judged from this reference-body, the time which elapses
between two strokes of the clock is not one second, but

$\dfrac{1}{\sqrt{1 - \dfrac{v^2}{c^2}}}$ seconds, *i.e.* a somewhat larger time. As a conse-

quence of its motion the clock goes more slowly than when at
rest. Here also the velocity c plays the part of an unattainable
limiting velocity.

THIRTEEN

Theorem of the
Addition of the Velocities.
The Experiment of Fizeau

Now in practice we can move clocks and measuring-rods only with velocities that are small compared with the velocity of light; hence we shall hardly be able to compare the results of the previous section directly with the reality. But, on the other hand, these results must strike you as being very singular, and for that reason I shall now draw another conclusion from the theory, one which can easily be derived from the foregoing considerations, and which has been most elegantly confirmed by experiment.

In Section 6 we derived the theorem of the addition of velocities in one direction in the form which also results from the hypotheses of classical mechanics. This theorem can also be deduced readily from the Galilei transformation (Section 11). In place of the man walking inside the carriage, we introduce a point moving relatively to the co-ordinate system K' in accordance with the equation

$$x = wt'.$$

43

44   *Relativity*

By means of the first and fourth equations of the Galilei transformation we can express x' and t' in terms of x and t, and we
then obtain

$$x = (v + w)t.$$

This equation expresses nothing else than the law of motion
of the point with reference to the system K (of the man with
reference to the embankment). We denote this velocity by
the symbol W, and we then obtain, as in Section 6,

$$W = v + w \qquad . \quad . \quad . \quad \text{(A)}.$$

But we can carry out this consideration just as well on the
basis of the theory of relativity. In the equation

$$x' = wt'$$

we must then express x' and t' in terms of x and t, making use
of the first and fourth equations of the *Lorentz transformation.*
Instead of the equation (A) we then obtain the equation

$$W = \frac{v + w}{1 + \dfrac{vw}{c^2}} \qquad . \quad . \quad . \quad \text{(B)},$$

which corresponds to the theorem of addition for velocities in
one direction according to the theory of relativity. The ques-

Theorem of the Addition of the Velocities. The Experiment of Fizeau 45

[18]

tion now arises as to which of these two theorems is the better in accord with experience. On this point we are enlightened by a most important experiment which the brilliant physicist Fizeau performed more than half a century ago, and which has been repeated since then by some of the best experimental physicists, so that there can be no doubt about its result. The experiment is concerned with the following question. Light travels in a motionless liquid with a particular velocity *w*. How quickly does it travel in the direction of the arrow in the tube *T* (see the accompanying diagram, Fig. 3) when the liquid above mentioned is flowing through the tube with a velocity *v*?

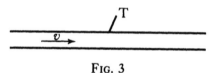

FIG. 3

In accordance with the principle of relativity we shall certainly have to take for granted that the propagation of light always takes place with the same velocity *w with respect to the liquid*, whether the latter is in motion with reference to other bodies or not. The velocity of light relative to the liquid and the velocity of the latter relative to the tube are thus known, and we require the velocity of light relative to the tube.

It is clear that we have the problem of Section 6 again before us. The tube plays the part of the railway embankment or of the co-ordinate system *K*, the liquid plays the part of the carriage or of the co-ordinate system *K'*, and finally, the light plays the part of the man walking along the carriage, or of the moving point in the present section. If we denote the velocity of the light relative to the tube by W, then this is given by the equation (A) or (B), according as the Galilei transformation or the Lorentz transformation corresponds to the facts. Experiment[1] decides in favour of equation (B) derived from the theory of relativity, and the agreement is, indeed, very exact. According to recent and most excellent measurements by Zeeman, the influence of the velocity of flow v on the propagation of light is represented by formula (B) to within one per cent.

[19]

Nevertheless we must now draw attention to the fact that a theory of this phenomenon was given by H. A. Lorentz long before the statement of the theory of relativity. This theory was of a purely electrodynamical nature, and was obtained by the use of particular hypotheses as to the electromagnetic structure of matter. This circumstance, however, does not in the least diminish the conclusiveness of the experiment as a crucial test in favour of the theory of relativity, for the electrodynamics of Maxwell-Lorentz, on which the original theory was based, in no way opposes the theory of relativity. Rather has the latter been developed from electrodynamics as an astoundingly simple combination of generalisation of the hypotheses, formerly independent of each other, on which electrodynamics was built.

[20]

---

[1] Fizeau found $W = w + v\left(1 - \dfrac{1}{n^2}\right)$, where $n = \dfrac{c}{w}$ is the index of refraction of the liquid. On the other hand, owing to the smallness of $\dfrac{vw}{c^2}$ as compared with 1, we can replace (B) in the first place by $W = (w + v\left(1 - \dfrac{vw}{c^2}\right)$, or to the same order of approximation by $w + v\left(1 - \dfrac{1}{n^2}\right)$, which agrees with Fizeau's result.

FOURTEEN

The Heuristic Value of the Theory of Relativity

Our train of thought in the foregoing pages can be epitomised in the following manner. Experience has led to the conviction that, on the one hand, the principle of relativity holds true and that on the other hand the velocity of transmission of light *in vacuo* has to be considered equal to a constant *c*. By uniting these two postulates we obtained the law of transformation for the rectangular co-ordinates x, y, z and the time t of the events which constitute the processes of nature. In this connection we did not obtain the Galilei transformation, but, differing from classical mechanics, the *Lorentz transformation*.

The law of transmission of light, the acceptance of which is justified by our actual knowledge, played an important part in this process of thought. Once in possession of the Lorentz transformation, however, we can combine this with the principle of relativity, and sum up the theory thus:

Every general law of nature must be so constituted that it is transformed into a law of exactly the same form when, instead of the space-time variables x, y, z, t of the original co-ordinate

47

system K, we introduce new space-time variables x', y', z', t' of a co-ordinate system K'. In this connection the relation between the ordinary and the accented magnitudes is given by the Lorentz transformation. Or in brief: General laws of nature are co-variant with respect to Lorentz transformations.

This is a definite mathematical condition that the theory of relativity demands of a natural law, and in virtue of this, the theory becomes a valuable heuristic aid in the search for general laws of nature. If a general law of nature were to be found which did not satisfy this condition, then at least one of the two fundamental assumptions of the theory would have been disproved. Let us now examine what general results the latter theory has hitherto evinced.

FIFTEEN

General Results of
the Theory

It is clear from our previous considerations that the (special) theory of relativity has grown out of electrodynamics and optics. In these fields it has not appreciably altered the predictions of theory, but it has considerably simplified the theoretical structure, *i.e.* the derivation of laws, and—what is incomparably more important—it has considerably reduced the number of independent hypotheses forming the basis of theory. The special theory of relativity has rendered the Maxwell-Lorentz theory so plausible, that the latter would have been generally accepted by physicists even if experiment had decided less unequivocally in its favour.

Classical mechanics required to be modified before it could come into line with the demands of the special theory of relativity. For the main part, however, this modification affects only the laws for rapid motions, in which the velocities of matter v are not very small as compared with the velocity of light. We have experience of such rapid motions only in the case of electrons and ions; for other motions the variations from the laws of classical mechanics are too small to make themselves evident in practice. We shall not consider the motion of stars until we come to speak of the general theory of relativity. In accordance with the theory of relativity the kinetic energy of a material point of mass m is no longer given by the well-known expression

$$m \frac{v^2}{2},$$

but by the expression

$$\frac{mc^2}{\sqrt{1 - \frac{v^2}{c^2}}}.$$

This expression approaches infinity as the velocity v approaches the velocity of light c. The velocity must therefore always remain less than c, however great may be the energies used to produce the acceleration. If we develop the expression for the kinetic energy in the form of a series, we obtain

[21]

$$mc^2 + m \frac{v^2}{2} = \frac{3}{8} m \frac{v^4}{c^2} + \ldots$$

When $\frac{v^2}{c^2}$ is small compared with unity, the third of these terms is always small in comparison with the second, which last is alone considered in classical mechanics. The first term

mc^2 does not contain the velocity, and requires no consideration if we are only dealing with the question as to how the energy of a point-mass depends on the velocity. We shall speak of its essential significance later.

The most important result of a general character to which the special theory of relativity has led is concerned with the conception of mass. Before the advent of relativity, physics recognised two conservation laws of fundamental importance, namely, the law of the conservation of energy and the law of the conservation of mass; these two fundamental laws appeared to be quite independent of each other. By means of the theory of relativity they have been united into one law. We shall now briefly consider how this unification came about, and what meaning is to be attached to it.

The principle of relativity requires that the law of the conservation of energy should hold not only with reference to a co-ordinate system K, but also with respect to every co-ordinate system K' which is in a state of uniform motion of translation relative to K, or, briefly, relative to every "Galileian" system of co-ordinates. In contrast to classical mechanics, the Lorentz transformation is the deciding factor in the transition from one such system to another.

By means of comparatively simple considerations we are led to draw the following conclusion from these premises, in conjunction with the fundamental equations of the electrodynamics of Maxwell: A body moving with the velocity v, which absorbs[1] an amount of energy E_0 in the form of radiation

[22]

---

[1] E_0 is the energy taken up, as judged from a co-ordinate system moving with the body.

without suffering an alteration in velocity in the process, has, as a consequence, its energy increased by an amount

$$\frac{E_0}{\sqrt{1 - \frac{v^2}{c^2}}}.$$

In consideration of the expression given above for the kinetic energy of the body, the required energy of the body comes out to be

$$\frac{\left(m + \frac{E_0}{c^2}\right)c^2}{\sqrt{1 - \frac{v^2}{c^2}}}.$$

Thus the body has the same energy as a body of mass $\left(m + \frac{E_0}{c^2}\right)$ moving with the velocity v. Hence we can say: If a body takes up an amount of energy E_0, then its inertial mass increases by an amount $\frac{E_0}{c^2}$; the inertial mass of a body is not a constant, but varies according to the change in the energy of the body. The inertial mass of a system of bodies can even be regarded as a measure of its energy. The law of the conservation of the mass of a system becomes identical with the law of the conservation of energy, and is only valid provided that the system neither takes up nor sends out energy. Writing the expression for the energy in the form

$$\frac{mc^2 + E_0}{\sqrt{1 - \dfrac{v^2}{c^2}}},$$

we see that the term mc^2, which has hitherto attracted our attention, is nothing else than the energy possessed by the body[1] before it absorbed the energy E_0.

A direct comparison of this relation with experiment is not possible at the present time (1920; see Note, p. 53), owing to the fact that the changes in energy E_0 to which we can subject a system are not large enough to make themselves perceptible

[23]   as a change in the inertial mass of the system. $\dfrac{E_0}{c^2}$ is too small in comparison with the mass m, which was present before the alteration of the energy. It is owing to this circumstance that classical mechanics was able to establish successfully the conservation of mass as a law of independent validity.

Let me add a final remark of a fundamental nature. The success of the Faraday-Maxwell interpretation of electromagnetic action at a distance resulted in physicists becoming convinced that there are no such things as instantaneous actions at a distance (not involving an intermediary medium) of the type of Newton's law of gravitation. According to the theory of relativity, action at a distance with the velocity of light always takes the place of instantaneous action at a distance or of action at a distance with an infinite velocity of transmission. This is connected with the fact that the velocity c plays a fundamental rôle in this theory. In Part II we shall see in what

---

[1] As judged from a co-ordinate system moving with the body.

54   *Relativity*

way this result becomes modified in the general theory of relativity.

NOTE.—With the advent of nuclear transformation processes, which result from the bombardment of elements by α-particles, protons, deuterons, neutrons or γ-rays, the equivalence of mass and energy expressed by the relation $E = mc^2$ has been amply confirmed. The sum of the reacting masses, together with the mass equivalent of the kinetic energy of the bombarding particle (or photon), is always greater than the sum of the resulting masses. The difference is the equivalent mass of the kinetic energy of the particles generated, or of the released electromagnetic energy (γ-photons). In the same way, the mass of a spontaneously disintegrating radioactive atom is always greater than the sum of the masses of the resulting atoms by the mass equivalent of the kinetic energy of the particles generated (or of the photonic energy). Measurements of the energy of the rays emitted in nuclear reactions, in combination with the equations of such reactions, render it possible to evaluate atomic weights to a high degree of accuracy.

R. W. L.

SIXTEEN

Experience and the
Special Theory of
Relativity

[24]   To what extent is the special theory of relativity supported by experience? This question is not easily answered for the reason already mentioned in connection
with the fundamental experiment of Fizeau. The special theory of relativity has crystallised out from the Maxwell-Lorentz
theory of electromagnetic phenomena. Thus all facts of experience which support the electromagnetic theory also support
the theory of relativity. As being of particular importance, I
mention here the fact that the theory of relativity enables us to
predict the effects produced on the light reaching us from the
fixed stars. These results are obtained in an exceedingly simple
manner, and the effects indicated, which are due to the relative
motion of the earth with reference to those fixed stars, are
found to be in accord with experience. We refer to the yearly
movement of the apparent position of the fixed stars resulting
from the motion of the earth round the sun (aberration), and to
the influence of the radial components of the relative motions
of the fixed stars with respect to the earth on the colour of the
light reaching us from them. The latter effect manifests itself

55

in a slight displacement of the spectral lines of the light transmitted to us from a fixed star, as compared with the position of
the same spectral lines when they are produced by a terrestrial
source of light (Doppler principle). The experimental argu-   [25]
ments in favour of the Maxwell-Lorentz theory, which are at
the same time arguments in favour of the theory of relativity,
are too numerous to be set forth here. In reality they limit the
theoretical possibilities to such an extent, that no other theory
than that of Maxwell and Lorentz has been able to hold its own
when tested by experience.

But there are two classes of experimental facts hitherto
obtained which can be represented in the Maxwell-Lorentz
theory only by the introduction of an auxiliary hypothesis,
which in itself—*i.e.* without making use of the theory of relativity—appears extraneous.

It is known that cathode rays and the so-called β-rays emitted by radioactive substances consist of negatively electrified
particles (electrons) of very small inertia and large velocity. By
examining the deflection of these rays under the influence of
electric and magnetic fields, we can study the law of motion of
these particles very exactly.

In the theoretical treatment of these electrons, we are faced
with the difficulty that electrodynamic theory of itself is unable to give an account of their nature. For since electrical
masses of one sign repel each other, the negative electrical
masses constituting the electron would necessarily be scat-
tered under the influence of their mutual repulsions, unless
there are forces of another kind operating between them, the

[26]   nature of which has hitherto remained obscure to us.[1] If we now assume that the relative distances between the electrical masses constituting the electron remain unchanged during the motion of the electron (rigid connection in the sense of classical mechanics), we arrive at a law of motion of the electron which does not agree with experience. Guided by purely for-

[27]   mal points of view, H. A. Lorentz was the first to introduce the hypothesis that the form of the electron experiences a contraction in the direction of motion in consequence of that motion, the contracted length being proportional to the expession $\sqrt{1 - \dfrac{v^2}{c^2}}$. This hypothesis, which is not justifiable by any electrodynamical facts, supplies us then with that particular law of motion which has been confirmed with great precision in recent years.

The theory of relativity leads to the same law of motion, without requiring any special hypothesis whatsoever as to the structure and the behaviour of the electron. We arrived at a similar conclusion of Section 13 in connection with the experiment of Fizeau, the result of which is foretold by the theory of relativity without the necessity of drawing on hypotheses as to the physical nature of the liquid.

The second class of facts to which we have alluded has reference to the question whether or not the motion of the earth in space can be made perceptible in terrestrial experiments. We have already remarked in Section 5 that all attempts of this

---

[1] The general theory of relativity renders it likely that the electrical masses of an electron are held together by gravitational forces.

nature led to a negative result. Before the theory of relativity was put forward, it was difficult to become reconciled to this negative result, for reasons now to be discussed. The inherited prejudices about time and space did not allow any doubt to arise as to the prime importance of the Galileian transformation for changing over from one body of reference to another. Now assuming that the Maxwell-Lorentz equations hold for a reference-body K, we then find that they do not hold for a reference-body K' moving uniformly with respect to K, if we assume that the relations of the Galileian transformation exist between the co-ordinates of K and K'. It thus appears that, of all Galileian co-ordinate systems, one (K) corresponding to a particular state of motion is physically unique. This result was interpreted physically by regarding K as at rest with respect to a hypothetical æther of space. On the other hand, all co-ordinate systems K' moving relatively to K were to be regarded as in motion with respect to the æther. To this motion of K' against the æther ("æther-drift" relative to K') were attributed the more complicated laws which were supposed to hold relative to K'. Strictly speaking, such an æther-drift ought also to be assumed relative to the earth, and for a long time the efforts of physicists were devoted to attempts to detect the existence of an æther-drift at the earth's surface.

In one of the most notable of these attempts Michelson devised a method which appears as though it must be decisive. Imagine two mirrors so arranged on a rigid body that the reflecting surfaces face each other. A ray of light requires a perfectly definite time T to pass from one mirror to the other and back again, if the whole system be at rest with respect to

[28]

the æther. It is found by calculation, however, that a slightly different time T' is required for this process, if the body, together with the mirrors, be moving relatively to the æther. And yet another point: it is shown by calculation that for a given velocity v with reference to the æther, this time T' is different when the body is moving perpendicularly to the planes of the mirrors from that resulting when the motion is parallel to these planes. Although the estimated difference between these two times is exceedingly small, Michelson and Morley performed an experiment involving interference in

[29]

which this difference should have been clearly detectable. But the experiment gave a negative result—a fact very perplexing to physicists. Lorentz and FitzGerald rescued the theory from this difficulty by assuming that the motion of the body relative

[30]

to the æther produces a contraction of the body in the direction of motion, the amount of contraction being just sufficient to compensate for the difference in time mentioned above. Comparison with the discussion in Section 12 shows that also from the standpoint of the theory of relativity this solution of the difficulty was the right one. But on the basis of the theory of relativity the method of interpretation is incomparably more satisfactory. According to this theory there is no such thing as a "specially favoured" (unique) co-ordinate system to occasion the introduction of the æther-idea, and hence there can be no æther-drift, nor any experiment with which to demonstrate it. Here the contraction of moving bodies follows from the two fundamental principles of the theory, without the introduction of particular hypotheses; and as the prime factor involved in this contraction we find, not the motion in itself,

60   *Relativity*

to which we cannot attach any meaning, but the motion with
respect to the body of reference chosen in the particular case
in point. Thus for a co-ordinate system moving with the earth
the mirror system of Michelson and Morley is not shortened,
but it *is* shortened for a co-ordinate system which is at rest
relatively to the sun.

SEVENTEEN

Minkowski's
Four-Dimensional Space

The non-mathematician is seized by a mysterious shuddering when he hears of "four-dimensional" things, by a feeling not unlike that awakened by thoughts of the occult. And yet there is no more common-place statement than that the world in which we live is a four-dimensional space-time continuum.

Space is a three-dimensional continuum. By this we mean that it is possible to describe the position of a point (at rest) by means of three numbers (co-ordinates) x, y, z, and that there is an indefinite number of points in the neighbourhood of this one, the position of which can be described by co-ordinates such as x_1, y_1, z_1, which may be as near as we choose to the respective values of the co-ordinates x, y, z of the first point. In virtue of the latter property we speak of a "continuum," and owing to the fact that there are three co-ordinates we speak of it as being "three-dimensional."

Similarly, the world of physical phenomena which was briefly called "world" by Minkowski is naturally four-dimensional in the space-time sense. For it is composed of individual events, each of which is described by four numbers, namely, three space co-ordinates x, y, z and a time co-ordinate, the time-value t. The "world" is in this sense also a continuum; for to every event there are as many "neighbouring" events (realised or at least thinkable) as we care to choose, the co-ordinates x_1, y_1, z_1, t_1 of which differ by an indefinitely small amount from those of the event x, y, z, t originally considered. That we have not been accustomed to regard the world in this sense as a four-dimensional continuum is due to the fact that in physics, before the advent of the theory of relativity, time played a different and more independent rôle, as compared with the space co-ordinates. It is for this reason that we have been in the habit of treating time as an independent continuum. As a matter of fact, according to classical mechanics, time is absolute, *i.e.* it is independent of the position and the condition of motion of the system of co-ordinates. We see this expressed in the last equation of the Galileian transformation ($t' = t$).

The four-dimensional mode of consideration of the "world" is natural on the theory of relativity, since according to this theory time is robbed of its independence. This is shown by the fourth equation of the Lorentz transformation:

$$t' = \frac{t - \frac{v}{c^2} x}{\sqrt{1 - \frac{v^2}{c^2}}}.$$

Moreover, according to this equation the time difference $\Delta t'$ of two events with respect to K' does not in general vanish,

even when the time difference Δt of the same events with
reference to K vanishes. Pure "space-distance" of two events
with respect to K results in "time-distance" of the same events
with respect to K'. But the discovery of Minkowski, which was
of importance for the formal development of the theory of
relativity, does not lie here. It is to be found rather in the fact
of his recognition that the four-dimensional space-time continuum of the theory of relativity, in its most essential formal
properties, shows a pronounced relationship to the three-
dimensional continuum of Euclidean geometrical space.[1] In
order to give due prominence to this relationship, however,
we must replace the usual time co-ordinate t by an imaginary
magnitude $\sqrt{-1} \cdot ct$ proportional to it. Under these conditions,
the natural laws satisfying the demands of the (special) theory
of relativity assume mathematical forms, in which the time
co-ordinate plays exactly the same rôle as the three space
co-ordinates. Formally, these four co-ordinates correspond exactly to the three space co-ordinates in Euclidean geometry. It
must be clear even to the non-mathematician that, as a consequence of this purely formal addition to our knowledge, the
theory perforce gained clearness in no mean measure.

These inadequate remarks can give the reader only a vague
notion of the important idea contributed by Minkowski. Without it the general theory of relativity, of which the fundamental ideas are developed in the following pages, would perhaps
have got no farther than its long clothes. Minkowski's work is
doubtless difficult of access to anyone inexperienced in math-

[31]

---

[1] Cf. the somewhat more detailed discussion in Appendix 2.

64   *Relativity*

ematics, but since it is not necessary to have a very exact grasp
of this work in order to understand the fundamental ideas of
either the special or the general theory of relativity, I shall
leave it here at present, and revert to it only towards the end
of Part II.

Part II

The General Theory of Relativity

EIGHTEEN

Special and General Principle of Relativity

The basal principle, which was the pivot of all our previous considerations, was the *special* principle of relativity, *i.e.* the principle of the physical relativity of all *uniform* motion. Let us once more analyse its meaning carefully.

It was at all times clear that, from the point of view of the idea it conveys to us, every motion must be considered only as a relative motion. Returning to the illustration we have frequently used of the embankment and the railway carriage, we can express the fact of the motion here taking place in the following two forms, both of whch are equally justifiable:

(*a*) The carriage is in motion relative to the embankment.

(*b*) The embankment is in motion relative to the carriage.

In (*a*) the embankment, in (*b*) the carriage, serves as the body of reference in our statement of the motion taking place. If it is simply a question of detecting or of describing the motion involved, it is in principle immaterial to what reference-body we refer the motion. As already mentioned, this is self-evident, but it must not be confused with the much

[32]

more comprehensive statement called "the principle of relativity," which we have taken as the basis of our investigations.

The principle we have made use of not only maintains that
we may equally well choose the carriage or the embankment
as our reference-body for the description of any event (for
this, too, is self-evident). Our principle rather asserts what
follows: If we formulate the general laws of nature as they are
obtained from experience, by making use of

      (*a*) the embankment as reference-body,

      (*b*) the railway carriage as reference-body,

then these general laws of nature (*e.g.* the laws of mechanics or
the law of the propagation of light *in vacuo*) have exactly the
same form in both cases. This can also be expressed as follows: For the *physical* description of natural processes, neither
of the reference-bodies K, K' is unique (lit. "specially marked
out") as compared with the other. Unlike the first, this latter
statement need not of necessity hold *a priori*; it is not contained in the conceptions of "motion" and "reference-body"
and derivable from them; only *experience* can decide as to its
correctness or incorrectness.

Up to the present, however, we have by no means maintained the equivalence of *all* bodies of reference K in connection with the formulation of natural laws. Our course was more
on the following lines. In the first place, we started out from
the assumption that there exists a reference-body K, whose
condition of motion is such that the Galileian law holds with
respect to it: A particle left to itself and sufficiently far removed from all other particles moves uniformly in a straight
line. With reference to K (Galileian reference-body) the laws

of nature were to be as simple as possible. But in addition to
K, all bodies of reference K' should be given preference in this
sense, and they should be exactly equivalent to K for the
formulation of natural laws, provided that they are in a state of
uniform rectilinear and non-rotary motion with respect to K; all
these bodies of reference are to be regarded as Galileian
reference-bodies. The validity of the principle of relativity
was assumed only for these reference-bodies, but not for others (*e.g.* those possessing motion of a different kind). In this
sense we speak of the *special* principle of relativity, or special
theory of relativity.

In contrast to this we wish to understand by the "general
principle of relativity" the following statement: All bodies of
reference K, K', etc., are equivalent for the description of
natural phenomena (formulation of the general laws of nature), whatever may be their state of motion. But before proceeding farther, it ought to be pointed out that this
formulation must be replaced later by a more abstract one, for
reasons which will become evident at a later stage.

Since the introduction of the special principle of relativity
has been justified, every intellect which strives after generalisation must feel the temptation to venture the step towards
the general principle of relativity. But a simple and apparently
quite reliable consideration seems to suggest that, for the
present at any rate, there is little hope of success in such an
attempt. Let us imagine ourselves transferred to our old friend
the railway carriage, which is travelling at a uniform rate. As
long as it is moving uniformly, the occupant of the carriage is
not sensible of its motion, and it is for this reason that he can

without reluctance interpret the facts of the case as indicating that the carriage is at rest, but the embankment in motion. Moreover, according to the special principle of relativity, this interpretation is quite justified also from a physical point of view.

If the motion of the carriage is now changed into a non-uniform motion, as for instance by a powerful application of the brakes, then the occupant of the carriage experiences a correspondingly powerful jerk forwards. The retarded motion is manifested in the mechanical behaviour of bodies relative to the person in the railway carriage. The mechanical behaviour is different from that of the case previously considered, and for this reason it would appear to be impossible that the same mechanical laws hold relatively to the non-uniformly moving carriage, as hold with reference to the carriage when at rest or in uniform motion. At all events it is clear that the Galileian law does not hold with respect to the non-uniformly moving carriage. Because of this, we feel compelled at the present juncture to grant a kind of absolute physical reality to non-uniform motion, in opposition to the general principle of relativity. But in what follows we shall soon see that this conclusion cannot be maintained.

NINETEEN

The Gravitational Field

If we pick up a stone and then let it go, why does it fall to the ground?" The usual answer to this question is: "Because it is attracted by the earth." Modern physics formulates the answer rather differently for the following reason. As a result of the more careful study of electromagnetic phenomena, we have come to regard action at a distance as a process impossible without the intervention of some intermediary medium. If, for instance, a magnet attracts a piece of iron, we cannot be content to regard this as meaning that the magnet acts directly on the iron through the intermediate empty space, but we are constrained to imagine—after the manner of Faraday—that the magnet always calls into being something physically real in the space around it, that something being what we call a "magnetic field." In its turn this magnetic field operates on the piece of iron, so that the latter strives to move towards the magnet. We shall not discuss here the justification for this incidental conception, which is indeed a somewhat arbitrary one. We shall only mention that with its aid electromagnetic phenomena can be theoretically represented much more satisfactorily than without it, and this applies particularly to the transmission of electromagnetic waves. The effects of gravitation also are regarded in an analogous manner.

The action of the earth on the stone takes place indirectly. The earth produces in its surroundings a gravitational field, which acts on the stone and produces its motion of fall. As we know from experience, the intensity of the action on a body   [33] diminishes according to a quite definite law, as we proceed farther and farther away from the earth. From our point of view this means: The law governing the properties of the gravitational field in space must be a perfectly definite one, in order correctly to represent the diminution of gravitational action with the distance from operative bodies. It is something like this: The body (*e.g.* the earth) produces a field in its immediate neighbourhood directly; the intensity and direction of the field at points farther removed from the body are thence determined by the law which governs the properties in space of the gravitational fields themselves.

In contrast to electric and magnetic fields, the gravitational field exhibits a most remarkable property, which is of fundamental importance for what follows. Bodies which are moving under the sole influence of a gravitational field receive an acceleration, *which does not in the least depend either on the material or on the physical state of the body*. For instance, a piece of lead and a piece of wood fall in exactly the same manner in a gravitational field (*in vacuo*), when they start off from rest or with the same initial velocity. This law, which holds most accurately, can be expressed in a different form in the light of the following consideration.

According to Newton's law of motion, we have

$$(\text{Force}) = (\text{inertial mass}) \times (\text{acceleration}),$$

where the "inertial mass" is a characteristic constant of the accelerated body. If now gravitation is the cause of the acceleration, we then have

$$(\text{Force}) = (\text{gravitational mass}) \times (\text{intensity of the gravitational field}),$$

where the "gravitational mass" is likewise a characteristic constant for the body. From these two relations follows:

$$(\text{acceleration}) = \frac{(\text{gravitational mass})}{(\text{inertial mass})} \, (\text{intensity of the gravitational field}).$$

If now, as we find from experience, the acceleration is to be independent of the nature and the condition of the body and always the same for a given gravitational field, then the ratio of the gravitational to the inertial mass must likewise be the same for all bodies. By a suitable choice of units we can thus [34] make this ratio equal to unity. We then have the following law: The *gravitational* mass of a body is equal to its *inertial* mass.

74   *Relativity*

It is true that this important law had hitherto been recorded
in mechanics, but it had not been *interpreted*. A satisfactory interpretation can be obtained only if we recognise the following
fact: *The same* quality of a body manifests itself according to circumstances as "inertia" or as "weight" (lit. "heaviness"). In
the following section we shall show to what extent this is actually the case, and how this question is connected with the
general postulate of relativity.

TWENTY

The Equality of Inertial and Gravitational Mass as an Argument for the General Postulate of Relativity

[35]

We imagine a large portion of empty space, so far removed from stars and other appreciable masses, that we have before us approximately the conditions required by the fundamental law of Galilei. It is then possible to choose a Galileian reference-body for this part of space (world), relative to which points at rest remain at rest and points in motion continue permanently in uniform rectilinear motion. As reference-body let us imagine a spacious chest resembling a room with an observer inside who is equipped with apparatus. Gravitation naturally does not exist for this observer. He must fasten himself with strings to the floor, otherwise the slightest impact against the floor will cause him to rise slowly towards the ceiling of the room.

To the middle of the lid of the chest is fixed externally a hook with rope attached, and now a "being" (what kind of a being is immaterial to us) begins pulling at this with a constant force. The chest together with the observer then begins to move "upwards" with a uniformly accelerated motion. In course of time their velocity will reach unheard-of values—

75

provided that we are viewing all this from another reference-body which is not being pulled with a rope.

But how does the man in the chest regard the process? The acceleration of the chest will be transmitted to him by the reaction of the floor of the chest. He must therefore take up this pressure by means of his legs if he does not wish to be laid out full length on the floor. He is then standing in the chest in exactly the same way as anyone stands in a room of a house on our earth. If he releases a body which he previously had in his hand, the acceleration of the chest will no longer be transmitted to this body, and for this reason the body will approach the floor of the chest with an accelerated relative motion. The observer will further convince himself *that the acceleration of the body towards the floor of the chest is always of the same magnitude, whatever kind of body he may happen to use for the experiment.*

Relying on his knowledge of the gravitational field (as it was discussed in the preceding section), the man in the chest will thus come to the conclusion that he and the chest are in a gravitational field which is constant with regard to time. Of course he will be puzzled for a moment as to why the chest does not fall in this gravitational field. Just then, however, he discovers the hook in the middle of the lid of the chest and the rope which is attached to it, and he consequently comes to the conclusion that the chest is suspended at rest in the gravitational field.

Ought we to smile at the man and say that he errs in his conclusion? I do not believe we ought to if we wish to remain consistent; we must rather admit that his mode of grasping the situation violates neither reason nor known mechanical laws.

Even though it is being accelerated with respect to the "Galileian space" first considered, we can nevertheless regard the
chest as being at rest. We have thus good grounds for extending the principle of relativity to include bodies of reference
which are accelerated with respect to each other, and as a
result we have gained a powerful argument for a generalised
postulate of relativity.

We must note carefully that the possibility of this mode of
interpretation rests on the fundamental property of the gravitational field of giving all bodies the same acceleration, or,
what comes to the same thing, on the law of the equality of
inertial and gravitational mass. If this natural law did not exist,
the man in the accelerated chest would not be able to interpret the behaviour of the bodies around him on the supposition of a gravitational field, and he would not be justified on
the grounds of experience in supposing his reference-body to
be "at rest."

Suppose that the man in the chest fixes a rope to the inner
side of the lid, and that he attaches a body to the free end of
the rope. The result of this will be to stretch the rope so that
it will hang "vertically" downwards. If we ask for an opinion
of the cause of tension in the rope, the man in the chest will
say: "The suspended body experiences a downward force in
the gravitational field, and this is neutralised by the tension of
the rope; what determines the magnitude of the tension of the
rope is the *gravitational mass* of the suspended body." On the
other hand, an observer who is poised freely in space will
interpret the condition of things thus: "The rope must perforce take part in the accelerated motion of the chest, and it

transmits this motion to the body attached to it. The tension of the rope is just large enough to effect the acceleration of the body. That which determines the magnitude of the tension of the rope is the *inertial mass* of the body." Guided by this example, we see that our extension of the principle of relativity implies the *necessity* of the law of the equality of inertial and gravitational mass. Thus we have obtained a physical interpretation of this law.

From our consideration of the accelerated chest we see that a general theory of relativity must yield important results on the laws of gravitation. In point of fact, the systematic pursuit of the general idea of relativity has supplied the laws satisfied by the gravitational field. Before proceeding farther, however, I must warn the reader against a misconception suggested by these considerations. A gravitational field exists for the man in the chest, despite the fact that there was no such field for the co-ordinate system first chosen. Now we might easily suppose that the existence of a gravitational field is always only an *apparent* one. We might also think that, regardless of the kind of gravitational field which may be present, we could always choose another reference-body such that *no* gravitational field exists with reference to it. This is by no means true for all gravitational fields, but only for those of quite special form. It is, for instance, impossible to choose a body of reference such that, as judged from it, the gravitational field of the earth (in its entirety) vanishes.

We can now appreciate why that argument is not convincing, which we brought forward against the general principle of relativity at the end of Section 18. It is certainly true that the

[36]

observer in the railway carriage experiences a jerk forwards as a result of the application of the brake, and that he recognises in this the non-uniformity of motion (retardation) of the carriage. But he is compelled by nobody to refer this jerk to a "real" acceleration (retardation) of the carriage. He might also interpret his experience thus: "My body of reference (the carriage) remains permanently at rest. With reference to it, however, there exists (during the period of application of the brakes) a gravitational field which is directed forwards and which is variable with respect to time. Under the influence of this field, the embankment together with the earth moves nonuniformly in such a manner that their original velocity in

[37]

the backwards direction is continuously reduced."

TWENTY-ONE

In What Respects
Are the Foundations of Classical
Mechanics and of the Special Theory
of Relativity Unsatisfactory?

We have already stated several times that classical mechanics starts out from the following law: Material particles sufficiently far removed from other material particles continue to move uniformly in a straight line or continue in a state of rest. We have also repeatedly emphasised that this fundamental law can only be valid for bodies of reference K which possess certain unique states of motion, and which are in uniform translational motion relative to each other. Relative to other reference-bodies K the law is not valid. Both in classical mechanics and in the special theory of relativity we therefore differentiate between reference-bodies K relative to which the recognised "laws of nature" can be said to hold, and reference-bodies K relative to which these laws do not hold.

But no person whose mode of thought is logical can rest satisfied with this condition of things. He asks: "How does it come that certain reference-bodies (or their states of motion) are given priority over other reference-bodies (or their states

[38]

[39]

[40]

80

of motion)? *What is the reason for this preference?* In order to show clearly what I mean by this question, I shall make use of a comparison.

I am standing in front of a gas range. Standing alongside of each other on the range are two pans so much alike that one may be mistaken for the other. Both are half full of water. I notice that steam is being emitted continuously from the one pan, but not from the other. I am surprised at this, even if I have never seen either a gas range or a pan before. But if I now notice a luminous something of bluish colour under the first pan but not under the other, I cease to be astonished, even if I have never before seen a gas flame. For I can only say that this bluish something will cause the emission of the steam, or at least *possibly* it may do so. If, however, I notice the bluish something in neither case, and if I observe that the one continuously emits steam whilst the other does not, then I shall remain astonished and dissatisfied until I have discovered some circumstance to which I can attribute the different behaviour of the two pans.

Analogously, I seek in vain for a real something in classical mechanics (or in the special theory of relativity) to which I can attribute the different behaviour of bodies considered with respect to the reference-systems K and K'.[1] Newton saw this objection and attempted to invalidate it, but without success. But E. Mach recognised it most clearly of all, and because of

[41]

---

[1] The objection is of importance more especially when the state of motion of the reference-body is of such a nature that it does not require any external agency for its maintenance, *e.g.* in the case when the reference-body is rotating uniformly.

82   *Relativity*

this objection he claimed that mechanics must be placed on a   [42]
new basis. It can only be got rid of by means of a physics
which is comformable to the general principle of relativity,
since the equations of such a theory hold for every body of
reference, whatever may be its state of motion.

TWENTY-TWO

A *Few Inferences from the General Principle of Relativity*

The considerations of Section 20 show that the general principle of relativity puts us in a position to derive properties of the gravitational field in a purely theoretical manner. Let us suppose, for instance, that we know the space-time "course" for any natural process whatsoever, as regards the manner in which it takes place in the Galileian domain relative to a Galileian body of reference K. By means of purely theoretical operations (*i.e.* simply by calculation) we are then able to find how this known natural process appears, as seen from a reference-body K' which is accelerated relatively to K. But since a gravitational field exists with respect to this new body of reference K', our consideration also teaches us how the gravitational field influences the process studied.

For example, we learn that a body which is in a state of uniform rectilinear motion with respect to K (in accordance with the law of Galilei) is executing an accelerated and in general curvilinear motion with respect to the accelerated reference-body K' (chest). This acceleration or curvature cor-

responds to the influence on the moving body of the gravitational field prevailing relatively to K'. It is known that a
gravitational field influences the movement of bodies in this
way, so that our consideration supplies us with nothing essentially new.

However, we obtain a new result of fundamental importance when we carry out the analogous consideration for a ray
of light. With respect to the Galileian reference-body K, such
a ray of light is transmitted rectilinearly with the velocity c. It
can easily be shown that the path of the same ray of light is
no longer a straight line when we consider it with reference to
the accelerated chest (reference-body K'). From this we conclude, *that, in general, rays of light are propagated curvilinearly
in gravitational fields*. In two respects this result is of great
importance.

In the first place, it can be compared with the reality. Although a detailed examination of the question shows that the
curvature of light rays required by the general theory of relativity is only exceedingly small for the gravitational fields at
our disposal in practice, its estimated magnitude for light rays
passing the sun at grazing incidence is nevertheless 1.7 sec-   [43]
onds of arc. This ought to manifest itself in the following way.
As seen from the earth, certain fixed stars appear to be in the
neighbourhood of the sun, and are thus capable of observation
during a total eclipse of the sun. At such times, these stars
ought to appear to be displaced outwards from the sun by an
amount indicated above, as compared with their apparent position in the sky when the sun is situated at another part of the
heavens. The examination of the correctness or otherwise of

[44] this deduction is a problem of the greatest importance, the early solution of which is to be expected of astronomers.[1]

In the second place our result shows that, according to the general theory of relativity, the law of the constancy of the velocity of light *in vacuo*, which constitutes one of the two fundamental assumptions in the special theory of relativity and to which we have already frequently referred, cannot claim any unlimited validity. A curvature of rays of light can only take place when the velocity of propagation of light varies with position. Now we might think that as a consequence of this, the special theory of relativity and with it the whole theory of relativity would be laid in the dust. But in reality this is not the case. We can only conclude that the special theory of relativity cannot claim an unlimited domain of validity; its results hold only so long as we are able to disregard the influences of gravitational fields on the phenomena (*e.g.* of light).

Since it has often been contended by opponents of the theory of relativity that the special theory of relativity is overthrown by the general theory of relativity, it is perhaps advisable to make the facts of the case clearer by means of an appropriate comparison. Before the development of electrodynamics the laws of electrostatics were looked upon as the laws of electricity. At the present time we know that electric fields can be derived correctly from electrostatic considerations only for the case, which is never strictly realised, in which the electrical masses are quite at rest relatively to each

---

[1] By means of the star photographs of two expeditions equipped by a Joint Committee of the Royal and Royal Astronomical Societies, the existence of the deflection of light demanded by theory was first confirmed during the solar eclipse of 29th May, 1919. (Cf. Appendix 3.)

other, and to the co-ordinate system. Should we be justified in saying that for this reason electrostatics is overthrown by the field-equations of Maxwell in electrodynamics? Not in the least. Electrostatics is contained in electrodynamics as a limiting case; the laws of the latter lead directly to those of the former for the case in which the fields are invariable with regard to time. No fairer destiny could be allotted to any physical theory, than that it should of itself point out the way to the introduction of a more comprehensive theory, in which it lives on as a limiting case.

[45]

In the example of the transmission of light just dealt with, we have seen that the general theory of relativity enables us to derive theoretically the influence of a gravitational field on the course of natural processes, the laws of which are already known when a gravitational field is absent. But the most attractive problem, to the solution of which the general theory of relativity supplies the key, concerns the investigation of the laws satisfied by the gravitational field itself. Let us consider this for a moment.

We are acquainted with space-time domains which behave (approximately) in a "Galileian" fashion under suitable choice of reference-body, *i.e.* domains in which gravitational fields are absent. If we now refer such a domain to a reference-body K' possessing any kind of motion, then relative to K' there exists a gravitational field which is variable with respect to space and time.[1] The character of this field will of course depend on the motion chosen for K'. According to the general theory of rel-

---

[1] This follows from a generalisation of the discussion in Section 20.

ativity, the general law of the gravitational field must be satisfied for all gravitational fields obtainable in this way. Even though by no means all gravitational fields can be produced in this way, yet we may entertain the hope that the general law of gravitation will be derivable from such gravitational fields of a special kind. This hope has been realised in the most beautiful manner. But between the clear vision of this goal and its actual realisation it was necessary to surmount a serious difficulty, and as this lies deep at the root of things, I dare not withhold it from the reader. We require to extend our ideas of the space-time continuum still farther.

TWENTY-THREE

Behaviour of Clocks
and Measuring-Rods on a
Rotating Body of Reference

Hitherto I have purposely refrained from speaking about the physical interpretation of space- and time-data in the case of the general theory of relativity. As a consequence, I am guilty of a certain slovenliness of treatment, which, as we know from the special theory of relativity, is far from being unimportant and pardonable. It is now high time that we remedy this defect; but I would mention at the outset, that this matter lays no small claims on the patience and on the power of abstraction of the reader.

We start off again from quite special cases, which we have frequently used before. Let us consider a space-time domain in which no gravitational field exists relative to a reference-body K whose state of motion has been suitably chosen. K is then a Galileian reference-body as regards the domain considered, and the results of the special theory of relativity hold relative to K. Let us suppose the same domain referred to a second body of reference K', which is rotating uniformly with respect to K. In order to fix our ideas, we shall imagine K' to be in the form of a plane circular disc, which rotates uniformly in its own plane about its centre. An observer who is sitting eccentrically on the disc K' is sensible of a force which acts outwards in a radial direction, and which would be interpreted as an effect of inertia (centrifugal force) by an observer who was at rest with respect to the original reference-body K. But the observer on the disc may regard his disc as a reference-body which is "at rest"; on the basis of the general principle of relativity he is justified in doing this. The force acting on himself, and in fact on all other bodies which are at rest relative to the disc, he regards as the effect of a gravitational field. Nevertheless, the space-distribution of this gravitational field is of a kind that would not be possible on Newton's theory of gravitation.[1] But since the observer believes in the general theory of relativity, this does not disturb him; he is quite in the right when he believes that a general law of gravitation can be formulated—a law which not only explains the motion of the stars correctly, but also the field of force experienced by himself.

The observer performs experiments on his circular disc with clocks and measuring-rods. In doing so, it is his intention to arrive at exact definitions for the signification of time- and space-data with reference to the circular disc K', these definitions being based on his observations. What will be his experience in this enterprise?

To start with, he places one of two identically constructed clocks at the centre of the circular disc, and the other on the

---

[1] The field disappears at the centre of the disc and increases proportionally to the distance from the centre as we proceed outwards.

edge of the disc, so that they are at rest relative to it. We now
ask ourselves whether both clocks go at the same rate from the
standpoint of the non-rotating Galileian reference-body K. As
judged from this body, the clock at the centre of the disc has
no velocity, whereas the clock at the edge of the disc is in
motion relative to K in consequence of the rotation. According
to a result obtained in Section 12, it follows that the latter
clock goes at a rate permanently slower than that of the clock
at the centre of the circular disc, *i.e.* as observed from K. It is
obvious that the same effect would be noted by an observer
whom we will imagine sitting alongside his clock at the centre
of the circular disc. Thus on our circular disc, or, to make the
case more general, in every gravitational field, a clock will go
more quickly or less quickly, according to the position in
which the clock is situated (at rest). For this reason it is not
possible to obtain a reasonable definition of time with the aid
of clocks which are arranged at rest with respect to the body
of reference. A similar difficulty presents itself when we attempt to apply our earlier definition of simultaneity in such a
case, but I do not wish to go any farther into this question.

Moreover, at this stage the definition of the space co-
ordinates also presents insurmountable difficulties. If the observer applies his standard measuring-rod (a rod which is short   [46]
as compared with the radius of the disc) tangentially to the
edge of the disc, then, as judged from the Galileian system,
the length of this rod will be less than 1, since, according to
Section 12, moving bodies suffer a shortening in the direction
of the motion. On the other hand, the measuring-rod will not
experience a shortening in length, as judged from K, if it is

applied to the disc in the direction of the radius. If, then, the observer first measures the circumference of the disc with his measuring-rod and then the diameter of the disc, on dividing the one by the other, he will not obtain as quotient the familiar number $\pi = 3 \cdot 14 \ldots$, but a larger number,[1] whereas of course, for a disc which is at rest with respect to K, this operation would yield π exactly. This proves that the propositions of Euclidean geometry cannot hold exactly on the rotating disc, nor in general in a gravitational field, at least if we attribute the length 1 to the rod in all positions and in every orientation. Hence the idea of a straight line also loses its meaning. We are therefore not in a position to define exactly the co-ordinates x, y, z relative to the disc by means of the method used in discussing the special theory, and as long as the co-ordinates and times of events have not been defined, we cannot assign an exact meaning to the natural laws in which these occur.

[49]   Thus all our previous conclusions based on general relativity would appear to be called in question. In reality we must make a subtle detour in order to be able to apply the postulate of general relativity exactly. I shall prepare the reader for this in the following paragraphs.

[47]

[48]

---

[1] Throughout this consideration we have to use the Galileian (non-rotating) system K as reference-body, since we may only assume the validity of the results of the special theory of relativity relative to K (relative to K' a gravitational field prevails).

Albert Einstein at home in Berlin, 1916.
(Einstein Archive, Hebrew University of Jerusalem)

TWENTY-FOUR

Euclidean and Non-Euclidean
Continuum

The surface of a marble table is spread out in front of me. I can get from any one point on this table to any other point by passing continuously from one point to a "neighbouring" one, and repeating this process a (large) number of times, or, in other words, by going from point to point without executing "jumps." I am sure the reader will appreciate with sufficient clearness what I mean here by "neighbouring" and by "jumps" (if he is not too pedantic). We express this property of the surface by describing the latter as a continuum.

Let us now imagine that a large number of little rods of equal length have been made, their lengths being small compared with the dimensions of the marble slab. When I say they are of equal length, I mean that one can be laid on any other without the ends overlapping. We next lay four of these little rods on the marble slab so that they constitute a quadrilateral figure (a square), the diagonals of which are equally long. To ensure the equality of the diagonals, we make use of a little testing-rod. To this square we add similar ones, each of which

[50]

has one rod in common with the first. We proceed in like manner with each of these squares until finally the whole marble slab is laid out with squares. The arrangement is such, that each side of a square belongs to two squares and each corner to four squares.

[51]

[52]

It is a veritable wonder that we can carry out this business without getting into the greatest difficulties. We only need to think of the following. If at any moment three squares meet at a corner, then two sides of the fourth square are already laid, and, as a consequence, the arrangement of the remaining two sides of the square is already completely determined. But I am now no longer able to adjust the quadrilateral so that its diagonals may be equal. If they are equal of their own accord, then this is an especial favour of the marble slab and of the little rods, about which I can only be thankfully surprised. We must needs experience many such surprises if the construction is to be successful.

If everything has really gone smoothly, then I say that the points of the marble slab constitute a Euclidean continuum with respect to the little rod, which has been used as a "distance" (line-interval). By choosing one corner of a square as "origin," I can characterize every other corner of a square with reference to this origin by means of two numbers. I only need state how many rods I must pass over when, starting from the origin, I proceed towards the "right" and then "upwards," in order to arrive at the corner of the square under consideration. These two numbers are then the "Cartesian co-ordinates" of this corner with reference to the "Cartesian co-ordinate system" which is determined by the arrangement of little rods.

By making use of the following modification of this abstract
experiment, we recognise that there must also be cases in
which the experiment would be unsuccessful. We shall suppose that the rods "expand" by an amount proportional to the
increase of temperature. We heat the central part of the marble slab, but not the periphery, in which case two of our little
rods can still be brought into coincidence at every position on
the table. But our construction of squares must necessarily
come into disorder during the heating, because the little rods
on the central region of the table expand, whereas those on
the outer part do not.

With reference to our little rods—defined as unit lengths—
the marble slab is no longer a Euclidean continuum, and we
are also no longer in the position of defining Cartesian co-
ordinates directly with their aid, since the above construction
can no longer be carried out. But since there are other things
which are not influenced in a similar manner to the little rods
(or perhaps not at all) by the temperature of the table, it is
possible quite naturally to maintain the point of view that the
marble slab is a "Euclidean continuum." This can be done in
a satisfactory manner by making a more subtle stipulation
about the measurement or the comparison of lengths.

But if rods of every kind (*i.e.* of every material) were to
behave *in the same way* as regards the influence of temperature
when they are on the variably heated marble slab, and if we
had no other means of detecting the effect of temperature
than the geometrical behaviour of our rods in experiments
analogous to the one described above, then our best plan
would be to assign the distance *one* to two points on the slab,

provided that the ends of one of our rods could be made to coincide with these two points; for how else should we define the distance without our proceeding being in the highest measure grossly arbitrary? The method of Cartesian co-ordinates must then be discarded, and replaced by another which does not assume the validity of Euclidean geometry for rigid bodies.[1] The reader will notice that the situation depicted here corresponds to the one brought about by the general postulate of relativity (Section 23).

---

[1] Mathematicians have been confronted with our problem in the following form. If we are given a surface (*e.g.* an ellipsoid) in Euclidean three-dimensional space, then there exists for this surface a two-dimensional geometry, just as much as for a plane surface. Gauss undertook the task of treating this two-dimensional geometry from first principles, without making use of the fact that the surface belongs to a Euclidean continuum of three dimensions. If we imagine constructions to be made with rigid rods *in the surface* (similar to that above with the marble slab), we should find that different laws hold for these from those resulting on the basis of Euclidean plane geometry. The surface is not a Euclidean continuum with respect to the rods, and we cannot define Cartesian co-ordinates *in the surface*. Gauss indicated the principles according to which we can treat the geometrical relationships in the surface, and thus pointed out the way to the method of Riemann of treating multi-dimensional, non-Euclidean *continua*. Thus it is that mathematicians long ago solved the formal problems to which we are led by the general postulate of relativity.

[53]

[54]

TWENTY-FIVE

Gaussian Co-ordinates

According to Gauss, this combined analytical and geometrical mode of handling the problem can be arrived at in the following way. We imagine a system of arbitrary curves (see Fig. 4) drawn on the surface of the table.

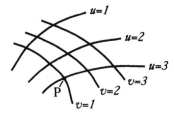

FIG. 4

These we designate as *u*-curves, and we indicate each of them by means of a number. The curves $u = 1$, $u = 2$ and $u = 3$ are drawn in the diagram. Between the curves $u = 1$ and $u = 2$ we must imagine an infinitely large number to be drawn, all of which correspond to real numbers lying between 1 and 2. We have then a system of *u*-curves, and this "infinitely dense" system covers the whole surface of the table. These *u*-curves must not intersect each other, and through each point of the surface one and only one curve must pass. Thus a perfectly definite value of u belongs to every point on the surface of the marble slab. In like manner we imagine a system of v-curves drawn on the surface. These satisfy the same conditions as the u-curves, they are provided with numbers in a corresponding manner, and they may likewise be of arbitrary shape. It follows that a value of u and a value of v belong to every point on the surface of the table. We call these two numbers the co-ordinates of the surface of the table (Gaussian co-ordinates). For example, the point P in the diagram has the Gaussian co-ordinates $u = 3$, $v = 1$. Two neighbouring points P and P' on the surface then correspond to the co-ordinates

$$P: \quad u, v$$
$$P': \quad u + du, v + dv,$$

where du and dv signify very small numbers. In a similar manner we may indicate the distance (line-interval) between P and P', as measured with a little rod, by means of the very small number ds. Then according to Gauss we have

$$ds^2 = g_{11}du^2 + 2g_{12}dudv + g_{22} dv^2,$$

where g_{11}, g_{12}, g_{22}, are magnitudes which depend in a perfectly definite way to u and v. The magnitudes g_{11}, g_{12} and g_{22} determine the behaviour of the rods relative to the u-curves and v-curves, and thus also relative to the surface of the table. For the case in which the points of the surface considered form a Euclidean continuum with reference to the measuring-rods, but only in this case, it is possible to draw the *u*-curves and *v*-curves and to attach numbers to them, in such a manner, that we simply have:

$$ds^2 = du^2 + dv^2.$$

Under these conditions, the *u*-curves and *v*-curves are straight lines in the sense of Euclidean geometry, and they are perpendicular to each other. Here the Gaussian co-ordinates are simply Cartesian ones. It is clear that Gauss co-ordinates are nothing more than an association of two sets of numbers with the points of the surface considered, of such a nature that numerical values differing very slightly from each other are associated with neighbouring points "in space."

So far, these considerations hold for a continuum of two dimensions. But the Gaussian method can be applied also to a continuum of three, four or more dimensions. If, for instance, a continuum of four dimensions be supposed available, we may represent it in the following way. With every point of the continuum we associate arbitrarily four numbers, x_1, x_2, x_3, x_4, which are known as "co-ordinates." Adjacent points correspond to adjacent values of the co-ordinates. If a distance ds is associated with the adjacent points P and P', this distance being measurable and well-defined from a physical point of view, then the following formula holds:

$$ds^2 = g_{11}dx_1{}^2 + 2g_{12}dx_1dx_2 \ldots + g_{44}dx_4{}^2,$$

where the magnitudes g_{11}, etc., have values which vary with the position in the continuum. Only when the continuum is a Euclidean one is it possible to associate the co-ordinates x_1 .. x_4 with the points of the continuum so that we have simply

$$ds^2 = dx_1{}^2 + dx_2{}^2 + dx_3{}^2 + dx_4{}^2.$$

In this case relations hold in the four-dimensional continuum which are analogous to those holding in our three-dimensional measurements.

However, the Gauss treatment for ds^2 which we have given above is not always possible. It is only possible when sufficiently small regions of the continuum under consideration may be regarded as Euclidean continua. For example, this obviously holds in the case of the marble slab of the table and local variation of temperature. The temperature is practically constant for a small part of the slab, and thus the geometrical behaviour of the rods is *almost* as it ought to be according to the rules of Euclidean geometry. Hence the imperfections of the construction of squares in the previous section do not show themselves clearly until this construction is extended over a considerable portion of the surface of the table.

We can sum this up as follows: Gauss invented a method for the mathematical treatment of continua in general, in which "size-relations" ("distances" between neighbouring points) are defined. To every point of a continuum are assigned as many numbers (Gaussian co-ordinates) as the continuum has dimensions. This is done in such a way, that only one meaning can be attached to the assignment, and that numbers (Gaus-

100   *Relativity*

sian co-ordinates) which differ by an indefinitely small amount
are assigned to adjacent points. The Gaussian co-ordinate system is a logical generalisation of the Cartesian co-ordinate
system. It is also applicable to non-Euclidean continua, but
only when, with respect to the defined "size" or "distance,"
small parts of the continuum under consideration behave more
nearly like a Euclidean system, the smaller the part of the
continuum under our notice.

TWENTY-SIX

The Space-Time Continuum of the Special Theory of Relativity Considered as a Euclidean Continuum

We are now in a position to formulate more exactly the idea of Minkowski, which was only vaguely indicated in Section 17. In accordance with the special theory of relativity, certain co-ordinate systems are given preference for the description of the four-dimensional, space-time continuum. We called these "Galileian co-ordinate systems." For these systems, the four co-ordinates x, y, z, t, which determine an event or—in other words—a point of the four-dimensional continuum, are defined physically in a simple manner, as set forth in detail in the first part of this book. For the transition from one Galileian system to another, which is moving uniformly with reference to the first, the equations of the Lorentz transformation are valid. These last form the basis for the derivation of deductions from the special theory of relativity, and in themselves they are nothing more than the expression of the universal validity of the law of transmission of light for all Galileian systems of reference.

Minkowski found that the Lorentz transformations satisfy the following simple conditions. Let us consider two neigh-

[55]

bouring events, the relative position of which in the four-dimensional continuum is given with respect to a Galileian reference-body *K* by the space co-ordinate differences *dx*, *dy*, *dz* and the time-difference *dt*. With reference to a second Galileian system we shall suppose that the corresponding differences for these two events are *dx′*, *dy′*, *dz′*, *dt′*. Then these magnitudes always fulfil the condition[1]

$$dx^2 + dy^2 + dz^2 - c^2dt^2 = dx'^2 + dy'^2 + dz'^2 - c^2dt'^2.$$

The validity of the Lorentz transformation follows from this condition. We can express this as follows: The magnitude

$$ds^2 = dx^2 + dy^2 + dz^2 - c^2dt^2,$$

which belongs to two adjacent points of the four-dimensional space-time continuum, has the same value for all selected (Galileian) reference-bodies. If we replace *x*, *y*, *z* $\sqrt{-1}$ *ct*, by x_1, x_2, x_3, x_4, we also obtain the result that

$$ds^2 = dx_1^2 + dx_2^2 + dx_3^2 + dx_4^2$$

is independent of the choice of the body of reference. We call the magnitude *ds* the "distance" apart of the two events or four-dimensional points.

---

[1] Cf. Appendices 1 and 2. The relations which are derived there for the co-ordinates themselves are valid also for co-ordinate *differences*, and thus also for co-ordinate differentials (indefinitely small differences).

Thus, if we choose as time-variable the imaginary variable $\sqrt{-1}\,ct$ instead of the real quantity t, we can regard the space-time continuum—in accordance with the special theory of relativity—as a "Euclidean" four-dimensional continuum, a result which follows from the considerations of the preceding section.

TWENTY-SEVEN

The Space-Time Continuum
of the General Theory of Relativity
Is Not a Euclidean Continuum

I n the first part of this book we were able to make use of
space-time co-ordinates which allowed of a simple and
direct physical interpretation, and which, according to
Section 26, can be regarded as four-dimensional Cartesian
co-ordinates. This was possible on the basis of the law of the
constancy of the velocity of light. But according to Section 21,
the general theory of relativity cannot retain this law. On the
contrary, we arrived at the result that according to this latter
theory the velocity of light must always depend on the co-
ordinates when a gravitational field is present. In connection
with a specific illustration in Section 23, we found that the
presence of a gravitational field invalidates the definition of
the co-ordinates and the time, which led us to our objective in
the special theory of relativity.

In view of the results of these considerations we are led to
the conviction that, according to the general principle of relativity, the space-time continuum cannot be regarded as a
Euclidean one, but that here we have the general case, corresponding to the marble slab with local variations of temper-

ature, and with which we made acquaintance as an example of a two-dimensional continuum. Just as it was there impossible to construct a Cartesian co-ordinate system from equal rods, so here it is impossible to build up a system (reference-body) from rigid bodies and clocks, which shall be of such a nature that measuring-rods and clocks, arranged rigidly with respect to one another, shall indicate position and time directly. Such was the essence of the difficulty with which we were confronted in Section 23.

But the considerations of Sections 25 and 26 show us the way to surmount this difficulty. We refer the four-dimensional space-time continuum in an arbitrary manner to Gauss coordinates. We assign to every point of the continuum (event) four numbers, x_1, x_2, x_3, x_4 (co-ordinates), which have not the least direct physical significance, but only serve the purpose of numbering the points of the continuum in a definite but arbitrary manner. This arrangement does not even need to be of such a kind that we must regard x_1, x_2, x_3 as "space" co-ordinates and x_4 as a "time" co-ordinate.

[56]

The reader may think that such a description of the world would be quite inadequate. What does it mean to assign to an event the particular co-ordinates x_1, x_2, x_3, x_4, if in themselves these co-ordinates have no significance? More careful consideration shows, however, that this anxiety is unfounded. Let us consider, for instance, a material point with any kind of motion. If this point had only a momentary existence without duration, then it would be described in space-time by a single system of values x_1, x_2, x_3, x_4. Thus its permanent existence must be characterised by an infinitely large number of such systems of values, the co-ordinate values of which are so close together as to give continuity; corresponding to the material point, we thus have a (uni-dimensional) line in the four-dimensional continuum. In the same way, any such lines in our continuum correspond to many points in motion. The only statements having regard to these points which can claim a physical existence are in reality the statements about their encounters. In our mathematical treatment, such an encounter is expressed in the fact that the two lines which represent the motions of the points in question have a particular system of co-ordinate values, x_1, x_2, x_3, x_4, in common. After mature consideration the reader will doubtless admit that in reality such encounters constitute the only actual evidence of a time-space nature with which we meet in physical statements.

When we were describing the motion of a material point relative to a body of reference, we stated nothing more than the encounters of this point with particular points of the reference-body. We can also determine the corresponding values of the time by the observation of encounters of the body with clocks, in conjunction with the observation of the encounter of the hands of clocks with particular points on the dials. It is just the same in the case of space-measurements by means of measuring-rods, as a little consideration will show.

[57]   The following statements hold generally: Every physical description resolves itself into a number of statements, each of which refers to the space-time coincidence of two events A and B. In terms of Gaussian co-ordinates, every such statement is expressed by the agreement of their four co-ordinates

x_1, x_2, x_3, x_4. Thus in reality, the description of the time-space continuum by means of Gauss co-ordinates completely replaces the description with the aid of a body of reference, without suffering from the defects of the latter mode of description; it is not tied down to the Euclidean character of the continuum which has to be represented.

TWENTY-EIGHT

Exact Formulation of the General Principle of Relativity

We are now in a position to replace the provisional formulation of the general principle of relativity given in Section 18 by an exact formulation. The form there used, "All bodies of reference K, K', etc., are equivalent for the description of natural phenomena (formulation of the general laws of nature), whatever may be their state of motion," cannot be maintained, because the use of rigid reference-bodies, in the sense of the method followed in the special theory of relativity, is in general not possible in space-time description. The Gauss co-ordinate system has to take the place of the body of reference. The following statement corresponds to the fundamental idea of the general principle of relativity: "*All Gaussian co-ordinate systems are essentially equivalent for the formulation of the general laws of nature.*"

We can state this general principle of relativity in still another form, which renders it yet more clearly intelligible than it is when in the form of the natural extension of the special principle of relativity. According to the special theory of relativity, the equations which express the general laws of nature pass over into equations of the same form when, by making use of the Lorentz transformation, we replace the space-time variables x, y, z, t, of a (Galileian) reference-body K by the space-time variables x', y', z', t', of a new reference-body K'. According to the general theory of relativity, on the other hand, by application of *arbitrary substitutions* of the Gauss variables x_1, x_2, x_3, x_4, the equations must pass over into equations of the same form; for every transformation (not only the Lorentz transformation) corresponds to the transition of one Gauss co-ordinate system into another.

If we desire to adhere to our "old-time" three-dimensional view of things, then we can characterise the development which is being undergone by the fundamental idea of the general theory of relativity as follows: The special theory of relativity has reference to Galileian domains, *i.e.* to those in which no gravitational field exists. In this connection a Galileian reference-body serves as body of reference, *i.e.* a rigid body the state of motion of which is so chosen that the Galileian law of the uniform rectilinear motion of "isolated" material points holds relatively to it.

Certain considerations suggest that we should refer the same Galileian domains to *non-Galileian* reference-bodies also. A gravitational field of a special kind is then present with respect to these bodies (cf. Sections 20 and 23).

In gravitational fields there are no such things as rigid bodies with Euclidean properties; thus the fictitious rigid body of reference is of no avail in the general theory of relativity. The motion of clocks is also influenced by gravitational fields, and in such a way that a physical definition of time which is made directly with the aid of clocks has by no means the same degree of plausibility as in the special theory of relativity.

For this reason non-rigid reference-bodies are used, which are as a whole not only moving in any way whatsoever, but which also suffer alterations in form *ad lib.* during their motion. Clocks, for which the law of motion is of any kind, however irregular, serve for the definition of time. We have to imagine each of these clocks fixed at a point on the non-rigid reference-body. These clocks satisfy only the one condition, that the "readings" which are observed simultaneously on adjacent clocks (in space) differ from each other by an indefinitely small amount. This non-rigid reference-body, which might appropriately be termed a "reference-mollusc," is in the main equivalent to a Gaussian four-dimensional co-ordinate system chosen arbitrarily. That which gives the "mollusc" a certain comprehensibility as compared with the Gauss co-ordinate system is the (really unjustified) formal retention of the separate existence of the space co-ordinates as opposed to the time co-ordinate. Every point on the mollusc is treated as a space-point, and every material point which is at rest relatively to it as at rest, so long as the mollusc is considered as reference-body. The general principle of relativity requires that all these molluscs can be used as reference-bodies with equal right and equal success in the formulation of the general laws of nature; the

Exact Formulation of the General Principle of Relativity   *111*

laws themselves must be quite independent of the choice of mollusc.

The great power possessed by the general principle of relativity lies in the comprehensive limitation which is imposed on the laws of nature in consequence of what we have seen above.

TWENTY-NINE

The Solution of the Problem of Gravitation on the Basis of the General Principle of Relativity

If the reader has followed all our previous considerations, he will have no further difficulty in understanding the methods leading to the solution of the problem of gravitation.

We start off from a consideration of a Galileian domain, *i.e.* a domain in which there is no gravitational field relative to the Galileian reference-body *K*. The behaviour of measuring-rods and clocks with reference to *K* is known from the special theory of relativity, likewise the behaviour of "isolated" material points; the latter move uniformly and in straight lines.

Now let us refer this domain to a random Gauss co-ordinate system or to a "mollusc" as reference-body *K'*. Then with respect to *K'* there is a gravitational field *G* (of a particular kind). We learn the behaviour of measuring-rods and clocks and also of freely-moving material points with reference to *K'* simply by mathematical transformation. We interpret this behaviour as the behaviour of measuring-rods, clocks and material points under the influence of the gravitational field *G*. Hereupon we introduce a hypothesis: that the influence of the gravitational field on measuring-rods, clocks and freely-moving material points continues to take place according to the same laws, even in the case where the prevailing gravitational field is *not* derivable from the Galileian special case, simply by means of a transformation of co-ordinates.

The next step is to investigate the space-time behaviour of the gravitational field *G*, which was derived from the Galileian special case simply by transformation of the co-ordinates. This behaviour is formulated in a law, which is always valid, no matter how the reference-body (mollusc) used in the description may be chosen.

This law is not yet the *general* law of the gravitational field, since the gravitational field under consideration is of a special kind. In order to find out the general law-of-field of gravitation we still require to obtain a generalisation of the law as found above. This can be obtained without caprice, however, by taking into consideration the following demands:

(*a*) The required generalisation must likewise satisfy the general postulate of relativity.

(*b*) If there is any matter in the domain under consideration, only its inertial mass, and thus according to Section 15 only its energy is of importance for its effect in exciting a field.

(*c*) Gravitational field and matter together must satisfy the law of the conservation of energy (and of impulse).

Finally, the general principle of relativity permits us to determine the influence of the gravitational field on the course of all those processes which take place according to known laws when a gravitational field is absent, *i.e.* which have already been fitted into the frame of the special theory of relativity. In this connection we proceed in principle according to the method which has already been explained for measuring-rods, clocks and freely-moving material points.

The theory of gravitation derived in this way from the general postulate of relativity excels not only in its beauty; nor in removing the defect attaching to classical mechanics which was brought to light in Section 21; nor in interpreting the empirical law of the equality of inertial and gravitational mass; but it has also already explained a result of observation in astronomy, against which classical mechanics is powerless.

[58]
[59]
[60]

If we confine the application of the theory to the case where the gravitational fields can be regarded as being weak, and in which all masses move with respect to the co-ordinate system with velocities which are small compared with the velocity of light, we then obtain as a first approximation the Newtonian theory. Thus the latter theory is obtained here without any particular assumption, whereas Newton had to introduce the hypothesis that the force of attraction between mutually attracting material points is inversely proportional to the square of the distance between them. If we increase the accuracy of the calculation, deviations from the theory of Newton make their appearance, practically all of which must nevertheless escape the test of observation owing to their smallness.

We must draw attention here to one of these deviations. According to Newton's theory, a planet moves round the sun in an ellipse, which would permanently maintain its position with respect to the fixed stars, if we could disregard the motion of the fixed stars themselves and the action of the other

planets under consideration. Thus, if we correct the observed   [61]
motion of the planets for these two influences, and if New-   [62]
ton's theory be strictly correct, we ought to obtain for the orbit
of the planet an ellipse, which is fixed with reference to the
fixed stars. This deduction, which can be tested with great
accuracy, has been confirmed for all the planets save one, with
the precision that is capable of being obtained by the delicacy
of observation attainable at the present time. The sole exception is Mercury, the planet which lies nearest the sun. Since
the time of Leverrier, it has been known that the ellipse
corresponding to the orbit of Mercury, after it has been corrected for the influences mentioned above, is not stationary
with respect to the fixed stars, but that it rotates exceedingly
slowly in the plane of the orbit and in the sense of the orbital
motion. The value obtained for this rotary movement of the
orbital ellipse was 43 seconds of arc per century, an amount
ensured to be correct to within a few seconds of arc. This
effect can be explained by means of classical mechanics only
on the assumption of hypotheses which have little probability,
and which were devised solely for this purpose.

On the basis of the general theory of relativity, it is found
that the ellipse of every planet round the sun must necessarily
rotate in the manner indicated above; that for all the planets,
with the exception of Mercury, this rotation is too small to be
detected with the delicacy of observation possible at the
present time; but that in the case of Mercury it must amount
to 43 seconds of arc per century, a result which is strictly in
agreement with observation.   [63]

Apart from this one, it has hitherto been possible to make   [64]

116   *Relativity*

[65]   only two deductions from the theory which admit of being
       tested by observation, to wit, the curvature of light rays by the
[66]   gravitational field of the sun,[1] and a displacement of the spec-
[67]   tral lines of light reaching us from large stars, as compared
[68]   with the corresponding lines for light produced in an analo-
[69]   gous manner terrestrially (*i.e.* by the same kind of atom).[2]
       These two deductions from the theory have both been con-
[70]   firmed.

---

[1] First observed by Eddington and others in 1919. (Cf. Appendix 3, pp. 145–151).
[2] Established by Adams in 1924. (Cf. p. 151).

Part III

Considerations on the Universe as a Whole

[71]

THIRTY

Cosmological Difficulties of Newton's Theory

Apart from the difficulty discussed in Section 21, there is a second fundamental difficulty attending classical celestial mechanics, which, to the best of my knowledge, was first discussed in detail by the astronomer Seeliger. If we ponder over the questions as to how the universe, considered as a whole, is to be regarded, the first answer that suggests itself to us is surely this: As regards space (and time) the universe is infinite. There are stars everywhere, so that the density of matter, although very variable in detail, is nevertheless on the average everywhere the same. In other words: However far we might travel through space, we should find everywhere an attenuated swarm of fixed stars of approximately the same kind and density.

This view is not in harmony with the theory of Newton. The latter theory rather requires that the universe should have a kind of centre in which the density of the stars is a maximum, and that as we proceed outwards from this centre the group-density of the stars should diminish, until finally, at great distances, it is succeeded by an infinite region of emp-

120   *Relativity*

tiness. The stellar universe ought to be a finite island in the
infinite ocean of space.[1]

[72]

This conception is in itself not very satisfactory. It is still
less satisfactory because it leads to the result that the light
emitted by the stars and also individual stars of the stellar
system are perpetually passing out into infinite space, never to
return, and without ever again coming into interaction with
other objects of nature. Such a finite material universe would
be destined to become gradually but systematically impover-
ished.

In order to escape this dilemma, Seeliger suggested a modification of Newton's law, in which he assumes that for great
distances the force of attraction between two masses diminishes more rapidly than would result from the inverse square
law. In this way it is possible for the mean density of matter
to be constant everywhere, even to infinity, without infinitely

[73]

large gravitational fields being produced. We thus free ourselves from the distasteful conception that the material universe ought to possess something of the nature of a centre. Of
course we purchase our emancipation from the fundamental
difficulties mentioned, at the cost of a modification and complication of Newton's law which has neither empirical nor

---

[1] *Proof*—According to the theory of Newton, the number of "lines of force" which come from
infinity and terminate in a mass m is proportional to the mass m. If, on the average, the mass
density p_0 is constant throughout the universe, then a sphere of volume V will enclose the average
mass p_0V. Thus the number of lines of force passing through the surface F of the sphere into its
interior is proportional to p_0V. For unit area of the surface of the sphere the number of lines of
force which enters the sphere is thus proportional to $p_0 \frac{V}{F}$ or to p_0R. Hence the intensity of the
field at the surface would ultimately become infinite with increasing radius R of the sphere, which
is impossible.

theoretical foundation. We can imagine innumerable laws which would serve the same purpose, without our being able to state a reason why one of them is to be preferred to the others; for any one of these laws would be founded just as little on more general theoretical principles as is the law of Newton.

THIRTY-ONE

The Possibility of a "Finite" and Yet "Unbounded" Universe

B ut speculations on the structure of the universe also move in quite another direction. The development of non-Euclidean geometry led to the recognition of the fact, that we can cast doubt on the *infiniteness* of our space without coming into conflict with the laws of thought or with experience (Riemann, Helmholtz). These questions have already been treated in detail and with unsurpassable lucidity by Helmholtz and Poincaré, whereas I can only touch on them briefly here.

[74]

In the first place, we imagine an existence in two-dimensional space. Flat beings with flat implements, and in particular flat rigid measuring-rods, are free to move in a *plane*. For them nothing exists outside of this plane: that which they observe to happen to themselves and to their flat "things" is the all-inclusive reality of their plane. In particular, the constructions of plane Euclidean geometry can be carried out by means of the rods, *e.g.* the lattice construction, considered in Section 24. In contrast to ours, the universe of these beings is two-dimensional; but, like ours, it extends to infinity. In their

[75]

The Possibility of a "Finite" and Yet "Unbounded" Universe   123

universe there is room for an infinite number of identical squares made up of rods, *i.e.* its volume (surface) is infinite. If these beings say their universe is "plane," there is sense in the statement, because they mean that they can perform the constructions of plane Euclidean geometry with their rods. In this connection the individual rods always represent the same distance, independently of their position.

Let us consider now a second two-dimensional existence, but this time on a spherical surface instead of on a plane. The flat beings with their measuring-rods and other objects fit exactly on this surface and they are unable to leave it. Their whole universe of observation extends exclusively over the surface of the sphere. Are these beings able to regard the geometry of their universe as being plane geometry and their rods withal as the realisation of "distance"? They cannot do this. For if they attempt to realise a straight line, they will obtain a curve, which we "three-dimensional beings" designate as a great circle, *i.e.* a self-contained line of definite finite length, which can be measured up by means of a measuring-rod. Similarly, this universe has a finite area that can be compared with the area of a square constructed with rods. The great charm resulting from this consideration lies in the recognition of the fact that *the universe of these beings is finite and yet has no limits.*

[76]

But the spherical-surface beings do not need to go on a world-tour in order to perceive that they are not living in a Euclidean universe. They can convince themselves of this on every part of their "world," provided they do not use too small a piece of it. Starting from a point, they draw "straight lines" (arcs of circles as judged in three-dimensional space) of equal length in all directions. They will call the line joining the free ends of these lines a "circle." For a plane surface, the ratio of the circumference of a circle to its diameter, both lengths being measured with the same rod, is, according to Euclidean geometry of the plane, equal to a constant value π, which is independent of the diameter of the circle. On their spherical surface our flat beings would find for this ratio the value

$$\pi = \frac{\sin\left(\dfrac{r}{R}\right)}{\left(\dfrac{v}{R}\right)},$$

i.e. a smaller value than π, the difference being the more considerable, the greater is the radius of the circle in comparison with the radius R of the "world-sphere." By means of this relation the spherical beings can determine the radius of their universe ("world"), even when only a relatively small part of their world-sphere is available for their measurements. But if this part is very small indeed, they will no longer be able to demonstrate that they are on a spherical "world" and not on a Euclidean plane, for a small part of a spherical surface differs only slightly from a piece of a plane of the same size.

Thus if the spherical-surface beings are living on a planet of which the solar system occupies only a negligibly small part of the spherical universe, they have no means of determining whether they are living in a finite or in an infinite universe,

The Possibility of a "Finite" and Yet "Unbounded" Universe   125

because the "piece of universe" to which they have access is in both cases practically plane, or Euclidean. It follows directly from this discussion, that for our sphere-beings the circumference of a circle first increases with the radius until the "circumference of the universe" is reached, and that it thenceforward gradually decreases to zero for still further increasing values of the radius. During this process the area of the circle continues to increase more and more, until finally it becomes equal to the total area of the whole "world-sphere."

Perhaps the reader will wonder why we have placed our "beings" on a sphere rather than on another closed surface. But this choice has its justification in the fact that, of all closed surfaces, the sphere is unique in possessing the property that all points on it are equivalent. I admit that the ratio of the circumference c of circle to its radius r depends on r, but for a given value of r it is the same for all points of the "world-sphere"; in other words, the "world-sphere" is a "surface of constant curvature."

[77]

To this two-dimensional sphere-universe there is a three-dimensional analogy, namely, the three-dimensional spherical space which was discovered by Riemann. Its points are likewise all equivalent. It possesses a finite volume, which is determined by its "radius" $(2\pi^2 R^3)$. Is it possible to imagine a spherical space? To imagine a space means nothing else than that we imagine an epitome of our "space" experience, *i.e.* of experience that we can have in the movement of "rigid" bodies. In this sense we *can* imagine a spherical space.

Suppose we draw lines or stretch strings in all directions

from a point, and mark off from each of these the distance γ
with a measuring-rod. All the free end-points of these lengths
lie on a spherical surface. We can specially measure up the
area (F) of this surface by means of a square made up of

[78]   measuring-rods. If the universe is Euclidean, then $F = 4\pi\gamma^2$; if
it is spherical, then F is always less than $4\pi\gamma^2$. With increasing
values of γ, F increases from zero up to a maximum value
which is determined by the "world-radius," but for still further increasing values of γ, the area gradually diminishes to
zero. At first, the straight lines which radiate from the starting
point diverge farther and farther from one another, but later

[79]   they approach each other, and finally they run together again
at a "counter-point" to the starting point. Under such conditions they have traversed the whole spherical space. It is easily
seen that the three-dimensional spherical space is quite analogous to the two-dimensional spherical surface. It is finite (*i.e.*
of finite volume), and has no bounds.

It may be mentioned that there is yet another kind of curved
space: "elliptical space." It can be regarded as a curved space
in which the two "counter-points" are identical (indistinguish-
able from each other). An elliptical universe can thus be considered to some extent as a curved universe possessing central
symmetry.

It follows from what has been said, that closed spaces without limits are conceivable. From amongst these, the spherical
space (and the elliptical) excels in its simplicity, since all
points on it are equivalent. As a result of this discussion, a
most interesting question arises for astronomers and physicists, and that is whether the universe in which we live is

The Possibility of a "Finite" and Yet "Unbounded" Universe   127

infinite, or whether it is finite in the manner of the spherical universe. Our experience is far from being sufficient to enable us to answer this question. But the general theory of relativity permits of our answering it with a moderate degree of certainty, and in this connection the difficulty mentioned in Section 30 finds its solution.

THIRTY-TWO

The Structure of Space
According to the General Theory
of Relativity

According to the general theory of relativity, the geo-metrical properties of space are not independent, but they are determined by matter. Thus we can draw conclusions about the geometrical structure of the universe only if we base our considerations on the state of the matter as being something that is known. We know from experience that, for a suitably chosen co-ordinate system, the velocities of the stars are small as compared with the velocity of transmission of light. We can thus as a rough approximation arrive at a conclusion as to the nature of the universe as a whole, if we treat the matter as being at rest.

We already know from our previous discussion that the behaviour of measuring-rods and clocks is influenced by grav-itational fields, *i.e.* by the distribution of matter. This in itself is sufficient to exclude the possibility of the exact validity of Euclidean geometry in our universe. But it is conceivable that our universe differs only slightly from a Euclidean one, and this notion seems all the more probable, since calculations show that the metrics of surrounding space is influenced only

128

The Structure of Space According to the General Theory of Relativity 129

to an exceedingly small extent by masses even of the magnitude of our sun. We might imagine that, as regards geometry, our universe behaves analogously to a surface which is irregularly curved in its individual parts, but which nowhere departs appreciably from a plane: something like the rippled surface of a lake. Such a universe might fittingly be called a quasi-Euclidean universe. As regards its space it would be infinite. But calculation shows that in a quasi-Euclidean universe the average density of matter would necessarily be *nil*. Thus such a universe could not be inhabited by matter everywhere; it would present to us that unsatisfactory picture which we portrayed in Section 30.

If we are to have in the universe an average density of matter which differs from zero, however small may be that difference, then the universe cannot be quasi-Euclidean. On the contrary, the results of calculation indicate that if matter be distributed uniformly, the universe would necessarily be spherical (or elliptical). Since in reality the detailed distribution of matter is not uniform, the real universe will deviate in individual parts from the spherical, *i.e.* the universe will be quasi-spherical. But it will be necessarily finite. In fact, the theory supplies us with a simple connection[1] between the space-expanse of the universe and the average density of matter in it.

[80]

---

[1] For the "radius" *R* of the universe we obtain the equation

$$R^2 = \frac{2}{\kappa\rho}.$$

The use of the C.G.S. system in this equation gives $\frac{2}{\kappa} = 108.10^{37}$; ρ is the average density of the matter and κ is a constant connected with the Newtonian constant of gravitational.

APPENDIX ONE

Simple Derivation of the
Lorentz Transformation

[SUPPLEMENTARY TO SECTION 11]

For the relative orientation of the co-ordinate systems indicated in Fig. 2, the x-axes of both systems permanently coincide. In the present case we can divide the problem into parts
by considering first only events which are localised on the
x-axis. Any such event is represented with respect to the coordinate system K by the abscissa x and the time t, and with
respect to the system K' by the abscissa x' and the time t'. We
require to find x' and t' when x and t are given.

A light-signal, which is proceeding along the positive axis of
x, is transmitted according to the equation

$$x = ct$$

or

$$x - ct = 0 \qquad . \quad . \quad . \quad (1).$$

Since the same light-signal has to be transmitted relative to K'
with the velocity c, the propagation relative to the system K'
will be represented by the analogous formula

131

$$x' - ct' = 0 \qquad . \quad . \quad . \quad (2).$$

Those space-time points (events) which satisfy (1) must also satisfy (2). Obviously this will be the case when the relation

$$(x' - ct' = \lambda(x - ct) \qquad . \quad . \quad . \quad (3),$$

is fulfilled in general, where λ indicates a constant; for, according to (3), the disappearance of $(x - ct)$ involves the disappearance of $(x' - ct')$.

If we apply quite similar considerations to light rays which are being transmitted along the negative x-axis, we obtain the condition

$$(x' + ct') = \mu(x + ct) \qquad . \quad . \quad . \quad (4).$$

By adding (or subtracting) equations (3) and (4), and introducing for convenience the constants a and b in place of the constants λ and μ, where

$$a = \frac{\gamma + \mu}{2}$$

and

$$b = \frac{\gamma - \mu}{2},$$

we obtain the equations

$$\left. \begin{array}{l} {}' = ax - bct \\ ct' = act - bx \end{array} \right\} \qquad \cdot \quad \cdot \quad \cdot \quad (5).$$

We should thus have the solution of our problem, if the constants a and b were known. These result from the following discussion.

For the origin of K' we have permanently $x' = 0$, and hence according to the first of the equations (5)

$$x = \frac{bc}{a} t.$$

If we call v the velocity with which the origin of K' is moving relative to K, we then have

$$v = \frac{bc}{a} \qquad \cdot \quad \cdot \quad \cdot \quad (6).$$

The same value v can be obtained from equations (5), if we calculate the velocity of another point of K' relative to K, or the velocity (directed towards the negative x-axis) of a point of K with respect to K'. In short, we can designate v as the relative velocity of the two systems.

Furthermore, the principle of relativity teaches us that, as judged from K, the length of a unit measuring-rod which is at rest with reference to K' must be exactly the same as the length, as judged from K', of a unit measuring-rod which is at rest relative to K. In order to see how the points of the x'-axis appear as viewed from K, we only require to take a "snapshot" of K' from K; this means that we have to insert a particular value of t (time of K), *e.g.* $t = 0$. For this value of t we then obtain from the first of the equations (5)

$$x' = ax.$$

Two points of the x'-axis which are separated by the distance $\Delta x' = 1$ when measured in the K' system are thus separated in our instantaneous photograph by the distance

$$\Delta x = \frac{1}{a} \qquad \cdot \quad \cdot \quad \cdot \quad (7).$$

But if the snapshot be taken from $K'(t' = 0)$, and if we eliminate t from the equations (5), taking into account the expression (6), we obtain

$$x' = a\left(1 - \frac{v^2}{c^2}\right)x.$$

From this we conclude that two points of the x-axis separated by the distance 1 (relative to K) will be represented on our snapshot by the distance

$$\Delta x' = a\left(1 - \frac{v^2}{c^2}\right) \qquad \cdot \quad \cdot \quad \cdot \quad (7a).$$

But from what has been said, the two snapshots must be identical; hence Δx in (7) must be equal to $\Delta x'$ in (7a), so that we obtain

$$a^2 = \frac{1}{1 - \dfrac{v^2}{c^2}} \qquad . \quad . \quad . \quad (7b).$$

The equations (6) and (7b) determine the constants a and b. By inserting the values of these constants in (5), we obtain the first and the fourth of the equations given in Section XI.

$$\left. \begin{array}{l} x' = \dfrac{x - vt}{\sqrt{1 - \dfrac{v^2}{c^2}}} \\[4ex] t' = \dfrac{t - \dfrac{v}{c^2} x}{\sqrt{1 - \dfrac{v^2}{c^2}}} \end{array} \right\} \qquad . \quad . \quad . \quad (8).$$

[81]   Thus we have obtained the Lorentz transformation for events on the x-axis. It satisfies the condition

$$x'^2 - c^2 t'^2 = x^2 - c^2 t^2 \qquad . \quad . \quad . \quad (8a).$$

The extension of this result, to include events which take place outside the x-axis, is obtained by retaining equations (8) and supplementing them by the relations

136   *Relativity*

$$\left.\begin{array}{c} y' = y \\ z' = z \end{array}\right\} \qquad \cdot \quad \cdot \quad \cdot \quad (9).$$

In this way we satisfy the postulate of the constancy of the velocity of light *in vacuo* for rays of light of arbitrary directions, both for the system K and for the system K'. This may be shown in the following manner.

We suppose a light-signal sent out from the origin of K at the time $t = 0$. It will be propagated according to the equation

$$r = \sqrt{x^2 + y^2 + z^2} = ct,$$

or, if we square this equation, according to the equation

$$x^2 + y^2 + z^2 - c^2 t^2 = 0 \quad \cdot \quad \cdot \quad \cdot \quad (10).$$

It is required by the law of propagation of light, in conjunction with the postulate of relativity, that the transmission of the signal in question should take place—as judged from K'—in accordance with the corresponding formula

$$r' = ct',$$

or,

$$x'^2 + y'^2 + z'^2 - c^2 t'^2 = 0 \quad \cdot \quad \cdot \quad (10a).$$

In order that equation (10a) may be a consequence of equation (10), we must have

$$x' + y'^2 + z'^2 - c^2 t'^2 = \sigma(x^2 + y^2 + z^2 - c^2 t^2) \qquad (11).$$

Since equation (8a) must hold for points on the x-axis, we thus have $\sigma = 1$. It is easily seen that the Lorentz transformation really satisfies equation (11) for $\sigma = 1$; for (11) is a consequence of (8a) and (9), and hence also of (8) and (9). We have thus derived the Lorentz transformation.

The Lorentz transformation represented by (8) and (9) still requires to be generalised. Obviously it is immaterial whether the axes of K' be chosen so that they are spatially parallel to those of K. It is also not essential that the velocity of translation of K' with respect to K should be in the direction of the x-axis. A simple consideration shows that we are able to construct the Lorentz transformation in this general sense from two kinds of transformations, viz. from Lorentz transformations in the special sense and from purely spatial transformations, which corresponds to the replacement of the rectangular co-ordinate system by a new system with its axes pointing in [82]   other directions.

Mathematically, we can characterise the generalised Lorentz transformation thus:

It expresses x', y', z', t', in terms of linear homogeneous functions of x, y, z, t, of such a kind that the relation

$$x'^2 + y'^2 + z'^2 - c^2 t'^2 = x^2 + y^2 + z^2 - c^2 t^2 \qquad (11a)$$

is satisfied identically. That is to say: If we substitute their expressions in x, y, z, t in place of x', y', z', t', on the left-hand side, then the left-hand side of (11a) agrees with the right-hand side.

APPENDIX TWO

Minkowski's Four-Dimensional Space ("World")

[SUPPLEMENTARY TO SECTION 17]

We can characterise the Lorentz transformation still more simply if we introduce the imaginary $\sqrt{-1} \cdot ct$ in place of t, as
time-variable. If, in accordance with this, we insert

$$x_1 = x$$
$$x_2 = y$$
$$x_3 = z$$
$$x_4 = \sqrt{-1} \cdot ct,$$

and similarly for the accented system K', then the condition
which is identically satisfied by the transformation can be
expressed thus:

$$x_1'^2 + x_2'^2 + x_3'^2 + x_4'^2 = x_1^2 + x_2^2 + x_3^2 + x_4^2 \quad (12).$$

That is, by the afore-mentioned choice of "co-ordinates,"
(11a) is transformed into this equation.

We see from (12) that the imaginary time co-ordinate x_4 enters into the condition of transformation in exactly the same way as the space co-ordinates x_1, x_2, x_3. It is due to this fact that, according to the theory of relativity, the "time" x_4 enters into natural laws in the same form as the space co-ordinates x_1, x_2, x_3.

A four-dimensional continuum described by the "co-ordinates" x_1, x_2, x_3, x_4, was called "world" by Minkowski, who also termed a point-event a "world-point." From a "happening" in three-dimensional space, physics becomes, as it were, an "existence" in the four-dimensional "world."

This four-dimensional "world" bears a close similarity to the three-dimensional "space" of (Euclidean) analytical geometry. If we introduce into the latter a new Cartesian co-ordinate system (x'_1, x'_2, x'_3) with the same origin, then x'_1, x'_2, x'_3, are linear homogeneous functions of x_1, x_2, x_3, which identically satisfy the equation

$$x_1'^2 + x_2'^2 + x_3'^2 = x_1{}^2 + x_2{}^2 + x_3{}^2.$$

The analogy with (12) is a complete one. We can regard Minkowski's "world" in a formal manner as a four-dimensional Euclidean space (with imaginary time co-ordinate); the Lorentz transformation corresponds to a "rotation" of the co-ordinate system in the four-dimensional "world."

APPENDIX THREE

*The Experimental Confirmation of the General
Theory of Relativity*   [83]

From a systematic theoretical point of view, we may imagine
the process of evolution of an empirical science to be a continuous process of induction. Theories are evolved and are
expressed in short compass as statements of a large number of
individual observations in the form of empirical laws, from
which the general laws can be ascertained by comparison.
Regarded in this way, the development of a science bears
some resemblance to the compilation of a classified catalogue.
It is, as it were, a purely empirical enterprise.

But this point of view by no means embraces the whole of   [84]
the actual process; for it slurs over the important part played
by intuition and deductive thought in the development of an
exact science. As soon as a science has emerged from its initial
stages, theoretical advances are no longer achieved merely by
a process of arrangement. Guided by empirical data, the in-
vestigator rather develops a system of thought which, in general, is built up logically from a small number of fundamental
assumptions, the so-called axioms. We call such a system of
thought a *theory*. The theory finds the justification for its ex-

141

istence in the fact that it correlates a large number of single observations, and it is just here that the "truth" of the theory lies.

Corresponding to the same complex of empirical data, there may be several theories, which differ from one another to a considerable extent. But as regards the deductions from the theories which are capable of being tested, the agreement between the theories may be so complete, that it becomes difficult to find any deductions in which the two theories differ from each other. As an example, a case of general interest is available in the province of biology, in the Darwinian theory of the development of species by selection in the struggle for existence, and in the theory of development which is based on the hypothesis of the hereditary transmission of acquired characters.

[85]

We have another instance of far-reaching agreement between the deductions from two theories in Newtonian mechanics on the one hand, and the general theory of relativity on the other. This agreement goes so far, that up to the present we have been able to find only a few deductions from the general theory of relativity which are capable of investigation, and to which the physics of pre-relativity days does not also lead, and this despite the profound difference in the fundamental assumptions of the two theories. In what follows, we shall again consider these important deductions, and we shall also discuss the empirical evidence appertaining to them which has hitherto been obtained.

The Experimental Confirmation of the General Theory of Relativity 143

(A) MOTION OF THE PERIHELION OF MERCURY

According to Newtonian mechanics and Newton's law of gravitation, a planet which is revolving round the sun would describe an ellipse round the latter, or, more correctly, round the common centre of gravity of the sun and the planet. In such a system, the sun, or the common centre of gravity, lies in one of the foci of the orbital ellipse in such a manner that, in the course of a planet-year, the distance sun-planet grows from a minimum to a maximum, and then decreases again to a minimum. If instead of Newton's law we insert a somewhat different law of attraction into the calculation, we find that, according to this new law, the motion would still take place in such a manner that the distance sun-planet exhibits periodic variations; but in this case the angle described by the line joining sun and planet during such a period (from perihelion—closest proximity to the sun—to perihelion) would differ from 360°. The line of the orbit would not then be a closed one but in the course of time it would fill up an annular part of the orbital plane, viz. between the circle of least and the circle of greatest distance of the planet from the sun.

[86]

According also to the general theory of relativity, which differs of course from the theory of Newton, a small variation from the Newton-Kepler motion of a planet in its orbit should take place, and in such a way, that the angle described by the radius sun-planet between one perihelion and the next should exceed that corresponding to one complete revolution by an amount given by

[87]

[88]

$$+ \frac{24\pi^3 a^2}{T^2 c^2 (1 - e^2)}.$$

(*N.B.*—One complete revolution corresponds to the angle 2π in the absolute angular measure customary in physics, and the above expression gives the amount by which the radius sun-planet exceeds this angle during the interval between one perihelion and the next.) In this expression a represents the major semi-axis of the ellipse, e its eccentricity, c the velocity of light, and T the period of revolution of the planet. Our result may also be stated as follows: According to the general theory of relativity, the major axis of the ellipse rotates round the sun in the same sense as the orbital motion of the planet. Theory requires that this rotation should amount to 43 seconds of arc per century for the planet Mercury, but for the other planets of our solar system its magnitude should be so small that it would necessarily escape detection.[1]

In point of fact, astronomers have found that the theory of Newton does not suffice to calculate the observed motion of Mercury with an exactness corresponding to that of the delicacy of observation attainable at the present time. After taking account of all the disturbing influences exerted on Mercury by the remaining planets, it was found (Leverrier—1859—and Newcomb—1895) that an unexplained perihelial movement of the orbit of Mercury remained over, the amount of which does not differ sensibly from the above-mentioned +43 sec-

[89]

---

[1] Especially since the next planet Venus has an orbit that is almost an exact circle, which makes it more difficult to locate the perihelion with precision.

The Experimental Confirmation of the General Theory of Relativity 145

onds of arc per century. The uncertainty of the empirical result amounts to a few seconds only.

(B) DEFLECTION OF LIGHT BY A GRAVITATIONAL FIELD

In Section 12 it has been already mentioned that according to the general theory of relativity, a ray of light will experience a curvature of its path when passing through a gravitational field, this curvature being similar to that experienced by the path of a body which is projected through a gravitational field. As a result of this theory, we should expect that a ray of light which is passing close to a heavenly body would be deviated towards the latter. For a ray of light which passes the sun at a distance of Δ sun-radii from its centre, the angle of deflection (*a*) should amount to   [90]

$$a = \frac{1.7 \text{ seconds of arc}}{\Delta}.$$

It may be added that, according to the theory, half of this deflection is produced by the Newtonian field of attraction of the sun, and the other half by the geometrical modification ("curvature") of space caused by the sun.

This result admits of an experimental test by means of the photographic registration of stars during a total eclipse of the sun. The only reason why we must wait for a total eclipse is because at every

[91]

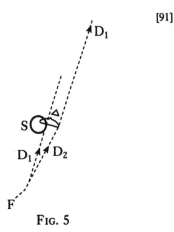

FIG. 5

other time the atmosphere is so strongly illuminated by the
light from the sun that the stars situated near the sun's disc are
invisible. The predicted effect can be seen clearly from the
accompanying diagram. If the sun (S) were not present, a star
which is practically infinitely distant would be seen in the
direction D_1, as observed from the earth. But as a consequence
of the deflection of light from the star by the sun, the star will
be seen in the direction D_2, *i.e.* at a somewhat greater distance
from the centre of the sun that corresponds to its real position.

In practice, the question is tested in the following way. The
stars in the neighbourhood of the sun are photographed during a solar eclipse. In addition, a second photograph of the
same stars is taken when the sun is situated at another position in the sky, *i.e.* a few months earlier or later. As compared
with the standard photograph, the positions of the stars on the
eclipse-photograph ought to appear displaced radially outwards (away from the centre of the sun) by an amount corresponding to the angle *a*.

We are indebted to the Royal Society and to the Royal
Astronomical Society for the investigation of this important
deduction. Undaunted by the war and by difficulties of both
a material and a psychological nature aroused by the war,
these societies equipped two expeditions—to Sobral (Brazil),
and to the island of Principe (West Africa)—and sent several
of Britain's most celebrated astronomers (Eddington, Cotting-
ham, Crommelin, Davidson), in order to obtain photographs
of the solar eclipse of 29th May, 1919. The relative discrep-
ancies to be expected between the stellar photographs ob-

The Experimental Confirmation of the General Theory of Relativity 147

tained during the eclipse and the comparison photographs amounted to a few hundredths of a millimetre only. Thus great accuracy was necessary in making the adjustments required for the taking of the photographs, and in their subsequent measurement.   [92]

The results of the measurements confirmed the theory in a thoroughly satisfactory manner. The rectangular components of the observed and of the calculated deviations of the stars (in seconds of arc) are set forth in the following table of results: [93]

Number of the Star.	First Co-ordinate.		Second Co-ordinate.	
	Observed.	Calculated.	Observed.	Calculated.
11	−0.19	−0.22	+0.16	+0.02
5	+0.29	+0.31	−0.46	−0.43
4	+0.11	+0.10	+0.83	+0.74
3	+0.20	+0.12	+1.00	+0.87
6	+0.10	+0.04	+0.57	+0.40
10	−0.08	+0.09	+0.35	+0.32
2	+0.95	+0.85	−0.27	−0.09

(c) DISPLACEMENT OF SPECTRAL LINES TOWARDS THE RED

In Section 23 it has been shown that in a system K' which is in rotation with regard to a Galileian system K, clocks of identical construction, and which are considered at rest with respect to the rotating reference-body, go at rates which are dependent on the positions of the clocks. We shall now examine this dependence quantitatively. A clock, which is situated at a distance γ from the centre of the disc, has a velocity relative to K which is given by

$$v = \omega\gamma$$

where ω represents the angular velocity of rotation of the disc K' with respect to K. If v_0 represents the number of ticks of the clock per unit time ("rate" of the clock) relative to K when the clock is at rest, then the "rate" of the clock (v) when it is moving relative to K with a velocity v, but at rest with respect to the disc, will, in accordance with Section 12, be given by

[94]

[95]

$$v = v_0\sqrt{1 - \frac{v^2}{c^2}},$$

or with sufficient accuracy by

$$v = v_0\left(1 - \frac{1}{2}\frac{v^2}{c^2}\right).$$

This expression may also be stated in the following form:

$$v = v_0\left(1 - \frac{1}{c^2}\frac{\omega^2\gamma^2}{2}\right).$$

If we represent the difference of potential of the centrifugal force between the position of the clock and the centre of the

The Experimental Confirmation of the General Theory of Relativity 149

disc by ϕ, *i.e.* the work, considered negatively, which must be performed on the unit of mass against the centrifugal force in order to transport it from the position of the clock on the rotating disc to the centre of the disc, then we have

$$\phi = - \frac{\omega^2 \gamma^2}{2}.$$

From this it follows that

$$\nu = \nu_0 \left(1 + \frac{\phi}{c^2} \right).$$

In the first place, we see from this expression that two clocks of identical construction will go at different rates when situated at different distances from the centre of the disc. This result is also valid from the standpoint of an observer who is rotating with the disc.

Now, as judged from the disc, the latter is in a gravitational field of potential ϕ, hence the result we have obtained will hold quite generally for gravitational fields. Furthermore, we can regard an atom which is emitting spectral lines as a clock, so that the following statement will hold:

An atom absorbs or emits light of a frequency which is dependent on the potential of the gravitational field in which it is situated.

The frequency of an atom situated on the surface of a heavenly body will be somewhat less than the frequency of an atom of the same element which is situated in free space (or

150   *Relativity*

on the surface of a smaller celestial body). Now $\phi = - K\dfrac{M}{\gamma}$, where K is Newton's constant of gravitation, and M is the mass of the heavenly body. Thus a displacement towards the red ought to take place for spectral lines produced at the surface of stars as compared with the spectral lines of the same element produced at the surface of the earth, the amount of this displacement being

[96]

$$\frac{\nu_0 - \nu}{\nu_0} = \frac{K}{c^2}\frac{M}{\gamma}.$$

For the sun, the displacement towards the red predicted by theory amounts to about two millionths of the wave-length. A trustworthy calculation is not possible in the case of the stars, because in general neither the mass M nor the radius γ are known.

It is an open question whether or not this effect exists, and at the present time (1920) astronomers are working with great zeal towards the solution. Owing to the smallness of the effect in the case of the sun, it is difficult to form an opinion as to its existence. Whereas Grebe and Bachem (Bonn), as a result of their own measurements and those of Evershed and Schwarz-
[97]   schild on the cyanogen bands, have placed the existence of
[98]   the effect almost beyond doubt, other investigators, particularly St. John, have been led to the opposite opinion in con-
[99]   sequence of their measurements.

Mean displacements of lines towards the less refrangible end of the spectrum are certainly revealed by statistical investiga-

The Experimental Confirmation of the General Theory of Relativity 151

tions of the fixed stars; but up to the present the examination of the available data does not allow of any definite decision being arrived at, as to whether or not these displacements are to be referred in reality to the effect of gravitation. The results of observation have been collected together, and discussed in detail from the standpoint of the question which has been engaging our attention here, in a paper by E. Freundlich entitled "Zur Prüfung der aligemeinen Relativitäts-Theorie" (*Die Naturwissenschaften*, 1919, No. 35, p. 520: Julius Springer, Berlin).

[100]

[101]

At all events, a definite decision will be reached during the next few years. If the displacement of spectral lines towards the red by the gravitational potential does not exist, then the general theory of relativity will be untenable. On the other hand, if the cause of the displacement of spectral lines be definitely traced to the gravitational potential, then the study of this displacement will furnish us with important information as to the mass of the heavenly bodies.

NOTE.—The displacement of spectral lines towards the red end of the spectrum was definitely established by Adams in 1924, by observations on the dense companion of Sirius, for which the effect is about thirty times greater than for the sun.

R. W. L.

A P P E N D I X   F O U R

The Structure of Space According to
the General Theory of Relativity

[SUPPLEMENTARY TO SECTION 32]

[102]

Since the publication of the first edition of this little book, our knowledge about the structure of space in the large ("cosmological problem") has had an important development, which ought to be mentioned even in a popular presentation of the subject.

My original considerations on the subject were based on two hypotheses:

1. There exists an average density of matter in the whole of space which is everywhere the same and different from zero.
2. The magnitude ("radius") of space is independent of time.

Both these hypotheses proved to be consistent, according to the general theory of relativity, but only after a hypothetical term was added to the field equations, a term which was not required by the theory as such nor did it seem natural from a theoretical point of view ("cosmological term of the field equations").

[103]

152

The Structure of Space According to the General Theory of Relativity 153

Hypothesis (2) appeared unavoidable to me at the time, since I thought that one would get into bottomless speculations if one departed from it.

However, already in the 'twenties, the Russian mathematician Friedman showed that a different hypothesis was natural from a purely theoretical point of view. He realized that it was possible to preserve hypothesis (1) without introducing the less natural cosmological term into the field equations of gravitation, if one was ready to drop hypothesis (2). Namely, the original field equations admit a solution in which the "world-radius" depends on time (expanding space). In that sense one can say, according to Friedman, that the theory demands an expansion of space.

A few years later Hubble showed, by a special investigation of the extra-galactic nebulae ("milky ways"), that the spectral lines emitted showed a red shift which increased regularly with the distance of the nebulae. This can be interpreted in regard to our present knowledge only in the sense of Doppler's principle, as an expansive motion of the system of stars in the large—as required, according to Friedman, by the field equations of gravitation. Hubble's discovery can, therefore, be considered to some extent as a confirmation of the theory.

There does arise, however, a strange difficulty. The interpretation of the galactic line-shift discovered by Hubble as an expansion (which can hardly be doubted from a theoretical point of view), leads to an origin of this expansion which lies "only" about 10^9 years ago, while physical astronomy makes it appear likely that the development of individual stars and

154   *Relativity*

systems of stars takes considerably longer. It is in no way known how this incongruity is to be overcome.

I further want to remark that the theory of expanding space, together with the empirical data of astronomy, permit no decision to be reached about the finite or infinite character of (three-dimensional) space, while the original "static" hypothesis of space yielded the closure (finiteness) of space.

APPENDIX FIVE

[104]   *Relativity and the Problem of Space*[1]

It is characteristic of Newtonian physics that it has to ascribe independent and real existence to space and time as well as to matter, for in Newton's law of motion the idea of acceleration appears. But in this theory, acceleration can only denote "acceleration with respect to space." Newton's space must thus be thought of as "at rest," or at least as "unaccelerated," in order that one can consider the acceleration, which appears in the law of motion, as being a magnitude with any meaning. Much the same holds with time, which of course likewise enters into the concept of acceleration. Newton himself and his most critical contemporaries felt it to be disturbing that one had to ascribe physical reality both to space itself as well as to its state of motion; but there was at that time no other alternative, if one wished to ascribe to mechanics a clear meaning.

---

[1] As with the original translation of this book in 1920, my old friend Emeritus Professor S. R. Milner, F.R.S. has again given me the benefit of his unique experience in this field, by reading the translation of this new appendix and making numerous suggestions for improvement. I am deeply grateful to him and to Professor A. G. Walker of the Mathematics Department of Liverpool University, who also read this appendix and offered various helpful suggestions.

                                                                    R. W. L.

155

156   *Relativity*

It is indeed an exacting requirement to have to ascribe physical reality to space in general, and especially to empty space. Time and again since remotest times philosophers have resisted such a presumption. Descartes argued somewhat on these lines: space is identical with extension, but extension is connected with bodies; thus there is no space without bodies and hence no empty space. The weakness of this argument lies primarily in what follows. It is certainly true that the concept extension owes its origin to our experiences of laying out or bringing into contact solid bodies. But from this it cannot be concluded that the concept of extension may not be justified in cases which have not themselves given rise to the formation of this concept. Such an enlargement of concepts can be justified indirectly by its value for the comprehension of empirical results. The assertion that extension is confined to bodies is therefore of itself certainly unfounded. We shall see later, however, that the general theory of relativity confirms Descartes' conception in a roundabout way. What brought Descartes to his remarkably attractive view was certainly the feeling that, without compelling necessity, one ought not to ascribe reality to a thing like space, which is not capable of being "directly experienced."[1]

The psychological origin of the idea of space, or of the necessity for it, is far from being so obvious as it may appear to be on the basis of our customary habit of thought. The old geometers deal with conceptual objects (straight line, point, surface), but not really with space as such, as was done later in

---

[1] This expression is to be taken *cum grano salis*.

analytical geometry. The idea of space, however, is suggested by certain primitive experiences. Suppose that a box has been constructed. Objects can be arranged in a certain way inside the box, so that it becomes full. The possibility of such arrangements is a property of the material object "box," something that is given with the box, the "space enclosed" by the box. This is something which is different for different boxes, something that is thought quite naturally at being independent of whether or not, at any moment, there are any objects at all in the box. When there are no objects in the box, its space appears to be "empty."

So far, our concept of space has been associated with the box. It turns out, however, that the storage possibilities that make up this box-space are independent of the thickness of the walls of the box. Cannot this thickness be reduced to zero, without the "space" being lost as a result? The naturalness of such a limiting process is obvious, and now there remains for our thought the space without the box, a self-evident thing, yet it appears to be so unreal if we forget the origin of this concept. One can understand that it was repugnant to Descartes to consider space as independent of material objects, a thing that might exist without matter.[1] (At the same time, this does not prevent him from treating space as a fundamental concept in his analytical geometry.) The drawing of attention to the vacuum in a mercury barometer has certainly disarmed

---

[1] Kant's attempt to remove the embarrassment by denial of the objectivity of space can, however, hardly be taken seriously. The possibilities of packing inherent in the inside space of a box are objective in the same sense as the box itself, and as the objects which can be packed inside it.

the last of the Cartesians. But it is not to be denied that, even at this primitive stage, something unsatisfactory clings to the concept of space, or to space thought of as an independent real thing.

The ways in which bodies can be packed into space (*e.g.* the box) are the subject of three-dimensional Euclidean geometry, whose axiomatic structure readily deceives us into forgetting that it refers to realisable situations.

If now the concept of space is formed in the manner outlined above, and following on from experience about the "filling" of the box, then this space is primarily a *bounded* space. This limitation does not appear to be essential, however, for apparently a larger box can always be introduced to enclose the smaller one. In this way space appears as something unbounded.

I shall not consider here how the concepts of the three-dimensional and the Euclidean nature of space can be traced back to relatively primitive experiences. Rather, I shall consider first of all from other points of view the rôle of the concept of space in the development of physical thought.

When a smaller box s is situated, relatively at rest, inside the hollow space of a larger box S, then the hollow space of s is a part of the hollow space of S, and the same "space," which contains both of them, belongs to each of the boxes. When s is in motion with respect to S, however, the concept is less simple. One is then inclined to think that s encloses always the same space, but a variable part of the space S. It then becomes necessary to apportion to each box its particular

398 SPECIAL AND GENERAL RELATIVITY

space, not thought of as bounded, and to assume that these two spaces are in motion with respect to each other.

Before one has become aware of this complication, space appears as an unbounded medium or container in which material objects swim around. But it must now be remembered that there is an infinite number of spaces, which are in motion with respect to each other. The concept of space as something existing objectively and independent of things belongs to pre-scientific thought, but not so the idea of the existence of an infinite number of spaces in motion relatively to each other. This latter idea is indeed logically unavoidable, but is far from having played a considerable rôle even in scientific thought.

But what about the psychological origin of the concept of time? This concept is undoubtedly associated with the fact of "calling to mind," as well as with the differentiation between sense experiences and the recollection of these. Of itself it is doubtful whether the differentiation between sense experience and recollection (or simple re-presentation) is something psychologically directly given to us. Everyone has experienced that he has been in doubt whether he has actually experienced something with his senses or has simply dreamt about it. Probably the ability to discriminate between these alternatives first comes about as the result of an activity of the mind creating order.

An experience is associated with a "recollection," and it is considered as being "earlier" in comparison with "present experiences." This is a conceptual ordering principle for recollected experiences, and the possibility of its accomplishment gives rise to the subjective concept of time, *i.e.* that concept of time which refers to the arrangement of the experiences of the individual.

What do we mean by rendering objective the concept of time? Let us consider an example. A person A ("I") has the experience "it is lightning." At the same time the person A also experiences such a behaviour of the person B as brings the behaviour of B into relation with his own experience "it is lightning." Thus it comes about that A associates with B the experience "it is lightning." For the person A the idea arises that other persons also participate in the experience "it is lightning." "It is lightning" is now no longer interpreted as an exclusively personal experience, but as an experience of other persons (or eventually only as a "potential experience"). In this way arises the interpretation that "it is lightning," which originally entered into the consciousness of an "experience," is now also interpreted as an (objective) "event." It is just the sum total of all events that we mean when we speak of the "real external world."

We have seen that we feel ourselves impelled to ascribe a temporal arrangement to our experiences, somewhat as follows. If β is later than α and γ later than β, then γ is also later than α ("sequence of experiences"). Now what is the position in this respect with the "events" which we have associated with the experiences? At first sight it seems obvious to assume that a temporal arrangement of events exists which agrees with the temporal arrangement of the experiences. In general, and unconsciously this was done, until sceptical doubts made themselves felt.[1] In order to arrive at the idea of an objective world, an additional constructive concept still is necessary: the event is localised not only in time, but also in space.

In the previous paragraphs we have attempted to describe how the concepts space, time and event can be put psychologically into relation with experiences. Considered logically, they are free creations of the human intelligence, tools of thought, which are to serve the purpose of bringing experiences into relation with each other, so that in this way they can be better surveyed. The attempt to become conscious of the empirical sources of these fundamental concepts should show to what extent we are actually bound to these concepts. In this way we become aware of our freedom, of which, in case of necessity, it is always a difficult matter to make sensible use.

We still have something essential to add to this stretch concerning the psychological origin of the concepts space-time-event (we will call them more briefly "space-like," in contrast to concepts from the psychological sphere). We have linked up the concept of space with experiences using boxes and the arrangement of material objects in them. Thus this formation of concepts already presupposes the concept of material objects (*e.g.* "boxes"). In the same way persons, who had to be introduced for the formation of an objective concept of time, also play the rôle of material objects in this connection. It appears to me, therefore, that the formation of the concept

---

[1] For example, the order of experiences in time obtained by acoustical means can differ from the temporal order gained visually, so that one cannot simply identify the time sequence of events with the time sequence of experiences.

of the material object must precede our concepts of time and space.

All these space-like concepts already belong to pre-scientific thought, along with concepts like pain, goal, purpose, etc. from the field of psychology. Now it is characteristic of thought in physics, as of thought in natural science generally, that it endeavours in principle to make do with "space-like" concepts *alone*, and strives to express with their aid all relations having the form of laws. The physicist seeks to reduce colours and tones to vibrations, the physiologist thought and pain to nerve processes, in such a way that the psychical element as such is eliminated from the causal nexus of existence, and thus nowhere occurs as an independent link in the causal associations. It is no doubt this attitude, which considers the comprehension of all relations by the exclusive use of only "space-like" concepts as being possible in principle, that is at the present time understood by the term "materialism" (since "matter" has lost its rôle as a fundamental concept).

Why is it necessary to drag down from the Olympian fields of Plato the fundamental ideas of thought in natural science, and to attempt to reveal their earthly lineage? Answer: In order to free these ideas from the taboo attached to them, and thus to achieve greater freedom in the formation of ideas or concepts. It is to the immortal credit of D. Hume and E. Mach that they, above all others, introduced this critical conception.

Science has taken over from pre-scientific thought the concepts space, time, and material object (with the important special case "solid body"), and has modified them and rendered them more precise. Its first significant accomplishment was the development of Euclidean geometry, whose axiomatic formulation must not be allowed to blind us to its empirical origin (the possibilities of laying out or juxtaposing solid bodies). In particular, the three-dimensional nature of space as well as its Euclidean character are of empirical origin (it can be wholly filled by like constituted "cubes").

The subtlety of the concept of space was enhanced by the discovery that there exist no completely rigid bodies. All bodies are elastically deformable and alter in volume with change in temperature. The structures, whose possible congruences are to be described by Euclidean geometry, cannot therefore be represented apart from physical concepts. But since physics after all must make use of geometry in the establishment of its concepts, the empirical content of geometry can be stated and tested only in the framework of the whole of physics.

In this connection atomistics must also be borne in mind, and its conception of finite divisibility; for spaces of subatomic extension cannot be measured up. Atomistics also compels us to give up, in principle, the idea of sharply and statically defined bounding surfaces of solid bodies. Strictly speaking, there are no *precise* laws, even in the macro-region, for the possible configurations of solid bodies touching each other.

In spite of this, no one thought of giving up the concept of space, for it appeared indispensable in the eminently satisfactory whole system of natural science. Mach, in the nineteenth century, was the only one who thought seriously of an elimination of the concept of space, in that he sought to replace it by the notion of the totality of the instantaneous distances between all material points. (He made this attempt in order to arrive at a satisfactory understanding of inertia).

The Field

In Newtonian mechanics, space and time play a dual rôle. First, they play the part of carrier or frame for things that happen in physics, in reference to which events are described by the space co-ordinates and the time. In principle, matter is thought of as consisting of "material points," the motions of which constitute physical happening. When matter is thought of as being continuous, this is done as it were provisionally in those cases where one does not wish to or cannot describe the discrete structure. In this case small parts (elements of volume) of the matter are treated similarly to material points, at least in so far as we are concerned merely with motions and not with occurrences which, at the moment, it is not possible or serves no useful purpose to attribute to motion (*e.g.* temperature changes, chemical processes). The second rôle of space and time was that of being an "inertial system." From all conceivable systems of reference, inertial systems were considered to be advantageous in that, with respect to them, the law of inertia claimed validity.

In this, the essential thing is that "physical reality," thought of as being independent of the subjects experiencing it, was conceived as consisting, at least in principle, of space and time on one hand, and of permanently existing material points, moving with respect to space and time, on the other. The idea of the independent existence of space and time can be expressed drastically in this way: If matter were to disappear, space and time alone would remain behind (as a kind of stage for physical happening).

The surmounting of this standpoint resulted from a development which, in the first place, appeared to have nothing to do with the problem of space-time, namely, the appearance of the *concept of field* and its final claim to replace, in principle, the idea of a particle (material point). In the framework of classical physics, the concept of field appeared as an auxiliary concept, in cases in which matter was treated as a continuum. For example, in the consideration of the heat conduction in a solid body, the state of the body is described by giving the temperature at every point of the body for every definite time. Mathematically, this means that the temperature T is represented as a mathematical expression (function) of the space co-ordinates and the time t (Temperature field). The law of heat conduction is represented as a local relation (differential equation), which embraces all special cases of the conduction of heat. The temperature is here a simple example of the concept of field. This is a quantity (or a complex of quantities), which is a function of the co-ordinates and the time. Another example is the description of the motion of a liquid. At every point there exists at any time a velocity, which is quantitatively described by its three "components" with respect to the axes of a co-ordinate system (vector). The components of the velocity at a point (field components), here also, are functions of the co-ordinates (x, y, z) and the time (t).

It is characteristic of the fields mentioned that they occur only within a ponderable mass; they serve only to describe a

state of this matter. In accordance with the historical development of the field concept, where no matter was available
there could also exist no field. But in the first quarter of the
nineteenth century it was shown that the phenomena of the
interference and motion of light could be explained with astonishing clearness when light was regarded as a wave-field,
completely analogous to the mechanical vibration field in an
elastic solid body. It was thus felt necessary to introduce a
field, that could also exist in "empty space" in the absence of
ponderable matter.

This state of affairs created a paradoxical situation, because,
in accordance with its origin, the field concept appeared to be
restricted to the description of states in the inside of a ponderable body. This seemed to be all the more certain, inas-
much as the conviction was held that every field is to be
regarded as a state capable of mechanical interpretation, and
this presupposed the presence of matter. One thus felt compelled, even in the space which had hitherto been regarded as
empty, to assume everywhere the existence of a form of matter, which was called "æther."

The emancipation of the field concept from the assumption
of its association with a mechanical carrier finds a place among
the psychologically most interesting events in the development of physical thought. During the second half of the nineteenth century, in connection with the researches of Faraday
and Maxwell, it became more and more clear that the description of electromagnetic processes in terms of field was vastly
superior to a treatment on the basis of the mechanical concepts of material points. By the introduction of the field con-

cept in electrodynamics, Maxwell succeeded in predicting the existence of electromagnetic waves, the essential identity of which with light waves could not be doubted, because of the equality of their velocity of propagation. As a result of this, optics was, in principle, absorbed by electrodynamics. *One* psychological effect of this immense success was that the field concept, as opposed to the mechanistic framework of classical physics, gradually won greater independence.

Nevertheless, it was at first taken for granted that electromagnetic fields had to be interpreted as states of the æther, and it was zealously sought to explain these states as mechanical ones. But as these efforts always met with frustration, science gradually became accustomed to the idea of renouncing such a mechanical interpretation. Nevertheless, the conviction still remained that electromagnetic fields must be states of the æther, and this was the position at the turn of the century.

The æther-theory brought with it the question: How does the æther behave from the mechanical point of view with respect to ponderable bodies? Does it take part in the motions of the bodies, or do its parts remain at rest relatively to each other? Many ingenious experiments were undertaken to decide this question. The following important facts should be mentioned in this connection: the "aberration" of the fixed stars in consequence of the annual motion of the earth, and the "Doppler effect," *i.e.* the influence of the relative motion of the fixed stars on the frequency of the light reaching us from them, for known frequencies of emission. The results of all these facts and experiments, except for one, the Michelson-

Morley experiment, were explained by H. A. Lorentz on the assumption that the æther does not take part in the motions of ponderable bodies, and that the parts of the æther have no relative motions at all with respect to each other. Thus the æther appeared, as it were, as the embodiment of a space absolutely at rest. But the investigation of Lorentz accomplished still more. It explained all the electromagnetic and optical processes within ponderable bodies known at that time, on the assumption that the influence of ponderable matter on the electric field—and conversely—is due solely to the fact that the constituent particles of matter carry electrical charges, which share the motion of the particles. Concerning the experiment of Michelson and Morley, H. A. Lorentz showed that the result obtained at least does not contradict the theory of an æther at rest.

In spite of all these beautiful successes the state of the theory was not yet wholly satisfactory, and for the following reasons. Classical mechanics, of which it could not be doubted that it holds with a close degree of approximation, teaches the equivalence of all inertial systems or inertial "spaces" for the formulation of natural laws, *i.e.* the invariance of natural laws with respect to the transition from one inertial system to another. Electromagnetic and optical *experiments* taught the same thing with considerable accuracy. But the foundation of electromagnetic *theory* taught that a particular inertial system must be given preference, namely that of the luminiferous æther at rest. This view of the theoretical foundation was much too unsatisfactory. Was there no modification that, like classical mechanics, would uphold the equivalence of inertial systems (special principle of relativity)?

The answer to this question is the special theory of relativity. This takes over from the theory of Maxwell–Lorentz the assumption of the constancy of the velocity of light in empty space. In order to bring this into harmony with the equivalence of inertial systems (special principle of relativity), the idea of the absolute character of simultaneity must be given up; in addition, the Lorentz transformations for the time and the space co-ordinates follow for the transition from one inertial system to another. The whole content of the special theory of relativity is included in the postulate: The laws of Nature are invariant with respect to the Lorentz transformations. The important thing of this requirement lies in the fact that it limits the possible natural laws in a definite manner.

What is the position of the special theory of relativity in regard to the problem of space? In the first place we must guard against the opinion that the four-dimensionality of reality has been newly introduced for the first time by this theory. Even in classical physics the event is localised by four numbers, three spatial co-ordinates and a time co-ordinate; the totality of physical "events" is thus thought of as being embedded in a four-dimensional continuous manifold. But the basis of classical mechanics this four-dimensional continuum breaks up objectively into the one-dimensional time and into three-dimensional spatial sections, only the latter of which contain simultaneous events. This resolution is the same for all inertial systems. The simultaneity of two definite events with reference to one inertial system involves the simultaneity of these events in reference to all inertial systems. This is what is meant when we say that the time of classical mechanics is absolute. According to the special theory of relativity it is otherwise. The sum total of events which are simultaneous with a selected event exist, it is true, in relation to a particular inertial system, but no longer independently of the choice of the inertial system. The four-dimensional continuum is now no longer resolvable objectively into sections, all of which contain simultaneous events; "now" loses for the spatially extended world its objective meaning. It is because of this that space and time must be regarded as a four-dimensional continuum that is objectively unresolvable, if it is desired to express the purport of objective relations without unnecessary conventional arbitrariness.

Since the special theory of relativity revealed the physical equivalence of all inertial systems, it proved the untenability of the hypothesis of an æther at rest. It was therefore necessary to renounce the idea that the electromagnetic field is to be regarded as a state of a material carrier. The field thus becomes an irreducible element of physical description, irreducible in the same sense as the concept of matter in the theory of Newton.

Up to now we have directed our attention to finding in what respect the concepts of space and time were *modified* by the special theory of relativity. Let us now focus our attention on those elements which this theory has taken over from classical mechanics. Here also, natural laws claim validity only when an inertial system is taken as the basis of space-time description. The principle of inertia and the principle of the constancy of the velocity of light are valid only with respect to an *inertial system*. The field-laws also can claim to have a meaning and validity only in regard to inertial systems. Thus, as in classical mechanics, space is here also an independent component in the representation of physical reality. If we imagine matter and field to be removed, inertial-space or, more accurately, this space together with the associated time remains behind. The four-dimensional structure (Minkowski-space) is thought of as being the carrier of matter and of the field. Inertial spaces, with their associated times, are only privileged four-dimensional co-ordinate systems, that are linked together by the linear Lorentz transformations. Since there exist in this four-dimensional structure no longer any sections which represent "now" objectively, the concepts of happening and becoming are indeed not completely suspended, but yet complicated. It appears therefore more natural to think of physical reality as a four-dimensional existence, instead of, as hitherto, the *evolution* of a three-dimensional existence.

This rigid four-dimensional space of the special theory of relativity is to some extent a four-dimensional analogue of H. A. Lorentz's rigid three-dimensional æther. For this theory also the following statement is valid: The description of physical states postulates space as being initially given and as existing independently. Thus even this theory does not dispel Descartes' uneasiness concerning the independent, or indeed, the *a priori* existence of "empty space." The real aim of the elementary discussion given here is to show to what extent these doubts are overcome by the general theory of relativity.

The Concept of Space in the General Theory of Relativity

This theory arose primarily from the endeavour to understand the equality of inertial and gravitational mass. We start out from an inertial system S_1, whose space is, from the physical point of view, empty. In other words, there exists in the part of space contemplated neither matter (in the usual sense) nor a field (in the sense of the special theory of relativity). With reference to S_1 let there be a second system of reference S_2 in uniform acceleration. Then S_2 is thus not an inertial system. With respect to S_2 every test mass would move with an acceleration, which is independent of its physical and chemical nature. Relative to S_2, therefore, there exists a state which, at least to a first approximation, cannot be distinguished from a gravitational field. The following concept is thus compatible with the observable facts: S_2 is also equivalent to an "inertial system"; but with respect to S_2 a (homogeneous) gravitational field is present (about the origin of which one does not worry in this connection). Thus when the gravitational field is included in the framework of the consideration, the inertial system loses its objective significance, assuming that this "principle of equivalence" can be extended to any relative motion whatsoever of the systems of reference. If it is possible to base a consistent theory on these fundamental ideas, it will satisfy of itself the fact of the equality of inertial and gravitational mass, which is strongly confirmed empirically.

Considered four-dimensionally, a non-linear transformation of the four co-ordinates corresponds to the transition from S_1 to S_2. The question now arises: What kind of non-linear transformations are to be permitted, or, how is the Lorentz transformation to be generalised? In order to answer this question, the following consideration is decisive.

We ascribe to the inertial system of the earlier theory this property: Differences in co-ordinates are measured by stationary "rigid" measuring rods, and differences in time by clocks at rest. The first assumption is supplemented by another, namely, that for the relative laying out and fitting together of measuring rods at rest, the theorems on "lengths" in Euclidean geometry hold. From the results of the special theory of relativity it is then concluded, by elementary considerations, that this direct physical interpretation of the co-ordinates is lost for systems of reference (S_2) accelerated relatively to inertial systems (S_1). But if this is the case, the co-ordinates now express only the order or rank of the "contiguity" and hence also the dimensional grade of the space, but do not express any of its metrical properties. We are thus led to extend the transformations to arbitrary continuous transformations.[1] This implies the general principle of relativity: Natural laws must be covariant with respect to arbitrary continuous transformations of the co-ordinates. This requirement (combined with that of the greatest possible logical simplicity of the laws) limits the natural laws concerned incomparably more strongly than the special principle of relativity.

---

[1] This inexact mode of expression will perhaps suffice here.

This train of ideas is based essentially on the field as an independent concept. For the conditions prevailing with respect to S_2 are interpreted as a gravitational field, without the question of the existence of masses which produce this field being raised. By virtue of this train of ideas it can also be grasped why the laws of the pure gravitational field are more directly linked with the idea of general relativity than the laws for fields of a general kind (when, for instance, an electromagnetic field is present). We have, namely, good ground for the assumption that the "field-free" Minkowski-space represents a special case possible in natural law, in fact, the simplest conceivable special case. With respect to its metrical character, such a space is characterised by the fact that $dx_1^2 + dx_2^2 + dx_3^2$ is the square of the spatial separation, measured with a unit gauge, of two infinitesimally neighbouring points of a three-dimensional "space-like" cross section (Pythagorean theorem), whereas dx_4 is the temporal separation, measured with a suitable time gauge, of two events with common (x_1, x_2, x_3). All this simply means that an objective metrical significance is attached to the quantity

$$ds^2 = dx_1^2 + dx_2^2 + dx_3^2 - dx_4^2 \qquad (1)$$

as is readily shown with the aid of the Lorentz transformations. Mathematically, this fact corresponds to the condition that ds^2 is invariant with respect to Lorentz transformations.

If now, in the sense of the general principle of relativity, this space (cf. eq. (1)) is subjected to an arbitrary continuous transformation of the co-ordinates, then the objectively significant quantity *ds* is expressed in the new system of co-ordinates by the relation

$$ds^2 = g_{ik} dx_i dx_k \qquad . \quad . \quad . \qquad (1a)$$

which has to be summed up over the indices *i* and *k* for all combinations 11, 12, . . . up to 44. The term g_{ik} now are not constants, but functions of the co-ordinates, which are determined by the arbitrarily chosen transformation. Nevertheless, the terms g_{ik} are not arbitrary functions of the new co-ordinates, but just functions of such a kind that the form (1a) can be transformed back into the form (1) by a continuous transformation of the four co-ordinates. In order that this may be possible, the function g_{ik} must satisfy certain general covariant equations of condition, which were derived by B. Riemann more than half a century before the formulation of the general theory of relativity ("Riemann condition"). According to the principle of equivalence, (1a) describes in general covariant form a gravitational field of a special kind, when the functions g_{ik} satisfy the Riemann condition.

It follows that the law for the pure gravitational field of a general kind must be satisfied when the Riemann condition is satisfied; but it must be weaker or less restricting than the Riemann condition. In this way the field law of pure gravitation is practically completely determined, a result which will not be justified in greater detail here.

We are now in a position to see how far the transition to the general theory of relativity modifies the concept of space. In accordance with classical mechanics and according to the special theory of relativity, space (space-time) has an existence independent of matter or field. In order to be able to describe at all that which fills up space and is dependent on the co-ordinates, space-time or the inertial system with its metrical properties must be thought of at once as existing, for otherwise the description of "that which fills up space" would have no meaning.[1] On the basis of the general theory of relativity, on the other hand, space as opposed to "what fills space," which is dependent on the co-ordinates, has no separate existence. Thus a pure gravitational field might have been described in terms of the g_{ik} (as functions of the co-ordinates), by solution of the gravitational equations. If we imagine the gravitational field, *i.e.* the functions g_{ik}, to be removed, there does not remain a space of the type (1), but absolutely *nothing*, and also no "topological space." For the functions g_{ik} describe not only the field, but at the same time also the topological and metrical structural properties of the manifold. A space of the type (1), judged from the standpoint of the general theory of relativity, is not a space without field, but a special case of the g_{ik} field, for which—for the co-ordinate system used, which in itself has no objective significance—the functions g_{ik} have values that do not depend on the co-ordinates. There is no such thing as an empty space, *i.e.* a space without field. Space-time does not claim existence on its own, but only as a structural quality of the field.

---

[1] If we consider that which fills space (e.g. the field) to be removed, there still remains the metric space in accordance with (1), which would also determine the inertial behaviour of a test body introduced into it.

Thus Descartes was not so far from the truth when he
believed he must exclude the existence of an empty space.
The notion indeed appears absurd, as long as physical reality
is seen exclusively in ponderable bodies. It requires the idea
of the field as the representative of reality, in combination
with the general principle of relativity, to show the true kernel
of Descartes' idea; there exists no space "empty of field."

Generalised Theory of Gravitation

The theory of the pure gravitational field on the basis of the
general theory of relativity is therefore readily obtainable, because we may be confident that the "field-free" Minkowski
space with its metric in conformity with (1) must satisfy the
general laws of field. From this special case the law of gravitation follows by a generalisation which is practically free from
arbitrariness. The further development of the theory is not so
unequivocally determined by the general principle of relativity; it has been attempted in various directions during the last
few decades. It is common to all these attempts, to conceive
physical reality as a field, and moreover, one which is a generalisation of the gravitational field, and in which the field law
is a generalisation of the law for the pure gravitational field.
After long probing I believe that I have now found[1] the most

---

[1] The generalisation can be characterised in the following way. In accordance with its derivation from empty "Minkowski space," the pure gravitational field of the functions g_{ik} has the property of symmetry given by $g_{ik} = g_{ki}$ ($g_{12} = g_{21}$, etc.). The generalised field is of the same kind, but without this property of symmetry. The derivation of the field law is completely analogous to that of the special case of pure gravitation.

natural form for this generalisation, but I have not yet been able to find out whether this generalised law can stand up against the facts of experience.

The question of the particular field law is secondary in the preceding general considerations. At the present time, the main question is whether a field theory of the kind here contemplated can lead to the goal at all. By this is meant a theory which describes exhaustively physical reality, including four-dimensional space, by a field. The present-day generation of physicists is inclined to answer this question in the negative. In conformity with the present form of the quantum theory, it believes that the state of a system cannot be specified directly, but only in an indirect way by a statement of the statistics of the results of measurement attainable on the system. The conviction prevails that the experimentally assured duality of nature (corpuscular and wave structure) can be realised only by such a weakening of the concept of reality. I think that such a far-reaching theoretical renunciation is not for the present justified by our actual knowledge, and that one should not desist from pursuing to the end the path of the relativistic field theory.

Translations of editorial notes

[4]. . . and later additions: "Appendix to the third edition. This year (1918) Springer-Verlag published an elaborate excellent textbook on the general theory of relativity, written by H. Weyl with the title *Space. Time. Matter*. It is to be warmly recommended to mathematicians and physicists."

. . . signed on November 9, 1920: "More than ever it is necessary in our hectic times to nurture those things which can bring people of a different language and nation closer to each other again. From this point of view it is of particular importance to facilitate the exchange of scientific endeavors even under these currently difficult conditions. I am glad that my booklet will now appear in the Russian language, the more so as Herr Itelson, whom I highly esteem, is guaranteed to provide an excellent translation. The author has often been scolded for saying his booklet is 'intelligible to all'; and therefore the Russian reader who encounters difficulties in comprehension should not get angry at himself or Herr Itelson. The really guilty one is no one other than the author himself."

. . . Czech translation (Einstein 1923): "I am glad that this little booklet, in which the basic ideas of the theory of relativity are presented without mathematical formalism, appears now in the national language of the country where I found the necessary contemplation to gradually give the general theory of relativity a more precise form, an endeavor whose basic idea I adopted already in 1908. In the quiet rooms of the Institute of Theoretical Physics at the German University of Prague, in Viničná ulice, I discovered in 1911 that the principle of equivalence demands a deflection of the light rays passing by the sun with observable magnitude—this without knowing that more than one hundred years ago a similar consequence had been anticipated from Newton's mechanics in combination with Newton's emission theory of light. I also discovered at Prague the still not completely confirmed consequence of the redshift of spectral lines. However, only after my return in 1912 to Zurich did I hit upon the decisive idea about the analogy between the mathematical problem connected with my theory and the theory of surfaces by Gauss—originally without knowledge of the research by Riemann, Ricci, and Levi-Civita. The latter research came to my attention only through my friend Grossmann in Zurich when I posed the problem to him only to find generally covariant tensors whose components depend only upon the derivatives of the coefficients of the quadratic fundamental invariant. Today it appears that we can clearly recognize the achievements and limitations of the theory. The theory provides deep insights into the physical nature of space, time, matter, and gravitation, but no adequate means to solve the problems of quanta and the atomic constitution of elementary electric structures that constitute matter." . . .

. . . The text is: "I added in a 4th appendix an exposition of my views on the problem of space in general and on the step-by-step modifications of our ideas about space under relativistic aspects. I wanted to show that space-time is not something to which we can attribute independent existence, disconnected from the objects proper of physical reality. Physical objects are not *within space*, but objects are rather *spatially extended*. The concept of 'empty space' thus loses its meaning."

[15] . . . was appended: "A simple derivation of the Lorentz transformation is given in the appendix."

[26] . . . was appended: "The general theory of relativity suggests the idea that the electric mass of an electron is held together by gravitational forces." . . .

[31] . . . was appended: "See the somewhat more elaborate exposition in the appendix."

[37] . . . final quotation marks: "It is the gravitational field also that gives the jerk to the observer." . . .

[44] . . . was appended: "The existence of the theoretically demanded deflection of light has been photographically documented at the solar eclipse on May 30, 1919, by the English astronomer Eddington." . . .

[47] . . . was appended: "For the whole consideration one has to use the Galilean (nonrotating) system K as a reference body, because we may assume the validity of the results of the special theory of relativity only with respect to K (relative to K' there is a gravitational field)."

[49] . . . in the hand of Ilse Einstein: "Note: This interpretation has often been opposed as unconvincing because not only the measuring rods but also the circular disk would suffer tangential contraction. This argument is not cogent because the rotating disk cannot be viewed as a Euclidean rigid body; such body would shatter when brought to rotation, according to the postulated reasons. In reality the disk plays no role in the whole consideration, but only the system—of rods at rest relative to each other—which rotates as a totality of radially and tangentially positioned rodlets."

[55] . . . was appended: "See in the appendix. The relations (11a) and (12), which are derived there for coordinates, apply also to differences of coordinates and, therefore, to coordinate differentials (as infinitely small differences)."

[56] . . . by the following one: "This assignment need not even be such that would

have to be viewed as 'spatial' coordinates and as a 'time-like' coordinate."

[59]. . . the paragraph: "The second result, the curvature of light rays in the gravitational field of the sun, has already been mentioned; the first one refers to the orbit of the planet Mercury."

[60]. . . the following text: "that is, the equations of the general theory of relativity."

[97]. . . the fourteenth edition: "Likewise Perot, based on his own observations."

[99]. . . was added: "that is to say they are not convinced by the evidence of the existing empirical material."

Doc. 43

Cosmological Considerations in the General Theory of Relativity [1]

This translation by W. Perrett and G. B. Jeffery is reprinted from H. A. Lorentz et al., *The Principle of Relativity* (Dover, 1952), pp. 175–188.

I T is well known that Poisson's equation
$$\nabla^2\phi = 4\pi K\rho \quad . \quad . \quad . \quad . \quad (1)$$
in combination with the equations of motion of a material point is not as yet a perfect substitute for Newton's theory of action at a distance. There is still to be taken into account the condition that at spatial infinity the potential ϕ tends toward a fixed limiting value. There is an analogous state of things in the theory of gravitation in general relativity. Here, too, we must supplement the differential equations by limiting conditions at spatial infinity, if we really have to regard the universe as being of infinite spatial extent. [2]

In my treatment of the planetary problem I chose these limiting conditions in the form of the following assumption : it is possible to select a system of reference so that at spatial infinity all the gravitational potentials $g_{\mu\nu}$ become constant. [3] But it is by no means evident *a priori* that we may lay down the same limiting conditions when we wish to take larger portions of the physical universe into consideration. In the following pages the reflexions will be given which, up to the present, I have made on this fundamentally important question.

§ 1. The Newtonian Theory

It is well known that Newton's limiting condition of the constant limit for ϕ at spatial infinity leads to the view that the density of matter becomes zero at infinity. For we imagine that there may be a place in universal space round about which the gravitational field of matter, viewed on a large scale, possesses spherical symmetry. It then follows from Poisson's equation that, in order that ϕ may tend to a

177

limit at infinity, the mean density ρ must decrease toward zero more rapidly than $1/r^2$ as the distance r from the centre increases.* In this sense, therefore, the universe according to Newton is finite, although it may possess an infinitely great total mass.

From this it follows in the first place that the radiation emitted by the heavenly bodies will, in part, leave the Newtonian system of the universe, passing radially outwards, to become ineffective and lost in the infinite. May not entire heavenly bodies fare likewise? It is hardly possible to give a negative answer to this question. For it follows from the assumption of a finite limit for ϕ at spatial infinity that a heavenly body with finite kinetic energy is able to reach spatial infinity by overcoming the Newtonian forces of attraction. By statistical mechanics this case must occur from time to time, as long as the total energy of the stellar system—transferred to one single star—is great enough to send that star on its journey to infinity, whence it never can return.

[4]

We might try to avoid this peculiar difficulty by assuming a very high value for the limiting potential at infinity. That would be a possible way, if the value of the gravitational potential were not itself necessarily conditioned by the heavenly bodies. The truth is that we are compelled to regard the occurrence of any great differences of potential of the gravitational field as contradicting the facts. These differences must really be of so low an order of magnitude that the stellar velocities generated by them do not exceed the velocities actually observed.

If we apply Boltzmann's law of distribution for gas molecules to the stars, by comparing the stellar system with a gas in thermal equilibrium, we find that the Newtonian stellar system cannot exist at all. For there is a finite ratio of densities corresponding to the finite difference of potential between the centre and spatial infinity. A vanishing of the density at infinity thus implies a vanishing of the density at the centre.

---

* ρ is the mean density of matter, calculated for a region which is large as compared with the distance between neighbouring fixed stars, but small in comparison with the dimensions of the whole stellar system.

It seems hardly possible to surmount these difficulties on the basis of the Newtonian theory. We may ask ourselves the question whether they can be removed by a modification of the Newtonian theory. First of all we will indicate a method which does not in itself claim to be taken seriously; it merely serves as a foil for what is to follow. In place of Poisson's equation we write

$$\nabla^2 \phi - \lambda \phi = 4\pi\kappa\rho \qquad . \qquad . \qquad . \quad (2)$$

where λ denotes a universal constant. If ρ_0 be the uniform density of a distribution of mass, then

$$\phi = -\frac{4\pi\kappa}{\lambda}\rho_0 \qquad . \qquad . \qquad . \quad (3)$$

is a solution of equation (2). This solution would correspond to the case in which the matter of the fixed stars was distributed uniformly through space, if the density ρ_0 is equal to the actual mean density of the matter in the universe. The solution then corresponds to an infinite extension of the central space, filled uniformly with matter. If, without making any change in the mean density, we imagine matter to be non-uniformly distributed locally, there will be, over and above the ϕ with the constant value of equation (3), an additional ϕ, which in the neighbourhood of denser masses will so much the more resemble the Newtonian field as $\lambda\phi$ is smaller in comparison with $4\pi\kappa\rho$. [5]

A universe so constituted would have, with respect to its gravitational field, no centre. A decrease of density in spatial infinity would not have to be assumed, but both the mean potential and mean density would remain constant to infinity. The conflict with statistical mechanics which we found in the case of the Newtonian theory is not repeated. With a definite but extremely small density, matter is in equilibrium, without any internal material forces (pressures) being required to maintain equilibrium.

§ 2. The Boundary Conditions According to the General Theory of Relativity

In the present paragraph I shall conduct the reader over the road that I have myself travelled, rather a rough and winding road, because otherwise I cannot hope that he will take much interest in the result at the end of the journey. The conclusion I shall arrive at is that the field equations of gravitation which I have championed hitherto still need a slight modification, so that on the basis of the general theory of relativity those fundamental difficulties may be avoided which have been set forth in § 1 as confronting the Newtonian theory. This modification corresponds perfectly to the transition from Poisson's equation (1) to equation (2) of § 1. We finally infer that boundary conditions in spatial infinity fall away altogether, because the universal continuum in respect of its spatial dimensions is to be viewed as a self-contained continuum of finite spatial (three-dimensional) volume.

The opinion which I entertained until recently, as to the limiting conditions to be laid down in spatial infinity, took its stand on the following considerations. In a consistent theory of relativity there can be no inertia *relatively to " space,"* but only an inertia of masses *relatively to one another*. If, therefore, I have a mass at a sufficient distance from all other masses in the universe, its inertia must fall to zero. We will try to formulate this condition mathematically.

According to the general theory of relativity the negative momentum is given by the first three components, the energy by the last component of the covariant tensor multiplied by $\sqrt{-g}$

$$m\sqrt{-g}\; g_{\mu a}\frac{dx_a}{ds} \quad . \quad . \quad . \quad . \quad (4)$$

where, as always, we set

$$ds^2 = g_{\mu\nu}dx_\mu dx_\nu \quad . \quad . \quad . \quad (5)$$

In the particularly perspicuous case of the possibility of choosing the system of co-ordinates so that the gravitational field at every point is spatially isotropic, we have more simply

$$ds^2 = -A(dx_1^2 + dx_2^2 + dx_3^2) + Bdx_4^2.$$

If, moreover, at the same time

$$\sqrt{-g} = 1 = \sqrt{A^3 B}$$

we obtain from (4), to a first approximation for small velocities;

$$m\frac{A}{\sqrt{B}}\frac{dx_1}{dx_4},\; m\frac{A}{\sqrt{B}}\frac{dx_2}{dx_4},\; m\frac{A}{\sqrt{B}}\frac{dx_3}{dx_4}$$

[6]

for the components of momentum, and for the energy (in the static case)

$$m\sqrt{B}.$$

From the expressions for the momentum, it follows that $m\dfrac{A}{\sqrt{B}}$ plays the part of the rest mass. As m is a constant peculiar to the point of mass, independently of its position, this expression, if we retain the condition $\sqrt{g\,-}\,=1$ at spatial infinity, can vanish only when A diminishes to zero, while B increases to infinity. It seems, therefore, that such a degeneration of the co-efficients $g_{\mu\nu}$ is required by the postulate of relativity of all inertia. This requirement implies that the potential energy $m\sqrt{B}$ becomes infinitely great at infinity. Thus a point of mass can never leave the system ; and a more detailed investigation shows that the same thing applies to light-rays. A system of the universe with such behaviour of the gravitational potentials at infinity would not therefore run the risk of wasting away which was mooted just now in connexion with the Newtonian theory.

I wish to point out that the simplifying assumptions as to the gravitational potentials on which this reasoning is based, have been introduced merely for the sake of lucidity. It is possible to find general formulations for the behaviour of the $g_{\mu\nu}$ at infinity which express the essentials of the question without further restrictive assumptions.

At this stage, with the kind assistance of the mathematician J. Grommer, I investigated centrally symmetrical, static gravitational fields, degenerating at infinity in the way mentioned. The gravitational potentials $g_{\mu\nu}$ were applied, and from them the energy-tensor $T_{\mu\nu}$ of matter was calculated on the basis of the field equations of gravitation. But here it proved that for the system of the fixed stars no boundary conditions of the kind can come into question at all, as was also rightly emphasized by the astronomer de Sitter recently.   [8]

For the contravariant energy-tensor $T^{\mu\nu}$ of ponderable matter is given by

$$T^{\mu\nu} = \rho\frac{dx_\mu}{ds}\frac{dx_\nu}{ds},$$

where ρ is the density of matter in natural measure. With

[7]

an appropriate choice of the system of co-ordinates the stellar velocities are very small in comparison with that of light. We may, therefore, substitute $\sqrt{g_{44}}\, dx_4$ for ds. This shows us that all components of $T^{\mu\nu}$ must be very small in comparison with the last component T^{44}. But it was quite impossible to reconcile this condition with the chosen boundary conditions. In the retrospect this result does not appear astonishing. The fact of the small velocities of the stars allows the conclusion that wherever there are fixed stars, the gravitational potential (in our case $\sqrt{B}$) can never be much greater than here on earth. This follows from statistical reasoning, exactly as in the case of the Newtonian theory. At any rate, our calculations have convinced me that such conditions of degeneration for the $g_{\mu\nu}$ in spatial infinity may not be postulated.

After the failure of this attempt, two possibilities next present themselves.

(a) We may require, as in the problem of the planets, that, with a suitable choice of the system of reference, the $g_{\mu\nu}$ in spatial infinity approximate to the values

$$\begin{array}{cccc} -1 & 0 & 0 & 0 \\ 0 & -1 & 0 & 0 \\ 0 & 0 & -1 & 0 \\ 0 & 0 & 0 & 1 \end{array}$$

(b) We may refrain entirely from laying down boundary conditions for spatial infinity claiming general validity; but at the spatial limit of the domain under consideration we have to give the $g_{\mu\nu}$ separately in each individual case, as hitherto we were accustomed to give the initial conditions for time separately.

The possibility (b) holds out no hope of solving the problem, but amounts to giving it up. This is an incontestable position, which is taken up at the present time by de Sitter.* But I must confess that such a complete resignation in this fundamental question is for me a difficult thing. I should not make up my mind to it until every effort to make headway toward a satisfactory view had proved to be vain.

Possibility (a) is unsatisfactory in more respects than one.

[10]

[9]

* de Sitter, Akad. van Wetensch. te Amsterdam, 8 Nov., 1916.

In the first place those boundary conditions pre-suppose a definite choice of the system of reference, which is contrary to the spirit of the relativity principle.   Secondly, if we adopt this view, we fail to comply with the requirement of the relativity of inertia.   For the inertia of a material point of mass m (in natural measure) depends upon the $g_{\mu\nu}$; but these differ but little from their postulated values, as given above, for spatial infinity.   Thus inertia would indeed be *influenced*, but would not be *conditioned* by matter (present in finite space).   If only one single point of mass were present, according to this view, it would possess inertia, and in fact an inertia almost as great as when it is surrounded by the other masses of the actual universe.   Finally, those statistical objections must be raised against this view which were mentioned in respect of the Newtonian theory.

From what has now been said it will be seen that I have not succeeded in formulating boundary conditions for spatial infinity.   Nevertheless, there is still a possible way out, without resigning as suggested under (*b*).   For if it were possible to regard the universe as a continuum which is *finite (closed) with respect to its spatial dimensions*, we should have no need at all of any such boundary conditions.   We shall proceed to show that both the general postulate of relativity and the fact of the small stellar velocities are compatible with the hypothesis of a spatially finite universe; though certainly, in order to carry through this idea, we need a generalizing modification of the field equations of gravitation.

§ 3. The Spatially Finite Universe with a Uniform Distribution of Matter

According to the general theory of relativity the metrical character (curvature) of the four-dimensional space-time continuum is defined at every point by the matter at that point and the state of that matter.   Therefore, on account of the lack of uniformity in the distribution of matter, the metrical structure of this continuum must necessarily be extremely complicated.   But if we are concerned with the structure only on a large scale, we may represent matter to ourselves as being uniformly distributed over enormous spaces, so that its density of distribution is a variable function which varies extremely slowly. Thus our procedure will somewhat resemble that of the geodesists who, by means of an ellipsoid, approximate to the shape of the earth's surface, which on a small scale is extremely complicated.

The most important fact that we draw from experience as to the distribution of matter is that the relative velocities of the stars are very small as compared with the velocity of light. So I think that for the present we may base our reasoning upon the following approximative assumption. There is a system of reference relatively to which matter may be looked upon as being permanently at rest. With respect to this system, therefore, the contravariant energy-tensor $T^{\mu\nu}$ of matter is, by reason of (5), of the simple form

[11]

$$\left.\begin{array}{cccc} 0 & 0 & 0 & 0 \\ 0 & 0 & 0 & 0 \\ 0 & 0 & 0 & 0 \\ 0 & 0 & 0 & \rho \end{array}\right\} \quad . \quad . \quad . \quad (6)$$

The scalar ρ of the (mean) density of distribution may be *a priori* a function of the space co-ordinates. But if we assume the universe to be spatially finite, we are prompted to the hypothesis that ρ is to be independent of locality. On this hypothesis we base the following considerations.

As concerns the gravitational field, it follows from the equation of motion of the material point

$$\frac{d^2x_\nu}{ds^2} + \{\alpha\beta, \nu\}\frac{dx_\alpha}{ds}\frac{dx_\beta}{ds} = 0$$

that a material point in a static gravitational field can remain at rest only when g_{44} is independent of locality. Since, further, we presuppose independence of the time co-ordinate x_4 for all magnitudes, we may demand for the required solution that, for all x_ν,

$$g_{44} = 1 \quad . \quad . \quad . \quad (7)$$

Further, as always with static problems, we shall have to set

$$g_{14} = g_{24} = g_{34} = 0 \quad . \quad . \quad (8)$$

It remains now to determine those components of the gravitational potential which define the purely spatial-geometrical relations of our continuum ($g_{11}, g_{12}, \ldots g_{33}$). From our assumption as to the uniformity of distribution of the masses generating the field, it follows that the curvature of the required space must be constant. With this distribution of mass, therefore, the required finite continuum of the x_1, x_2, x_3, with constant x_4, will be a spherical space.

We arrive at such a space, for example, in the following way. We start from a Euclidean space of four dimensions, ξ_1, ξ_2, ξ_3 ξ_4, with a linear element $d\sigma$; let, therefore,

$$d\sigma^2 = d\xi_1^2 + d\xi_2^2 + d\xi_3^2 + d\xi_4^2 . \quad . \quad . \quad (9)$$

In this space we consider the hyper-surface

$$R^2 = \xi_1^2 + \xi_2^2 + \xi_3^2 + \xi_4^2, \quad . \quad . \quad . \quad (10)$$

where R denotes a constant. The points of this hyper-surface form a three-dimensional continuum, a spherical space of radius of curvature R.

The four-dimensional Euclidean space with which we started serves only for a convenient definition of our hyper-surface. Only those points of the hyper-surface are of interest to us which have metrical properties in agreement with those of physical space with a uniform distribution of matter. For the description of this three-dimensional continuum we may employ the co-ordinates ξ_1, ξ_2, ξ_3 (the projection upon the hyper-plane $\xi_4 = 0$) since, by reason of (10), ξ_4 can be expressed in terms of ξ_1, ξ_2, ξ_3. Eliminating ξ_4 from (9), we obtain for the linear element of the spherical space the expression

$$\left.\begin{array}{c} d\sigma^2 = \gamma_{\mu\nu} d\xi_\mu d\xi_\nu \\ \gamma_{\mu\nu} = \delta_{\mu\nu} + \dfrac{\xi_\mu \xi_\nu}{R^2 - \rho^2} \end{array}\right\} \quad . \quad . \quad . \quad (11)$$

where $\delta_{\mu\nu} = 1$, if $\mu = \nu$; $\delta_{\mu\nu} = 0$, if $\mu \neq \nu$, and $\rho^2 = \xi_1^2 + \xi_2^2 + \xi_3^2$. The co-ordinates chosen are convenient when it is a question of examining the environment of one of the two points $\xi_1 = \xi_2 = \xi_3 = 0$.

Now the linear element of the required four-dimensional space-time universe is also given us. For the potential $g_{\mu\nu}$, both indices of which differ from 4, we have to set

$$g_{\mu\nu} = -\left(\delta_{\mu\nu} + \frac{x_\mu x_\nu}{R^2 - (x_1^2 + x_2^2 + x_3^2)}\right) \quad . \quad (12)$$

[12]

which equation, in combination with (7) and (8), perfectly defines the behaviour of measuring-rods, clocks, and light-rays.

§ 4. On an Additional Term for the Field Equations of Gravitation

My proposed field equations of gravitation for any chosen system of co-ordinates run as follows :—

$$
\left.
\begin{aligned}
G_{\mu\nu} &= - \kappa(T_{\mu\nu} - \tfrac{1}{2}g_{\mu\nu}T), \\
G_{\mu\nu} &= - \frac{\partial}{\partial x_a}\{\mu\nu,\,a\} + \{\mu a,\,\beta\}\,\{\nu\beta,\,a\} \\
&\quad + \frac{\partial^2 \log\sqrt{-g}}{\partial x_\mu \partial x_\nu} - \{\mu\nu,\,a\}\frac{\partial \log\sqrt{-g}}{\partial x_a}
\end{aligned}
\right\} \quad (13)
$$

The system of equations (13) is by no means satisfied when we insert for the $g_{\mu\nu}$ the values given in (7), (8), and (12), and for the (contravariant) energy-tensor of matter the values indicated in (6). It will be shown in the next paragraph how this calculation may conveniently be made. So that, if it were certain that the field equations (13) which I have hitherto employed were the only ones compatible with the postulate of general relativity, we should probably have to conclude that the theory of relativity does not admit the hypothesis of a spatially finite universe.

However, the system of equations (14) allows a readily suggested extension which is compatible with the relativity postulate, and is perfectly analogous to the extension of Poisson's equation given by equation (2). For on the left-hand side of field equation (13) we may add the fundamental tensor $g_{\mu\nu}$, multiplied by a universal constant, $-\lambda$, at present unknown, without destroying the general covariance. In place of field equation (13) we write

$$
G_{\mu\nu} - \lambda g_{\mu\nu} = - \kappa(T_{\mu\nu} - \tfrac{1}{2}g_{\mu\nu}T) \quad . \quad . \quad (13a)
$$

This field equation, with λ sufficiently small, is in any case also compatible with the facts of experience derived from the solar system. It also satisfies laws of conservation of momentum and energy, because we arrive at (13a) in place of (13) by introducing into Hamilton's principle, instead of the scalar of Riemann's tensor, this scalar increased by a universal constant; and Hamilton's principle, of course, guarantees the validity of laws of conservation. It will be shown in § 5 that field equation (13a) is compatible with our conjectures on field and matter.

§ 5. Calculation and Result

Since all points of our continuum are on an equal footing, it is sufficient to carry through the calculation for *one* point, e.g. for one of the two points with the co-ordinates

$$x_1 = x_2 = x_3 = x_4 = 0.$$

Then for the $g_{\mu\nu}$ in (13a) we have to insert the values

$$
\begin{array}{cccc}
-1 & 0 & 0 & 0 \\
0 & -1 & 0 & 0 \\
0 & 0 & -1 & 0 \\
0 & 0 & 0 & 1
\end{array}
$$

wherever they appear differentiated only once or not at all. We thus obtain in the first place

$$G_{\mu\nu} = \frac{\partial}{\partial x_1}[\mu\nu, 1] + \frac{\partial}{\partial x_2}[\mu\nu, 2] + \frac{\partial}{\partial x_3}[\mu\nu, 3] + \frac{\partial^2 \log \sqrt{-g}}{\partial x_\mu \partial x_\nu}.$$

From this we readily discover, taking (7), (8), and (13) into account, that all equations (13*a*) are satisfied if the two relations

$$-\frac{2}{R^2} + \lambda = -\frac{\kappa\rho}{2}, \quad -\lambda = -\frac{\kappa\rho}{2},$$

or

$$\lambda = \frac{\kappa\rho}{2} = \frac{1}{R^2} \quad . \quad . \quad . \quad . \quad (14)$$

are fulfilled.

Thus the newly introduced universal constant λ defines both the mean density of distribution ρ which can remain in equilibrium and also the radius R and the volume $2\pi^2R^3$ of spherical space. The total mass M of the universe, according to our view, is finite, and is in fact

$$M = \rho \cdot 2\pi^2R^3 = 4\pi^2\frac{R}{\kappa} = \pi^2\sqrt{\frac{32}{\kappa^3\rho}} \quad . \quad . \quad (15)$$

[15]

Thus the theoretical view of the actual universe, if it is in correspondence with our reasoning, is the following. The curvature of space is variable in time and place, according to the distribution of matter, but we may roughly approximate to it by means of a spherical space.   At any rate, this view is logically consistent, and from the standpoint of the general theory of relativity lies nearest at hand ; whether, from the standpoint of present astronomical knowledge, it is tenable, will not here be discussed.   In order to arrive at this consistent view, we admittedly had to introduce an extension of the field equations of gravitation which is not justified by our actual knowledge of gravitation.   It is to be emphasized, however, that a positive curvature of space is given by our results, even if the supplementary term is not introduced. That term is necessary only for the purpose of making possible a quasi-static distribution of matter, as required by the fact of the small velocities of the stars.

[16]

Doc. 44

Reply to the Plaintiff's Written Statement of 27 December 1916

Not translated for this volume.

Doc. 45

[p. 82]

On the Quantum Theorem of Sommerfeld and Epstein

by A. Einstein

(Presented at the session of May 11)

(cf. above, p. 79)

§1. Previous Formulation. There is hardly any more doubt that the quantum condition for periodic mechanical systems with one degree of freedom is (after SOMMERFELD and DEBYE)

[1]

$$\int p\,dq = \int p\frac{dq}{dt}\,dt = nh. \tag{1}$$

The integral is to be extended here over one full period of movement; q denotes the coordinate, p the associated coordinate of momentum of the system. SOMMERFELD's work on the theory of spectra proves with certainty that in systems with several degrees of freedom several quantum conditions have to take the place of this *single* quantum condition; in general as many (l) as the system has degrees of freedom. These l conditions are, according to SOMMERFELD,

[2]

$$\int p_i\,dq_i = n_i h. \tag{2}$$

As this formulation is not independent of the choice of coordinates, it can only be correct for a distinct choice of coordinates. System (2) represents a distinct statement about the movement only after this choice has been made and if the q_i are periodic functions of time.

[3]
[4]

Further principal progress is due to EPSTEIN (and SCHWARZSCHILD). The former based his rule for the selection of the SOMMERFELD coordinates q_i on JACOBI's theorem, which states in a well-known manner: Let $H(H(q_i p_i))$ be the HAMILTONian function of the q_i and p_i and t, which occurs in the canonical equations

$$\dot{p}_i = -\frac{\partial H}{\partial q_i} \tag{3}$$

$$\dot{q}_i = \frac{\partial H}{\partial p_i}, \tag{4}$$

and which is identical to the energy function if it does not explicitly contain the time   [p. 83]
t.[1] If $J(t, q_1 \ldots q_l, \alpha_1 \ldots \alpha_l)$ is a complete integral of the HAMILTON-JACOBI partial
differential equations

$$\frac{\partial J}{\partial t} + H\left(q_i, \frac{\partial J}{\partial q_i}\right) = 0, \tag{5}$$

the solution of the canonical equations is

$$\frac{\partial J}{\partial \alpha_i} = \beta_i \tag{6}$$

$$\frac{\partial J}{\partial q_i} = p_i. \tag{7}$$

If H does not explicitly depend upon time—as we shall assume in the following—we
can solve (5) with

$$J = J^* - ht, \tag{8}$$

where h represents a constant and J^* no longer depends explicitly upon the time t.
Then (5), (6), and (7) are replaced by the equations

$$H\left(q_i, \frac{\partial J^*}{\partial q_i}\right) = h \tag{5a}$$

$$\left.\begin{array}{l}\dfrac{\partial J^*}{\partial \alpha_i} = \beta_i \\[2mm] \dfrac{\partial J^*}{\partial h} = t - t_0\end{array}\right\} \tag{6a}$$

$$\frac{\partial J^*}{\partial q_i} = p_i, \tag{7a}$$

where, however, the first of the equations (6a) represents only $l-1$ equations, and   [5] {1}
where α_l is replaced by the constant h, and β_l by the constant $-t_0$.

According to EPSTEIN, the coordinates q_i are to be chosen such that a complete
integral of (5a) exists in the form of

---

[1]Because in this case one has $\dfrac{dH}{dt} = \displaystyle\sum_i \frac{\partial H}{\partial q_i}\dot{q}_i + \sum_i \frac{\partial H}{\partial p_i}\dot{p}_i = 0.$

$$J^* = \sum_i J_i(q_i),$$

(8a)

where J_i depends upon q_i but is independent of the other q. SOMMERFELD's quantum conditions (2) shall then be valid for these coordinates q_i if the q_i are periodic functions of t.

[p. 84]   Notwithstanding the great successes that have been achieved by the SOM-MERFELD-EPSTEIN extension of the quantum theorem for systems of several degrees of freedom, it still remains unsatisfying that one has to depend on the separation of variables, due to (8), because it probably has nothing to do with the quantum problem per se. In the following we would like to suggest a minor modification of the SOMMERFELD-EPSTEIN condition and thereby avoid this deficiency. I will briefly point

[6]   out the basic idea and explain it later in more detail.

§2. Modified Formulation. While *pdq* is an invariant with systems of one degree of freedom (i.e., it is independent of the choice of the coordinates q), the individual products $p_i dq_i$ in a system of several degrees of freedom, on the other hand, are not invariants; therefore, the quantum condition (2) has no invariant meaning. Invariant is only the sum $\sum_i p_i dq_i$ extended over all l degrees of freedom. In order to derive from this sum a multiplicity of invariant quantum conditions, one can proceed as follows. Look at the p_i as functions of q_i. Or speaking geometrically, one can view p_i as a vector (of "covariant" character) within an l-dimensional space of the q_i. If I draw any closed curve inside this q_i-space (and it need by no means be an "orbital path" of the mechanical system) then its line integral

$$\int \sum p_i dq_i$$

(9)

is an invariant. If the p_i are any functions of the q_i, then each closed curve will, in general, provide a different value for the integral (9). However, if the vector p_i is derivable from a potential J^*, i.e., if the following relations hold true

$$\frac{\partial p_i}{\partial q_k} - \frac{\partial p_k}{\partial q_i} = 0,$$

(10)

respectively,

$$p_i = \frac{\partial J^*}{\partial q_i},$$

(10a)

then the integral (9) has the same value for all closed curves that can be continuously transformed into one another. And the integral (9) vanishes for all curves that can be

[p. 85]   contracted to a point by a continuous modification. However, if the space of the q_i

considered is multiply-connected, then there do exist closed curves that cannot be shrunk to a point by continuous deformation. If then J^* is not a single-valued (but an ∞-valued) function of the q_i, integral (9) will, in general, differ from zero for such curves. Nevertheless, there will be a finite number of closed paths in q-space into which all closed lines in this space can be reduced by continuous processes. In this sense, the finite number of conditions

$$\int \sum_i p_i dq_i = n_i h \qquad (11)$$

can be prescribed as quantum conditions. In my opinion they have to replace the quantum conditions (2). We will have to expect that the number of equations (10) that cannot be reduced into each other will be equal to the number of degrees of freedom of the system. If this number is smaller, we are confronted with a case of "degeneracy" (*Entartung*).

The basic idea suggested above (intentionally full of gaps) will be explained in more detail in the following.

§3. Descriptive Derivation of the HAMILTON-JACOBI Differential Equation. When a point P of the coordinate space is given with the coordinates Q_i and with the associated velocity, i.e., the associated momentum coordinates P_i, then its movement is completely determined by the canonical equations (3) and (4).[2] Every point on the orbital curve L thus has a certain velocity, i.e., the p_i on L are distinct functions of q_i. If one imagines in an $(l-1)$-dimensional "surface" of the coordinate space every point P given by its Q_i and P_i, then to every point belongs a motion with an orbital curve L in the coordinate space. These orbits fill the coordinate space (or part of it) continuously, provided the P_i on the surface are continuous functions of the Q_i. There will be an orbital curve through every point (q_i) of the coordinate space; and, [p. 86] therefore, there will be distinct momentum coordinates p_i associated with this point. Thus, a vector field p_i is given for the coordinate space. We consider it our task to find the law of this vector field.

Considering the p_i as functions of the q_i in the canonical system of equations (3), we have to replace the left-hand sides by

$$\sum_k \frac{\partial p_i}{\partial q_k} \frac{dq_k}{dt}$$

and this, according to (4), can be written as

---

[2]It shall be assumed that H does not explicitly depend upon time t.

$$\sum_k \frac{\partial p_i}{\partial q_k} \frac{\partial H}{\partial p_k}.$$

Thus, we get in place of (3)

$$\frac{\partial H}{\partial q_i} + \sum_k \frac{\partial H}{\partial p_k} \frac{\partial p_i}{\partial q_k} = 0. \tag{12}$$

This is a system of l linear differential equations which must be satisfied by the p_k as functions of the q_k.

We ask now if there are vector fields for which a potential J^* exists, i.e., which satisfy the conditions (10) and (10a). In this case (12), due to (10), takes the form

$$\frac{\partial H}{\partial q_i} + \sum_k \frac{\partial H}{\partial p_k} \frac{\partial p_k}{\partial q_i} = 0.$$

This equation says that H is independent of the q_i. Therefore, fields of potentials of the desired kind do exist, and their potential J^* satisfies the HAMILTON-JACOBI equation (5a), or J satisfies (5), respectively.

Thus, it has been shown that equations (3) can be replaced by (7a) and (5a), or by (7) and (5), respectively. We still want to show that the equation system (4) is satisfied by (6a) or (6), even though it is of no consequence for the following deliberations. Equations (4) form a system of total differential equations for the determination of the q_i as functions of time, after the integration of (5a) and [p. 87] expressing the p_i as functions of the q_i due to (7a). According to the theory of differential equations of the first order, this system of total differential equations is equivalent to the partial differential equation

$$\sum_k \frac{\partial H}{\partial p_k} \frac{\partial \phi}{\partial q_k} + \frac{\partial \phi}{\partial t} = 0. \tag{13}$$

The latter is satisfied by

$$\phi = \frac{\partial J}{\partial \alpha_i}$$

if J is a complete integral of (5). Because, if the value of ϕ is inserted into the left-hand side of (13), one obtains under due consideration of (7)

$$\sum_k \frac{\partial H}{\partial \left(\frac{\partial J}{\partial q_k} \right)} \frac{\partial^2 J}{\partial q_k \partial \alpha_i} + \frac{\partial^2 J}{\partial t \partial \alpha_i}$$

or

$$\frac{\partial}{\partial \alpha_i}\left\{H\left(q_k, \frac{\partial J}{\partial q_k}\right) + \frac{\partial J}{\partial t}\right\},$$

and these are quantities that vanish due to (5). From this follows that equations (4) are integrated by (6), (6a), respectively.

§4. The p_i-field of a Single Orbit. We come now to a very essential point which I carefully avoided mentioning during the preliminary sketch of the basic idea in §2. In our deliberations in §3 we imagined the p_i-field to be generated by $(l-1)$-fold infinitely many, mutually independent movements that were illustrated by exactly as many orbital curves in q_i-space. But now we imagine that we follow the undisturbed movement of a single system for an infinitely long time and the associated orbital curve to be (figuratively) drawn into the q_i-space. Here two cases can occur.

1. There exists a part of q_i-space such that in the course of time the orbital curve comes arbitrarily close to every point of this (l-dimensional) domain.

2. The orbital curve in its entirety fits into a continuum of fewer than l dimensions. A special case of this category is a movement with an exactly closed orbit.

Case 1 is the general one; the cases 2 result as specializations from 1. As an example of case 1 we think of the movement of a material point under the action of [p. 88] a central force, described by two coordinates which fix the position of the point in the orbital plane (e.g., polar coordinates r and ϕ). Case 2 occurs when the law of attraction is exactly proportional to $\dfrac{1}{r^2}$ and when the deviations from a KEPLERian motion—which the theory of relativity demands—are neglected. The path of the orbit is then closed and its points constitute a continuum of one dimension. When seen in three-dimensional space, the central movement is always a movement of type 2 because the entire orbital curve can be fit into a continuum of two dimensions. In a three-dimensional view, central movement has to be perceived as a special case of movement, defined by a more complicated force law (e.g., the movement studied by EPSTEIN in the theory of the STARK effect).

The following consideration relates to the general case 1. Consider an element $d\tau$ of the q_i-space. The orbital curve of the motion under consideration passes infinitely often through $d\tau$. There is a system (p_i) of momentum coordinates for each one of such crossings. A priori, two types of orbits are possible, obviously of fundamentally different characteristics.

Type a): The p_i-systems repeat such that only a *finite* number of p_i-systems belong to $d\tau$. The process of motion can be represented with the p_i as single-valued functions of the q_i.

Type b): There are infinitely many p_i-systems at the location under consideration. In this case the p_i *cannot* be represented as functions of the q_i.

One notices immediately that type b) excludes the quantum condition we formulated in §2. On the other hand, classical statistical mechanics deals essentially *only* with type b); because only in this case is the microcanonic ensemble of *one* system equivalent to the time ensemble.[3]

[p. 89]   In summarizing we can say: The application of the quantum condition (11) demands that there exist orbits such that a *single* orbit determines the p_i-field for which a potential J^* exists.

{2}   §5. *The "Rationalized Coordinate Space."* It has already been noted that the p_i are, in general, multivalued functions of the q_i. As a simple example we look again at the planar revolution of a point that is under the attracting force of a fixed center. The point moves such that its distance r from the attracting center oscillates periodically between a minimal value r_1 and a maximal value r_2. Looking at a point of the space of the q_i, i.e., a point in the ring-shaped surface bounded by the radii r_1 and r_2, one realizes that the orbital curve, in the course of time, will pass infinitely often and arbitrarily close to this point, or—to phrase it a little imprecisely—will pass through it. But the radial component of velocity has different signs, depending upon whether the passing occurs on an orbital segment of increasing or decreasing r; and thus, the p_i are two-valued functions of the q_ν.

The associated inconvenience for the imagination is best resolved by the well-known method that RIEMANN has introduced to the theory of functions. Imagine the ring-shaped surface to be doubled such that two congruent ring-shaped leaves are on

top of each other. We imagine the orbital parts with positive $\dfrac{dr}{dt}$ in the top-most ring,

those with negative $\dfrac{dr}{dt}$ drawn onto the lower ring, with the associated vector system

of the p_ν. We think of the two leaves as connected along the circular lines because the orbit must proceed from one ring-shaped leaf to the other whenever the orbital curve touches one of the two limiting circles. It is easily understood that the p_ν from both leaves agree along these circles. Interpreted on this double surface, the p_ν are
[7]   not only continuous functions of the q_ν but also univalent functions—this represents the value of this concept.

---

[3]The microcanonic ensemble contains systems which possess, with given q_i, still arbitrarily given p_i (commensurable with the energy values).

There are obviously two types of closed curves on this double surface; they can neither be shrunk to a point by continuous deformation, nor can one type be reduced [p. 90] to the other. Fig. 1 below depicts an example for each of the two types (L_1 and L_2). The parts of a line on the lower leaf are drawn with dashes. All other closed curves on the double surface can either be contracted into a point by continuous deformation or, by the same process, can be mapped onto the types L_1 or L_2 with one or several revolutions. The quantum theorem (11) would here have to be applied to path types L_1 and L_2.

Obviously, these considerations are to be generalized to all movements that satisfy the condition of §4. Phase space has to be imagined split into a number of

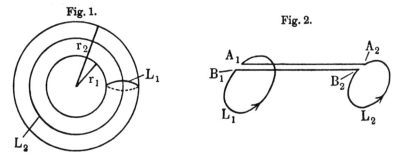

Fig. 1.   Fig. 2.

"tracts" that are connected along $(l - 1)$-dimensional "surfaces" such that—when interpreted in the so-constructed manifold the p_i are univalent and continuous functions (also at transitions from one tract to another tract). We shall call this auxiliary geometrical construction the "rationalized phase space." The quantum theorem (11) shall apply to all lines that are closed within the rationalized coordinate space.

In order to attach a precise meaning to the quantum theorem in this formulation, the integral $\int \Sigma p_i dq_i$, extended over the rationalized q_i-space, must have the same value for all those closed curves, which can be continuously deformed into one another. The proof for this follows the customary scheme entirely. In the rationalized q_i-space, let L_1 and L_2 (see the schematic in Fig. 2) be closed curves that can continuously be deformed into one another under preservation of the indicated directional orientation. The contour in the figure is then a closed curve that can be [p. 91] continuously shrunk into a point. From this follows, due to (10), that the line integral extended over the entire trail vanishes. If one considers furthermore that the integrals extended over the infinitely close connecting lines $\overline{A_1 A_2}$ and $\overline{B_1 B_2}$ are equal, because of the single-valuedness of the p_i in the rationalized q_i-space, one has the result that the integrals extended over L_1 and L_2 are equal.

The potential J^* is infinitely multivalued also in the rationalized q_i-space; but according to the quantum theorem this multivaluedness is the simplest one imaginable. Because, if J^* is the value of the potential of *one* point in the rationalized q_i-space, the other values are $J^* + nh$ where n is an integer.

Supplement in proof. Further thinking showed that the second condition for the applicability of formula (11) in §4 is always automatically fulfilled, i.e., the theorem holds: If a movement produces a p_i-field, then this necessarily has a potential J^*.

According to JACOBI's theorem, every movement can be derived from a complete integral J^* of (5a). At any rate, there exists at least one function J^* of the q_i from which the momentum coordinates p_i of a system of movements under consideration can be computed for every point of an orbital curve by means of the equations

$$p_i = \frac{\partial J^*}{\partial q_i}.$$

We now have to remember that J^* has been obtained by means of a partial differential equation, i.e., through a prescription of how the function J^* is to be continued in the q_i-space. Therefore, if we want to know how J^* changes for a system in the course of its movement, we have to envision how J^* extends along an orbital curve (and its neighborhood) according to the differential equation. If the orbit after a certain (rather long) time returns within close proximity to a point P through which the orbital curve has passed before, then $\dfrac{\partial J^*}{\partial q_i}$ provides the momentum coordinate for both times if we integrate J^* continuously along the entire piece of

[p. 92] orbital curve between both points. It is by no means to be expected that this continuation leads to a return to previous values of the $\dfrac{\partial J^*}{\partial q_i}$; instead, one should, in general, anticipate finding a totally different system of the p_i every time the configuration of the q_i under consideration is approximately attained again in the course of movement. Thus, it is absolutely impossible to represent the p_i as functions of the q_i for a motion that *continues indefinitely*. But if the p_i—or resp. a finite number of value systems of these quantities—repeat under repetition of the coordinate configuration, the $\dfrac{\partial J^*}{\partial q_i}$ are expressible as functions of the q_i for an indefinitely continuing motion. Therefore, if a p_i-field exists for an indefinitely continuing motion, then there also exists an associated potential J^*.

Consequently we can say the following: If there exist l integrals for the 21 equations of motion in the form

$$R_k(q_i, p_i) = \text{const}, \tag{14}$$

where the R_k are algebraic functions of the p_i, then $\Sigma p_i dq_i$ is always a complete   {3}
differential, provided the p_i are expressed in terms of the q_i with the help of (14).

The quantum condition demands that the integral $\int_i \Sigma p_i dq_i$, when extended over an
irreducible curve, is a multiple of h. This quantum condition coincides with the one
by SOMMERFELD-EPSTEIN when, specifically, each p_i depends only upon the
associated q_i.

If there exist fewer than l integrals of type (14), as is the case, for example,
according to POINCARÉ in the three-body problem, then the p_i are not expressible by
the q_i and the quantum condition of SOMMERFELD-EPSTEIN fails also in the slightly
generalized form that has been given here.

Additional notes by translator

{1}   "α_n" and "β_n" have been corrected here to "α_l" and "β_l."

{2}   The German word "rationell" is an economic concept, quite remote from
      rational numbers, though also derived from the Latin word "ratio." As will be
      seen later, Einstein has a geometric rationalization in mind in order to facilitate
      a more concrete and descriptive understanding of an otherwise abstract entity
      of physics.

{3}   The index "i" in "dq_i" was missing.

Doc. 46
Review of Hermann von Helmholtz, *Two Lectures on Goethe*

Not translated for this volume.

Doc. 47

A Derivation of Jacobi's Theorem   [1]

by A. Einstein

The canonical equations of dynamics

$$\frac{dp_i}{dt} = - \frac{\partial H}{\partial q_i} \tag{1}$$

$$\frac{dq_i}{dt} = \frac{\partial H}{\partial p_i}, \tag{2}$$

where H is, in the most general case, a function of the coordinates q_i, the momenta p_i and time t, can be integrated—as is well known—according to HAMILTON-JACOBI by determining a function J of the q_i and time t as a solution of the partial differential equation

$$\frac{\partial J}{\partial t} + \overline{H} = 0. \tag{3}$$

$\overline{H}$ is here obtained from H by replacing the p_i in H by the derivatives $\frac{\partial J}{\partial q_i}$. If J is a complete integral of these equations with the constants of integration α_i, then the system (1), (2) of the canonical equations is generally integrated by the equations

$$\frac{\partial J}{\partial q_i} = p_i \tag{4}$$

$$\frac{\partial J}{\partial \alpha_i} = \beta_i. \tag{5}$$

The more detailed textbooks on dynamics verify by calculation that satisfying (3),   [2]
(4), and (5) has the consequence of satisfying the canonical equations (1), (2).
However, I do not know of a natural way, free of surprising tricks of the trade, where
one begins at the canonical equations and arrives at the HAMILTON-JACOBI system
(3), (4), (5). Such a way is given in the following.

If I give for a distinct time t_0 the coordinates q_i^0 and the associated momenta p_i^0
of the system, then its motion is determined by (1) and (2). I represent this motion   [p. 607]
as the movement of a point in the n-dimensional space of the coordinates q_i. If I
imagine at time t_0 the momenta p_i^0 given for all points (q_i) of the coordinate space
by means of the equations (1) and (2) of the corresponding system such that the p_i^0
are continuous functions of the q_i, then these initial conditions determine the

movements of all points due to (1) and (2). We call the essence of all these motions a "field of flow" (*Strömungsfeld*).

Instead of describing the field of flow in the sense of (1) and (2) by giving coordinates and momenta of each point of the system as functions of time, I can also think of the state of motion—as measured by the p_i—to be given at every point (q_i) as a function of time, such that the q_i and t are viewed as independent variables. Both kinds of description correspond exactly to those in hydrodynamics based on the LAGRANGEan or EULERian equations of motions for fluids, respectively.

In terms of the second kind of description I have to replace the left-hand side of (1) by

$$\frac{\partial p_i}{\partial t} + \sum_\nu \frac{\partial p_i}{\partial q_\nu} \frac{dq_\nu}{dt}$$

or, according to (2), by

$$\frac{\partial p_i}{\partial t} + \sum_\nu \frac{\partial H}{\partial p_\nu} \frac{\partial p_i}{\partial q_\nu}.$$

Therefore, we have, according to (1), the system of equations

$$\frac{\partial p_i}{\partial t} + \frac{\partial H}{\partial q_i} + \sum_\nu \frac{\partial H}{\partial p_\nu} \frac{\partial p_i}{\partial q_\nu} = 0. \tag{6}$$

The $\dfrac{\partial H}{\partial q_i}$ and the $\dfrac{\partial H}{\partial p_i}$ are known functions of the q_i, the p_i, and t. Consequently, (6) is the system of partial differential equations with which the components p_i of the momentum vector of the field of flow comply.

The question suggests itself if there are fields of flow in which the momentum vector has a potential such as to satisfy the conditions

$$\frac{\partial p_i}{\partial q_k} - \frac{\partial p_k}{\partial q_i} = 0 \tag{7}$$

$$p_i = \frac{\partial J}{\partial q_i}. \tag{7a}$$

[p. 608] If (7) is satisfied, (6) takes the form

$$\frac{\partial p_i}{\partial t} + \left(\frac{\partial H}{\partial q_i} + \sum_\nu \frac{\partial H}{\partial p_\nu} \frac{\partial p_\nu}{\partial q_i} \right) = 0.$$

The second term is the complete derivation of H with respect to the coordinate q_i. If $\overline{H}$ denotes that function of q_i and time t which results from H when the p_i in H are expressed in terms of q_i and t, one gets

$$\frac{\partial p_i}{\partial t} + \frac{\partial \overline{H}}{\partial q_i} = 0$$

or, introducing the potential function J from (7a)

$$\frac{\partial}{\partial q_i}\left(\frac{\partial J}{\partial t} + \overline{H}\right) = 0.$$

These equations are satisfied if one demands for J the differential equation

$$\frac{\partial J}{\partial t} + \overline{H} = 0,$$

which is nothing else than the HAMILTONian equation (3). In conjunction with (7a) it solves the equations (6) of the field of flow.

The equations (5) are obtained in the following manner. If J is a complete integral with the arbitrary constants α_i, then (3) remains valid if α_i is replaced by $\alpha_i + d\alpha_i$ in J. Hence,

$$\frac{\partial^2 J}{\partial t \partial \alpha_i} + \sum_\nu \frac{\partial H}{\partial p_\nu} \frac{\partial^2 J}{\partial q_\nu \partial \alpha_i} = 0$$

must hold. Because of (2) we can write instead

$$\left(\frac{\partial}{\partial t} + \sum_\nu \frac{dq_\nu}{dt} \frac{\partial}{\partial q_\nu}\right)\left(\frac{\partial J}{\partial \alpha_i}\right) = 0.$$

But the operator in parentheses is identical with the operator $\left(\dfrac{d}{dt}\right)$, a time derivative in the sense of the "LAGRANGEan" description. Therefore, $\dfrac{\partial J}{\partial \alpha_i}$ remains constant for

one system during its motion and, consequently, an equation system of the form (5) must hold for the movement of a point in the system.

Doc. 48
Marian von Smoluchowski

Not translated for this volume.

Doc. 49

The Nightmare   [1]

by Albert Einstein   [Reprint prohibited]

I consider the final secondary school exam* that follows the normal school education *not only unnecessary but even harmful.* I consider it unnecessary because the teachers of a school are without doubt able to judge the maturity of a young man who attends the school for several years. The teachers' impression of a student derived during the school years, together with the usual numerous papers from assignments—which every student has to complete—are a succinctly complete and better basis on which to judge the student than any carefully executed examination.

I consider the final secondary school exam harmful for two reasons. The fear of exams as well as the *large mass of topical subjects that have to be assimilated by memorization harm the health of many young men to a considerable degree.* This fact is too well known to need to be verified in detail. But I will nevertheless mention the well known fact that many men in the most varied professions have been plagued, into their later years, by nightmares whose origins trace back to the final secondary school exam. And these are men who have stood up to their responsibilities in life and can by no means be counted among the "neurasthenics."

The final secondary school exam is furthermore harmful because *it lowers the level of teaching in the last school years.* Instead of an exclusively substance-oriented occupation with the individual subjects, one too often finds a lapse into a shallow drilling of the students for the exam. Instead of in-depth teaching one gets a more or less *showmanship exercise* designed to give the class a certain luster in front of the examiners.

Therefore, away with the final secondary school exam!

**Translator's note.* Readers who are not familiar with the German educational system of the past should not misinterpret Einstein's article as "humor." He criticizes here the traditional German *Abitur* exam that was used from the Empire to well after World War II. Passing the exam qualified the candidate for study at any German university without any further entrance exam, and in any department. The exam lasted five or six consecutive days, with written exams, uninterrupted, from 8 AM to noon; usually oral exams followed in the afternoon from 2 to 6 PM. It covered all major and thus mandatory subjects such as mathematics, physics, chemistry, German, and two of four languages: French, English, Latin, Greek. The only grades used from school-year exams were those in biology, geography, history, and religion. All exam questions or topics were selected by the Ministry of Culture. A government official delivered the exams in sealed envelopes daily, breaking the seal in front of the class and teachers. The grading, however, was done locally for each school. Oral exams were individual with at least two teachers present in order to prevent "irregularities."